HIGH DEFINITION TELEVISION

HIGH DEFINITION TELEVISION

Hi-Vision Technology

by

NHK Science and Technical Research Laboratories

Prepared under the Auspices of

NHK (The Japan Broadcasting Corporation)

Translated by James G. Parker

TechSearch International, Inc.

VNR VAN NOSTRAND REINHOLD

New York

Softcover reprint of the hardcover 1st edition 1993

Library of Congress Catalog Card Number 92-17071

ISBN-13: 978-1-4684-6538-9 e-ISBN-13: 978-1-4684-6536-5
DOI: 10.1007/978-1-4684-6536-5

Van Nostrand Reinhold
115 Fifth Avenue
New York, New York 10003

Chapman and Hall
2-6 Boundary Row
London, SE1 8HN, England

Thomas Nelson Australia
102 Dodds Street
South Melbourne 3205
Victoria, Australia

Nelson Canada
1120 Birchmount Road
Scarborough, Ontario MIK 5G4, Canada

16 15 14 13 12 11 10 9 8 7 6 5 4 3 2 1

Library of Congress Cataloging-in-Publication Data

Haibijon gijutsu. English
High definition television : hi-vision technology / NHK Science and Technical Research Laboratories : prepared under the auspices of NHK (the Japan Broadcasting Corporation) : translated by James G. Parker.
p. cm.
Translation of: Haibijon gijutsu.
Includes index.
ISBN-13: 978-1-4684-6538-9
1. High definition television. I. Nihon Hōsō Kyōkai. Hōsō Gijutsu Kenkyūjc. II. Title.
TK6679.H3513 1992
621.388—dc20 92-17071
CIP

Contents

Foreword

Hi-Vision is a new television system that Japan is the first to propose to the world. It has long been in development by NHK (the Japanese Broadcasting Corporation). The term Hi-Vision itself is becoming well-known worldwide.

NHK has been involved in the research and development of a high-definition television system for almost twenty years. Over this period, the project has moved from basic visual, auditory and psychological research to the development of experimental and broadcast quality equipment. With practical implementation near at hand, a considerable amount of equipment is now already on the market. Furthermore, efforts are underway to commercialize the technology by improving the performance of household and broadcast systems and establishing an international standard.

Experiments and exhibitions conducted toward this end include a direct satellite broadcast via the BS-2 satellite, international relay transmissions using communications satellites and Intelsat satellites, a UHF terrestrial broadcast in the United States, and exhibits at many expositions. These events have earned high praise both in Japan and abroad, thereby further promoting the commercialization of Hi-Vision.

In addition to the obvious application in broadcasting, Hi-Vision has applications in cable television, packaged media such as videotapes, movies, printing, education, medicine, and many other industrial fields. Practical applications have already begun in some of these areas.

In view of these developments, it is significant that a book that systematically deals with Hi-Vision technologies is being published. Until now there has not been any publication that adequately dealt with Hi-Vision technologies, and students and engineers interested in the subject have had to sift through numerous journals and papers.

Believing that there was now a need to systematically present the results of a quarter century of research and development, the NHK Science and Technical Research Laboratories decided to compile this volume. Each section has been written by the research staff members directly involved in the project and knowledgeable in the latest developments. We encourage people interested in Hi-Vision technology to read this book.

The path of research and development is a never-ending one as each new development leads on to an even newer one. Even as we anticipate the future developments in Hi-Vision technology, we look forward to conducting research for the post-Hi-Vision era and to supporting the unending development of broadcasting.

Masahiko Ohkawa
Managing Director
NHK (Japan Broadcasting Corporation)

Preface

1. THE EMERGENCE OF HI-VISION

The term Hi-Vision stands for a next-generation television system developed by NHK, and is a contraction of high-definition television. This system was designed with a higher resolution and larger screen than conventional television so as to create a sense of realism or telepresence. It has 1125 scan lines, or twice the 525 lines of conventional television, and the screen aspect ratio has been widened from the current 3:4 to 9:16.

The current standard NTSC (National Television System Committee) television format being used in Japan was developed and first adopted in the United States in 1953. Although it represented the most advanced technology at the time, the image quality of current television is lower than that of motion pictures and photographs. The low resolution is particularly noticeable on recently available display screens which are 30 inches and larger in size. Meanwhile, developments in electronics since NTSC was adopted have been quite remarkable in areas ranging from semiconductor technology to satellite technology, and have provided the basic elements for improving the quality of television images.

At the NHK Science and Technical Laboratories, we have been conducting research on next-generation television technology ever since the diffusion of color television in the 1960s. Since then, the research and development has resulted in the practical utilization of television cameras, VCRs, displays, and other broadcast quality production equipment. In addition, a band compression method for Hi-Vision signal transmission known as MUSE (Multiple Sub-Nyquist Encoding) was developed, and the possibility of Hi-Vision satellite broadcasting was confirmed in transmission experiments with the BS-2b direct broadcast satellite. In late 1991, Hi-Vision broadcasting was commenced on an experimental basis using broadcast satellite BS-3, which was launched in 1990.

As the image quality of Hi-Vision is comparable to that of 35mm film, it is being developed for video media applications in areas such as motion pictures, printing, education, and medicine. The application of Hi-Vision technology will extend far beyond broadcasting as it is combined in new ways with other video media.

2. THE OBJECTIVES OF HI-VISION

Table P1 shows that the Hi-Vision standard has twice as many scan lines as conventional television and a wider screen aspect ratio of 9:16

Table P1. Hi-Vision and conventional television standards.

	Hi-Vision	Conventional NTSC TV
No. of scan lines	1125	525
Aspect ratio (screen height to width ratio)	9:16 (3:5.33)	9:12 (3:4)
Interlace ratio	2:1	2:1
Field frequency	60 Hz	59.94 Hz
Video signal bandwidth	20 MHz	4.2 MHz
Audio signal modulation method	PCM	FM

(or 3:5.33 versus 3:4 for the old standard), all of which results in a pixel count about five times larger than conventional television. As will be explained in Chapter 1, the standard was set to conform to human visual characteristics in a manner appropriate for a next-generation television system.

Based on the coarseness of the scan lines, current television is best viewed from a distance of six to seven times the height of the screen. The screen thus covers a field of view of only 10°, and so the impact on the viewer is rather small. In Hi-Vision, the field of view was set at 30° to give the viewer more realism, and as Figure P.1 shows, the optimal viewing distance was set at three times the screen height. The wider angle of view reduces the influence of the frame around the image and thereby increases the realism of the image. Under these conditions, 1125 scan lines are needed to bring the resolution of the scanning line structure below the perception of the human eye.

The larger the screen size, the more desirable is a wide screen aspect ratio. Since Hi-Vision was intended for large displays, an aspect ratio of 9 : 16 was chosen. This ratio is close to that of the motion picture industry's standard Vista Vision screen.

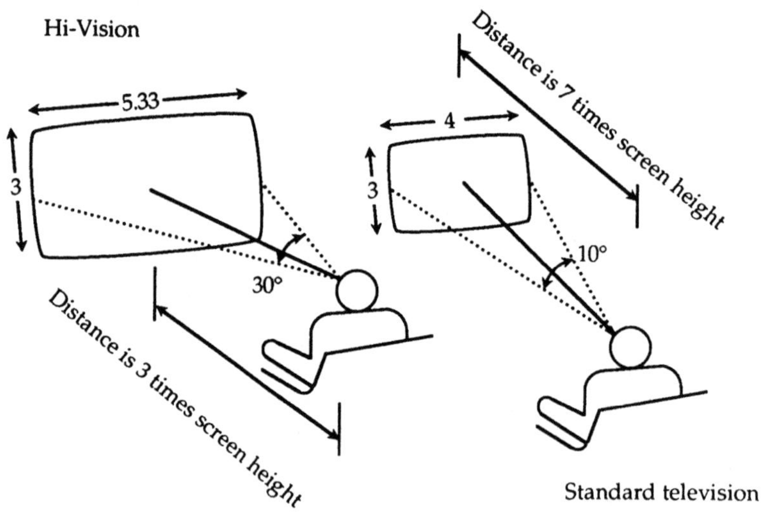

FIGURE P1. Comparison of viewing conditions for Hi-Vision and standard television.

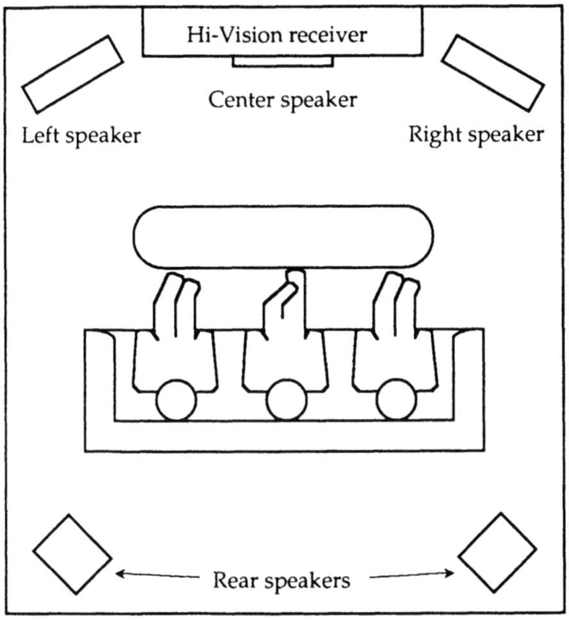

FIGURE P2. Basic speaker arrangement for 3-1 format 4-channel stereo system.

The audio format for Hi-Vision is a 3-1 mode 4-channel format as shown in Figure P.2. A center channel improves the orientation of the audio image with respect to the screen, which is useful for large screen displays. PCM (Pulse Code Modulation) was adopted for audio transmission.

The 1125-line, 60 Hz Hi-Vision studio standard is presently being used in Japan, the United States, and Canada, and an effort is underway at the CCIR (Comité Consultatif Internationale des Radio Communications) to adopt it as a worldwide standard. However, this effort toward a unified standard is meeting resistance from European nations, who use a 625-line, 50 Hz format and are opposed to a field frequency of 60 Hz, and more generally from groups that have economic concerns regarding the introduction of Japanese technology.

3. TECHNOLOGIES SUPPORTING HI-VISION

While Hi-Vision is a new television format, its principles are not completely different from those of conventional television. However, it is made possible by the wide range of recent advances in electronic technologies. A system-wide advance can be effected only by advances in many individual technologies such as the improvement of the optoelectric conversion film of a camera tube, development of a VCR magnetic head for high density recording, development of (Super High Frequency) band transmission technology for broadcasting, and development of a large screen display for household receivers.

Figure P.3 shows the technologies that form the basis of Hi-Vision. While the figure shows Hi-Vision technology broken up into the four areas of image pickup, recording, transmission, and display, it should be noted that Hi-Vision was made possible not by any particular breakthrough comparable to superconductivity, but by the accumulation of a wide range of technologies that includes LSI (Large Scale Integration) technology and satellite transmission technology, which were not available 30 years ago. At the foundation are microelectronics and digital technologies, which made possible the

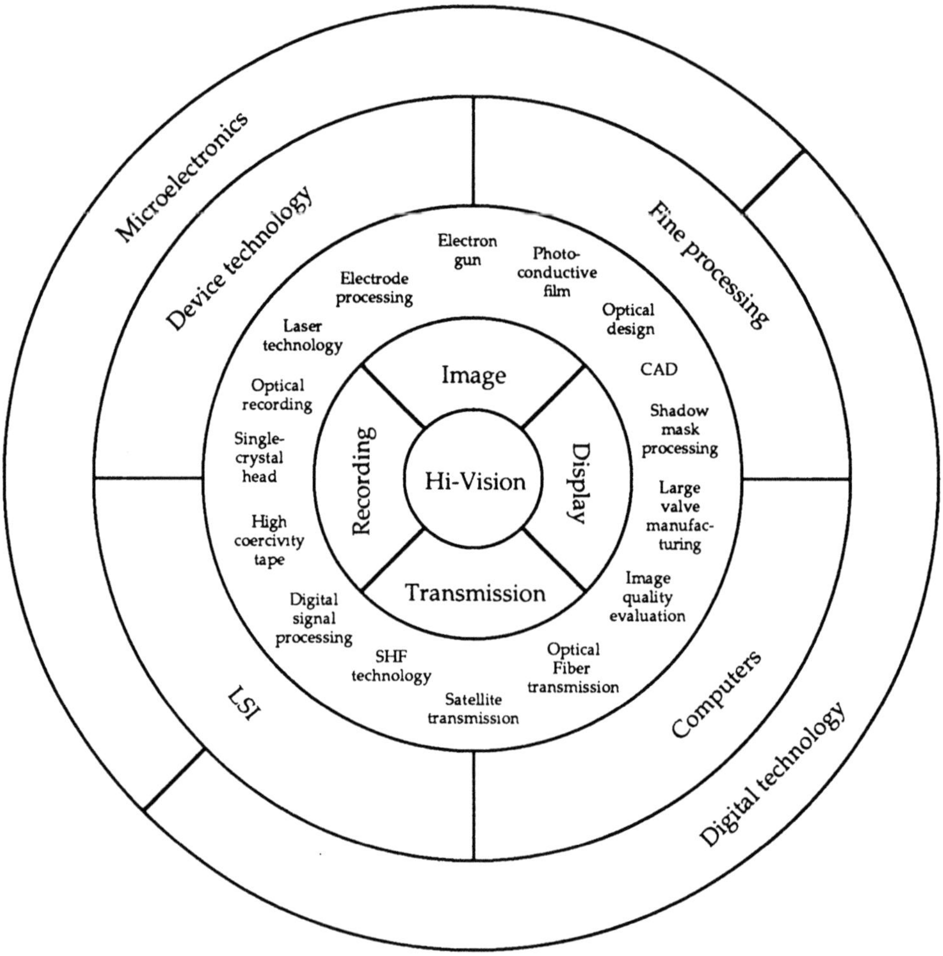

FIGURE P3. Technologies supporting Hi-Vision.

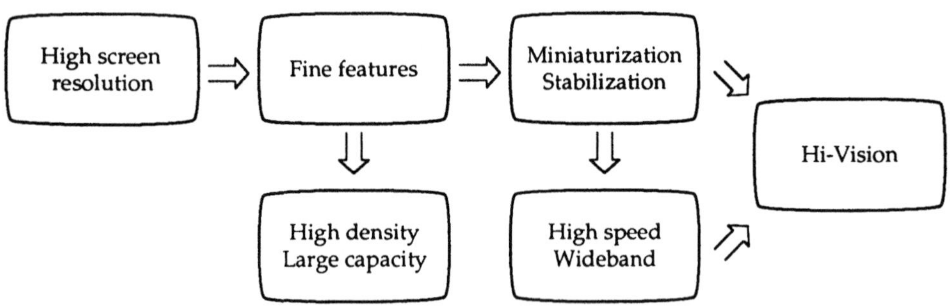

FIGURE P4. Technologies leading to Hi-Vision.

trend toward higher resolution and speed shown in Figure P.4. In this book we will discuss the many technologies represented in Hi-Vision.

4. THE FUTURE OF HI-VISION TECHNOLOGY

While Hi-Vision program production, transmission, and reception equipment have already been manufactured, research and development continues to take their capabilities to higher levels.

1. Increasing Camera Sensitivity

One of these areas of development is the Hi-Vision camera. Because the f-stop on a Hi-Vision camera must be increased two stops to obtain the same depth of field as a conventional camera, the camera needs to be four times as sensitive. This was considered impossible with present camera tubes until the recent discovery of the avalanche effect in optoelectric conversion films, which increases the quantum efficiency of the optoelectric conversion tenfold. Development of cameras having this type of camera tube is under way.

2. Digitization of VTRs

The extensive use of digital image processing in Hi-Vision makes the development of digital signal recording and transmission capabilities a high priority. The data recording rate of VTRs must be 1.2 Gbits/second. Prototype machines will soon be able to record for one hour with a 12-inch reel of metal particle tape, and standards for the VTR have already been established.*

3. Development of a Hi-Vision Receiver

Hi-Vision is intended to be shown on screens one meter in width. Presently displays available for practical use are of the rear projector variety, with red, green, and blue CRT projectors used to project the image onto a screen through a lens.

In the future, displays are expected to be flat panels that can be hung on the wall. The most promising type of flat panel is the plasma display, which uses a plasma discharge to cause phosphors to glow. A 20-inch plasma display has been developed at the NHK Laboratories. This display uses a plasma discharge to excite the phosphors in each pixel with ultraviolet light. A 30-inch panel is currently under development.

A band compression method for Hi-Vision broadcasting known as MUSE (Multiple Sub-Nyquist Sampling Encoding) has been developed. As relatively complex signal processing is required at the receiver, LSIs are currently being developed with the aim of reducing the cost of a MUSE receiver to a level comparable to conventional receivers in time for regular Hi-Vision broadcasting.

Hi-Vision will be broadcast by satellite. Advantages of satellite broadcasting include instant nationwide coverage, the reception of high quality images unhindered by ghosting, and the fact that present DBS (Direct Broadcast Satellite) antennas can be used for reception. The band compression method MUSE is expected to find applications beyond broadcasting in areas such as cable television, videotapes, and video disks.

5. THE NEW VIDEO ERA OPENED BY HI-VISION

Because it rivals 35mm film in image quality, Hi-Vision is being used in movie production, where Hi-Vision images are converted to 35mm motion picture film, as well as in converting movies to Hi-Vision for presentation in video theaters. Another application already in use is video printing, where Hi-Vision images are printed directly onto paper without using film.

In this way, Hi-Vision is expected to be at the center of the coming video culture because of its versatility in mixing different video media. This trend conforms with advances in other video media such as layout scanners, Computer Type Setting (CTS) and DeskTop Publishing (DTP)

*Digital VTRs are currently being used in Hi-Vision program production.

in the printing field, optical storage media for offices, and the complete digitization of image databases. Other applications include the fusion of Hi-Vision with computer graphics, the combination of Hi-Vision with optical disks for video exhibits in libraries and museums, educational applications, and the use of audio visual catalogs in the distribution sector.

New media combinations will invade the home as well, for example in the form of a video disk with a collection of images and an accompanying narration soundtrack that would put a book to shame. With the development of new video media such as this, the high resolution, large screen Hi-Vision receiver will not only be receiving broadcasts and playing back videotapes, but also performing a vital function as a video data terminal.

Junichi Ishida

HIGH DEFINITION TELEVISION

1

Hi-Vision Standards

Kengo Ohgushi, Junji Kumada, Taiji Nishizawa, Tetsuo Mitsuhashi

1.1 BASIC PARAMETERS FOR HI-VISION

Research in television image quality under the current standard format has been going on for about half a century. Much research and experience have been gained during that time on the relationship between physical factors such as signal-to-noise ratio or signal bandwidth and image quality. However, practically no investigations have undertaken the advanced study of visual and auditory psychological effects as carefully as has Hi-Vision research. Thus in the development of the Hi-Vision system, numerous psychological experiments and studies were conducted to establish the basis for the fundamental parameters.

In this chapter, we will discuss how the basic Hi-Vision parameters were established and introduce the various visual psychological experiments that were conducted. The audio system is discussed in Section 1.5, and so will be mentioned in passing here. The term image quality in the following discussions refers to the total impression received from a screen.

1.1.1 Image Quality Factors and Measures

Many theories have been proposed regarding the psychological factors (psychological response to viewing an image) and physical factors (physical characteristics that influence image quality) that affect television image quality. Table 1.1 is an example from Hiwatashi.[1] As the table shows, space and time factors affect image quality, but the final quality, as shown in Figure 1.1, is determined by an overall evaluation of the combination of these psychological factors.

Up until now, methods for improving image quality have directly tied together the study of physical factors to the final overall image quality, without studying psychological factors along the way. But since psychological effects are important in Hi-Vision, understanding the factors behind them is basic and vitally important.

To determine what psychological factors are important to image quality, we conducted a study using SD (Semantic Differential) method. The result was that eight factors including strength, beauty, brightness, and texture explained 75% of the variation in image quality, and the sense of telepresence was found to be important. Further, in studying the correspondence between these psychological factors and the physical factors such as the system parameters of screen size, brightness, and pixel count, we found that each physical factor affected several psychological factors, and further, that the overall image quality decision was strongly correlated to each psychological factor. For example, the subjec-

TABLE 1.1. Factors affecting image quality.

	PHYSICAL FACTORS		PSYCHOLOGICAL FACTORS		Example of
	Optical factors	Electrical factors	Sensitivity/perception level	Cognitive/emotional level	evaluation
Screen configuration	Screen size, shape, aspect ratio, nature of light	Bandwidth	Optimal viewing distance; peripheral and central vision	Telepresence, impact, three dimensionality, fatigue	Sharpness
2-D spatial factors: Linear	Geometric distortion Blur, Halation, Development effect } OTF, MTF	Scan distortion (polarization distortion, bandwidth) (High band), (Middle band) } Frequency characteristic (spectrum) Outline compensation	Optical illusion Vision, resolution Spatial frequency characteristic of sight Mach effect	Recognition rate, readability	Sharpness
Planar: Shading	Brightness (max, avg.) Contrast, Half tone } Conversion characteristic (gamma)	Signal level Dynamic range Conversion characteristic (gamma)	Brightness perception, brightness function adaptation, contrast effect	Brightness	Gray scale reproduction
Color	Leakage light, indoor light Spectral characteristic Color additive and subtractive methods	Color signal, DG and DP	Color difference, Color phase, Saturation } Color perception ⇓ {Color adaptation, Color contrast, Area effect} Other	Color harmony Color preference Color memory	Color tone reproduction
3-D spatial factor: Depth	Binocular type 3-dimensional display	(Bandwidth)	3-dimensional perception	3-dimensionality	
Time factors--flicker, blinking	Frame rate, flicker, residual image	Interlace	C.F.F., time frequency characteristic, Broca-Sulzer effect, Bartley effect, subjective color		
Time-space factors--change in shape, motion	Residual image	Interframe correlation	Graphic residual effect, motion residual effect, motion vision (real and hypothetical)		
Noise and other factors	Granularity; moire; dust, scratches; screen jitter; double images, transfer	Random noise (shot and thermal noise); periodic noise (hum,sine wave); pulse noise; jitter; ghost	Noise evaluation curve for brightness and frequency	Tolerability	Level of interference

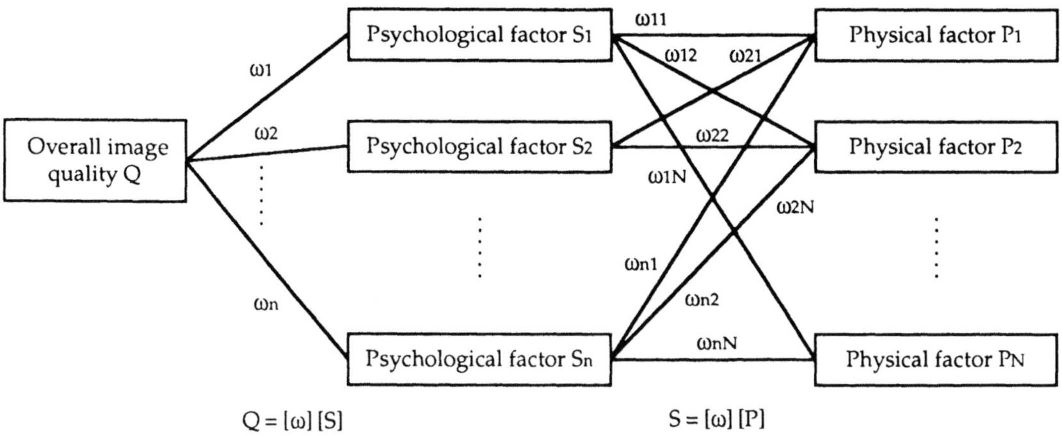

FIGURE 1.1. Factors determining overall image quality.

tive evaluation results in Figure 1.2 show how screen area affects the psychological factors of telepresence, impact on the viewer, three dimensionality, balance, and the overall evaluation of attractiveness. The figure shows that large screens greatly increase the effect of three dimensionality and impact.[2]

From these results, the Hi-Vision image quality evaluation was broken down into seven categories, as shown in Table 1.2. The special feature of this measure is that since most people tend to avoid the extreme categories, in order to remove the contingency that the effect of image quality improvements will not be correctly evaluated, an anchoring category corresponding to the best image imaginable of "perfect—no further improvement needed" was implemented. In this way, highly reliable eval-

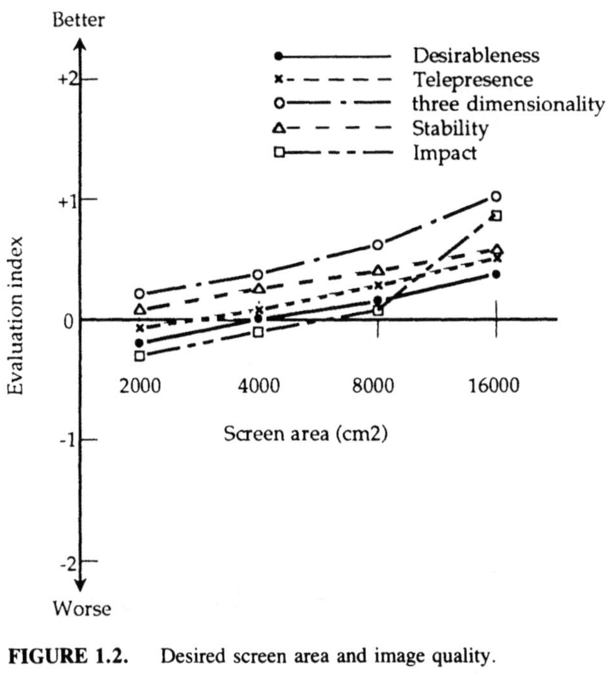

FIGURE 1.2. Desired screen area and image quality.

TABLE 1.2. Subjective evaluation scale.

Image quality scale	General purpose TV image scale (CCIR Rec. 500)	Hi-Vision
7		Ideal Hi-Vision image
6		Excellent
5	Excellent	Very good
4	Good	Somewhat good
3	Fair	Somewhat bad
2	Poor	Very bad
1	Bad	Extremely bad

uation data could be obtained. The image quality of an 8″ × 10″ slide would definitely be a category 7, while NTSC TV would be about a category 3.

1.1.2 Screen Format and Viewing Distance

The term screen format is used to group together the parameters related to the display of images such as screen size and shape. An analysis of psychological factors shows that the screen format is closely related to telepresence and impact on the viewer. Since the screen format is closely related to viewing distance, it will also be dealt with in this section.

Similar to the experience of watching a wide screen movie, when the display area is enlarged and the volume of visual information presented in the viewer's field of vision is increased, the viewing conditions are not as restricted by the frame around the image. The screen itself seems to disappear, the image acquires depth, and the viewer experiences telepresence, or a sense of being there. Large screens are especially important in heightening this psychological effect.

However, when the actual viewing conditions in a home are considered, screen size and viewing distance naturally are restricted. Thus with Hi-Vision we began by quantifying telepresence, and conducted our investigation as described below.

(1) Visual Angle and Viewing Distance

Viewing distance can be described in absolute terms as the distance (L) between the viewer and the screen, or in relative terms as a multiple of the screen height. Ordinarily, use of the relative distance is preferred. The relationship between the relative distance D and the viewer's field of vision angle Θ, as shown in the following equation, is that as D decreases Θ increases.

$$\theta = 2\tan^{-1}\{1/(2D3)\} \qquad (1)$$

A number of experiments were conducted to determine the value of D. Figure 1.3 shows the method proposed to quantify the relationship between telepresence and visual angle. In other words, if the viewer looks at a hemispherical screen that covers practically his entire field of vision, he will be drawn into the image and unconsciously lean forward. The degree of the

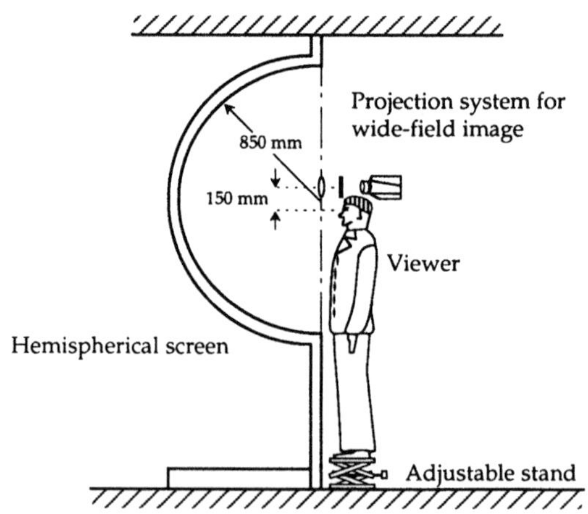

FIGURE 1.3. Apparatus to measure the effect of a wide field of vision.

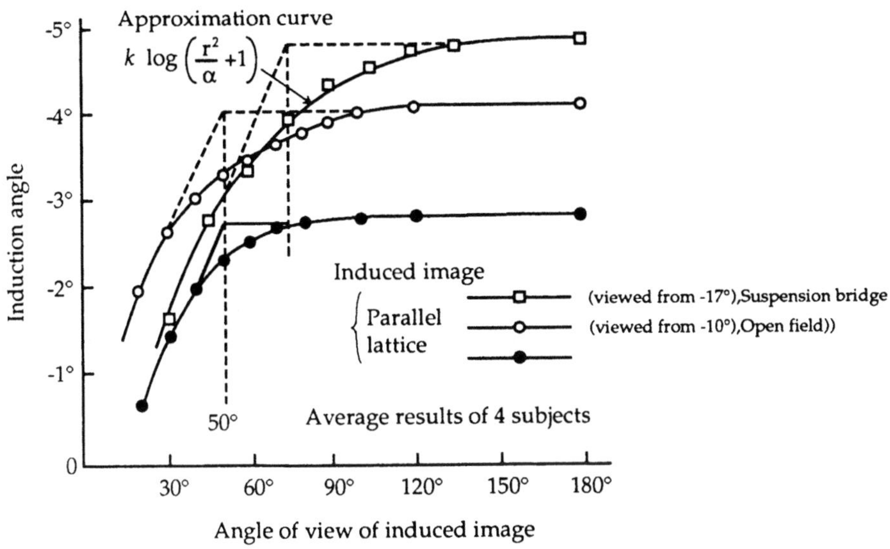

FIGURE 1.4. Subjective coordinate axis induction effect of observed view angle.

lean (called the inducement angle) thus becomes an indicator of telepresence. Measurement results of the inducement angle with respect to visual angle are shown in Figure 1.4.[3] Telepresence begins to occur at a visual angle of about 20°, and at 30° becomes quite conspicuous.

Further, in studying the viewer's desired viewing distance using an image with sufficient resolution, the result was that the optimal distance was 2 to 3H(H= screen height, with a visual angle of 20° to 30°). The overall viewing experience actually deteriorated at closer distances because viewers would be overwhelmed by the screen or not be able to see the entire screen at once. (Figure 1.5).

In contrast to the above experiments with still

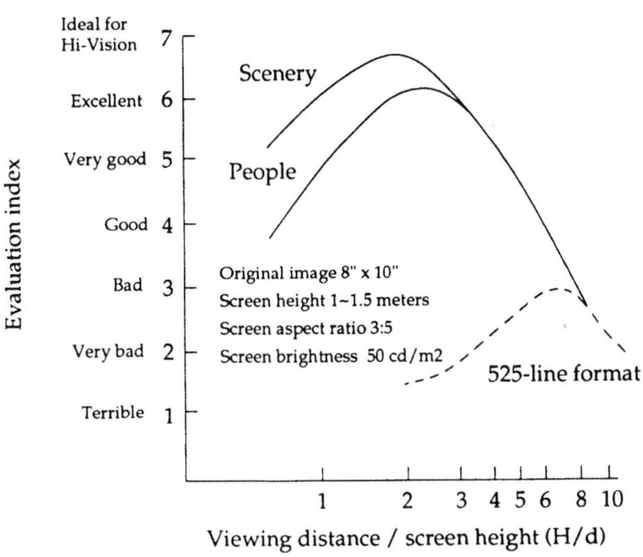

FIGURE 1.5. Image quality as a function of viewing distance.

images, experiments using moving images indicated that the visual angle was too wide for viewers, who complained of dizziness and instead preferred a viewing distance of 4*H* (visual angle about 19°). The dizziness, of course, varied depending on the amount of motion in the material being viewed. With regard to the viewing distance for Hi-Vision, due to the importance of creating the sense of telepresence by enlarging the visual angle, we considered achieving a balance with the still image effect mentioned above, and decided on 3*H*, or a visual angle of about 20°, as the standard viewing distance.

(2) Screen shape

Screen shape refers to a screen's aspect ratio and curvature. While the shape is closely related to the screen's size, we will discuss screen size in the next section.

We conducted a subjective evaluation for the most desirable aspect ratio for Hi-Vision using slides with various aspect ratios. The results are shown in Figure 1.6.[4] Regardless of the image on the screen or the screen's surface area, the optimal aspect ratio was found to be 3:5, followed by 3:6. Moreover, there was a tendency for the desired aspect ratio to increase as the screen area increased. Thus the appeal of movie screens with large aspect ratios is understandable.

Furthermore, in studying the nature of the human field of vision, we found that the effective field of vision where one's vision is superior and information is instantly accepted is, as Figure 1.7 shows, 30° horizontally and about 20° vertically, for an aspect ratio of about 1:1.5.[5]

From these psychological and physiological studies, we determined that a Hi-Vision aspect ratio of about 3:5 was optimal. This corresponds to a viewing distance of 3*H*, a horizontal visual angel of 30° and a vertical visual angle of 20°.

On the other hand, the aspect ratios used in movies covers quite a broad range, from about 1:1.3 to 1:2.7. While these aspect ratios are intended to take advantage of the large screen and high image quality, their psychological effect is not necessarily clear. Currently, the most commonly used aspect ratio is Vista Vision's 1:1.8. The SMPTE (Society of Motion Picture and Television Engineers) has recommended an aspect ratio of 9:16 for movie film because it is most commonly used. This is equivalent to 3:5.33, and is very close to the psychologically and physiologically desirable aspect ratio of 3:5 described above.

Regarding screen curvature, while it is recognized that an appropriate curvature increases the sense of depth and naturalness compared to a flat screen, the optimal degree of curvature varies greatly depending on the image on the screen. Furthermore, the curvature complicates the manufacturing of CRTs. Thus we decided that a flat screen was appropriate.

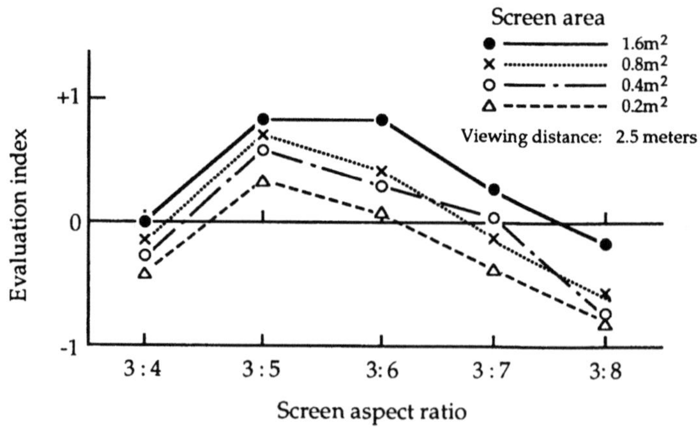

FIGURE 1.6. Screen aspect ratio perspective.

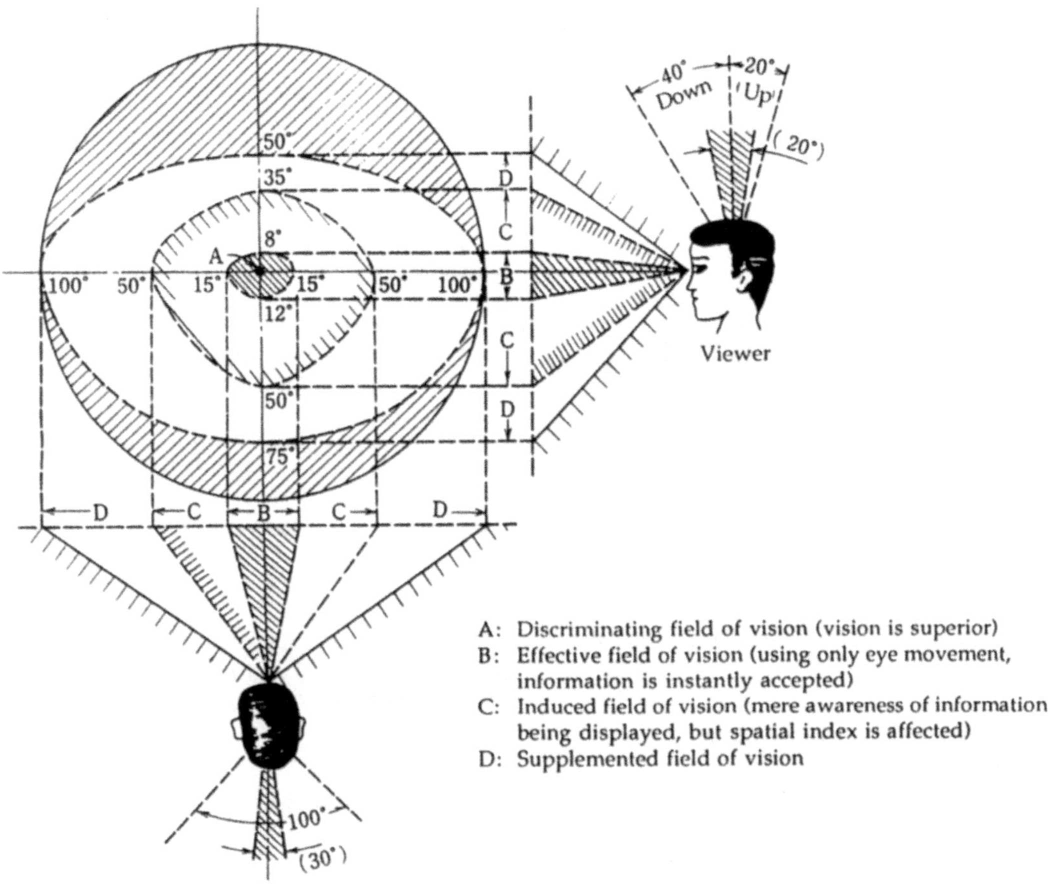

FIGURE 1.7. Field of vision and acceptance of information.

Based on the above studies, we decided to make the Hi-Vision screen shape flat with an aspect ratio of 9:16.

(3) Screen size

In the discussion so far, we have determined the desired visual angle, or in other words the relative screen size. However, as Figure 1.8 shows, even with the same visual angle, the larger the absolute screen size is, the greater is the telepresence experienced.[3] The absolute screen size can be calculated from Equation 1.1 if the absolute viewing distance is determined.

At the minimum viewing distance viewers tend to experience eye fatigue. Just as when one reads a book for a long time, when one continuously watches an object from a short distance, the strain on the eye's focusing muscles causes eye fatigue. To avoid this, a viewing distance of at least two meters is desirable. Furthermore, the eye's ability to focus, which allows it to detect depth information, decreases at distances greater than two meters, and even a flat image can appear to be three dimensional.

On the other hand, the greatest factor affecting the maximum viewing distance is the size of the room. Needless to say, a viewing distance that is too large is not desirable in terms of efficient use of room space. If we assume that the display will be placed in a room about 15 square meters in size, then three meters would be about the maximum viewing distance.

Based on these considerations, we decided on a standard viewing distance of 2.5 meters. With a horizontal visual angle of 30° and an

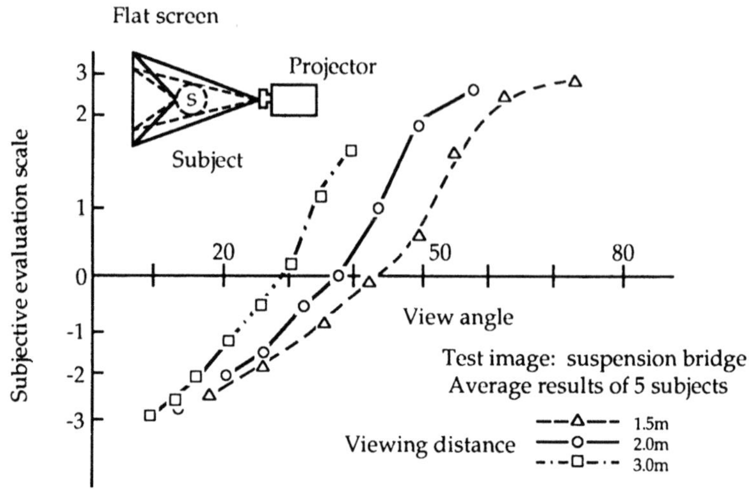

FIGURE 1.8. Subjective evaluation of a sense of realism experienced from a large screen display.

aspect ratio of 9:16, the standard screen size is 0.75m × 1.3m.

(4) Display conditions—contrast and maximum brightness

Under normal conditions, the desirable contrast ratio for a television image is 30 to 50. On the other hand, for a movie a contrast ratio of at least 100 is recommended, and experience has shown that a certain amount of high contrast is necessary when high image quality is desired. Thus a contrast value of 50 is considered desirable for Hi-Vision.

In studying the ordinary TV viewing conditions of households, the ambient lighting was found to be about 54 lx on the display screen, which converts to about 3 cd/m^2 in terms of screen surface brightness. Thus if the minimum brightness is 3 cd/cm^2, the maximum brightness needs to be 90 to 150 cd/m^2.

One visual characteristic related to maximum brightness is glare. Glare depends not only on brightness but on the size of the light source and the amount of ambient light. If the contrast ratio of the field of vision within 4° around the screen is no more than 50, glare can be permitted up to 500 cd/m^2. If we estimate a leeway of about 50% under various use conditions in the home, then a maximum brightness of about 250 cd/m^2 can be tolerated.

Based on the above considerations, an image contrast value of 50 and maximum brightness of 150 to 200 cd/m^2 are desirable.

1.1.3 Scanning Method

The scanning method includes the number of scanning lines, frames per second, interlace, and image bandwidth. It is the most basic aspect of a television system.

(1) Number of Scanning Lines

The number of scanning lines must be set not only to determine the vertical resolution but also to prevent the occurrence of unsightly scanning line artifacts that look like blinds over the image. In order to ensure sufficient vertical resolution and to remove scanning line artifacts, the number of scanning lines must be increased enough to be undiscernable.

The minimum size that can be seen by a person with normal vision covers a visual angle of one minute. Thus the necessary number of scanning lines at a viewing distance of $3H$ (20° visual angle) is about 1,000 lines.

In order to curtail the signal bandwidth, interlace scanning is used. Figure 1.9 shows the evaluation results for artifacts from 2:1 interlace versus progressive scanning. At a viewing dis-

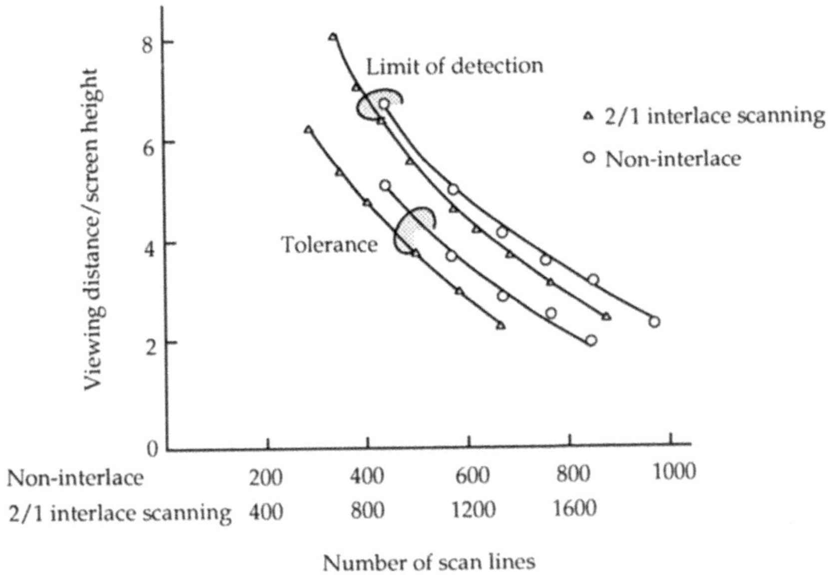

FIGURE 1.9. Viewing distance at which scanning lines are not visible.

tance of $3H$, 2:1 interlace has a permissible limit of 1,100 scanning lines. Moreover, 2:1 interlace produces about the same amount of artifacts as progressive scanning having six-tenths the number of scanning lines per frame.[6] Thus the bandwidth reduction effect of 2:1 interlace is about 1/1.2. Higher interlace ratios such as 3:1 or 5:1 are not practical, as they produce more pronounced artifacts peculiar to interlacing such as visual pairing, interline flicker, and crawling.

On the other hand, as the vertical resolution is determined by the number of scanning lines per frame, there is no difference between 2:1 interlace and progressive scanning. For this reason, when Hi-Vision images are recorded on film, the interlace method has an advantage in obtaining the same resolution for still images.

In addition to the above considerations regarding image quality, another factor in determining the number of scanning lines is convertibility with current television formats. In other words, with regard to both the 525 lines for NTSC and 625 lines for PAL (Phase Alternation by Line) and SECAM (Sequential á Memoire Color Television System), it would be desirable to have the simplest ratio of integers possible. The number of lines that best meets this condition in the neighborhood of 1,100 is 1,125, the ratio to 525 being 15:7 and to 625 being 9:5.

(2) Frames per second

The number of frames per second is determined based on flicker and the reproducibility of moving images. While flicker is mainly determined by screen brightness, it is also related to visual angle. In Figure 1.10, the relationship between the location on the retina and the number of frames per second at which flicker is no longer visible (this limit expressed in frequency is known as CFF or Critical Flicker Frequency) is plotted for various brightness levels of the index.[7] Since the CFF is higher (flicker is more visible) in the peripheral portions of the retina than in the central region, this point must be considered when determining the field frequency. When the viewing distance is three times the height of the screen and the eye is looking at the center of the screen, the whole screen is within $\pm 15°$ from the center of the retina, but when the eye is looking at the left (or right) edge, the opposite edge is 30° away. This situation is indicated by the dotted lines in the figure. At a brightness of 150 to 2100 cd/m^2, a field frequency of 50Hz is below the CFF and will not eliminate flicker, but 60Hz will.

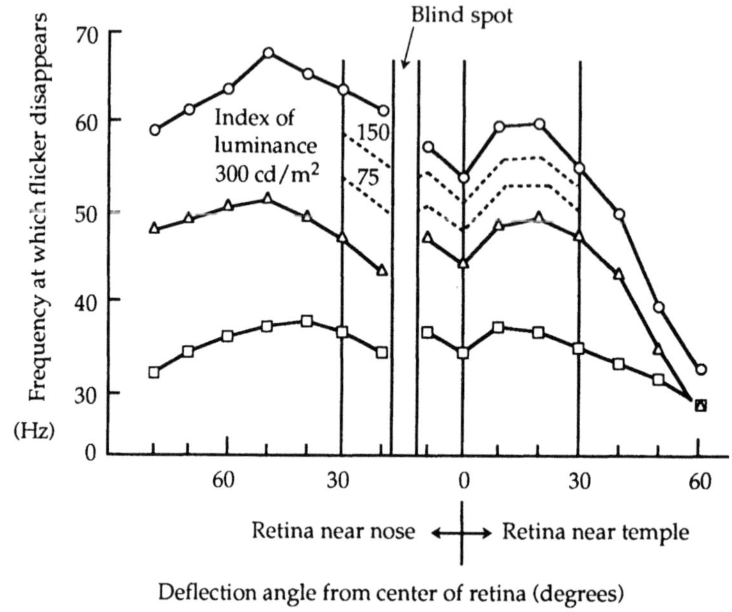

Size of index: visual angle 16°, Background light: 2/3 of index luminance, Number of subjects: 3

FIGURE 1.10. Flicker characteristics of the human eye.

As for reproducing movement, as Figure 1.11 shows, at 60 fields per second, movements of up to 24° per second can be reproduced smoothly.[8] This is about as fast as the visual system can follow an object, and so a field frequency of 60 fields per second is both necessary and sufficient. Further, as Figure 1.12 shows regarding the resolution of moving images, the faster the movement is, the lower is the MTF of vision.[9] While sufficient for ordinary images, for slow motion and other special effects requiring higher moving image resolution, the same methods are used as in current format cameras, such as the use of shutters.

As with the number of scanning lines, convertibility of the field frequency with current

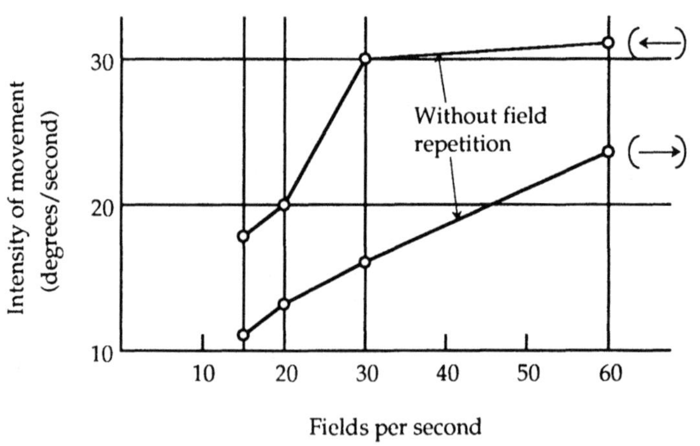

FIGURE 1.11. Critical maximum speed at which a window pattern appears to move smoothly.

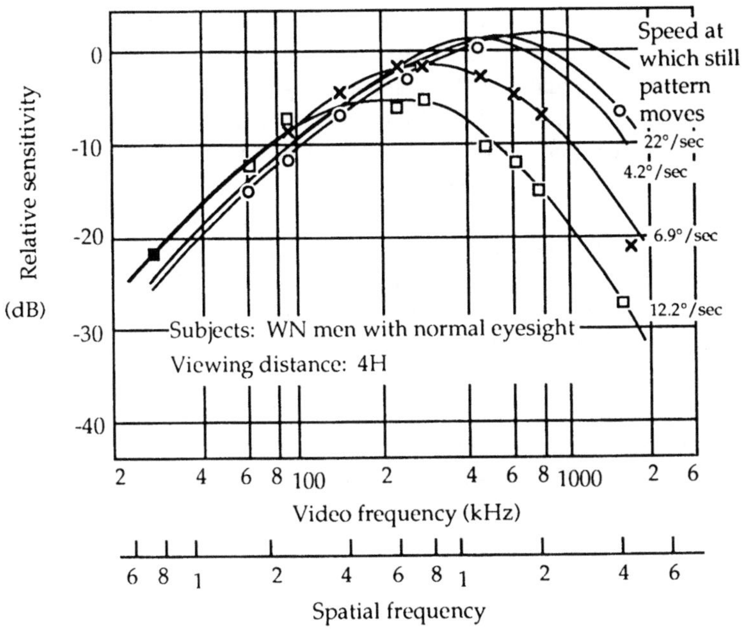

FIGURE 1.12. Spatial frequency characteristic of vision for motion space sine wave.

television formats is important. However, unlike converting the number of scanning lines, field frequency conversion requires not simply an integer ratio, but the smallest common multiple between the 60Hz and NTSC and 50Hz of PAL and SECAM. While 300Hz may be a common multiple, it is an unrealistic value.

For this reason, a field frequency of 60Hz was chosen based on considerations of visual characteristics.

(3) Image Signal Bandwidth

The image signal bandwidth is determined by the necessary horizontal resolution (number of

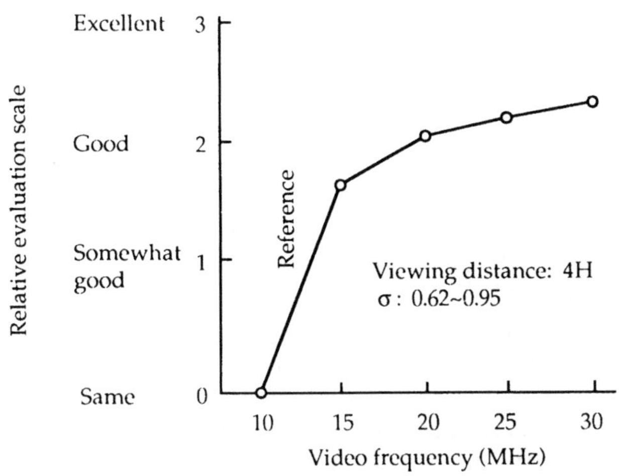

FIGURE 1.13. Relationship between video signal frequency bandwidth and image quality evaluation.

pixels) and field frequency. With regard to resolution, while an image is supposed to be sharpest when each pixel's height and width are equal (that is, when the horizontal and vertical resolution are the same), when we look at experimental results on the optimal pixel shape (the ratio between horizontal and vertical resolution, or the height and width of a pixel), we find that a rather large inequality is permissible.

Figure 1.13 shows the evaluation results of changing the image signal bandwidth of an image with 1,125 scanning lines.[10] The evaluation value starts to level off at a bandwidth of about 20 MHz (horizontal resolution of about 600 lines), and is practically flat by 30 MHz. Thus a signal bandwidth of at least 20 MHz is necessary.

Incidentally, the spatial frequency characteristics of the visual system differ for luminance and chrominance signals. Figure 1.14 shows the spatial frequency characteristics for the red-green axis, which is most sensitive to variations in the luminance signal and chromaticity, and for the yellow-blue axis, which is least sensitive.[11] It can be seen that for an ordinary image without any particularly saturated parts, sharpness is mainly determined by the luminance signal, and the 20 MHz mentioned above can be construed as the luminance signal. As for the bandwidth of the chrominance signal, judging from the same figure, it is one-third to one-fourth of the luminance signal.

Figure 1.15 shows the evaluation results of the chrominance signal bandwidth.[12] For the R-Y and B-Y combinations (similar to the yellow-blue and red-green axes mentioned above), 7.0 MHz and 5.5 MHz respectively were found to be sufficient.

Figure 1.16 compares the image quality of television and film.[10] Hi-Vision, with 1,125 lines, surpassed 35mm movie film and was found to be comparable to slides. As an electronic imaging system to replace film, the Hi-Vision system has the basic ingredients for future widespread use.

The basic parameters for the Hi-Vision image signal are shown in Table 1.3. The scanning format has 1,125 lines, 30 frames per second, 2:1 interlace, and 60 Hz field frequency. The screen format has an aspect ratio of 16:9 and a standard viewing distance of 3H.

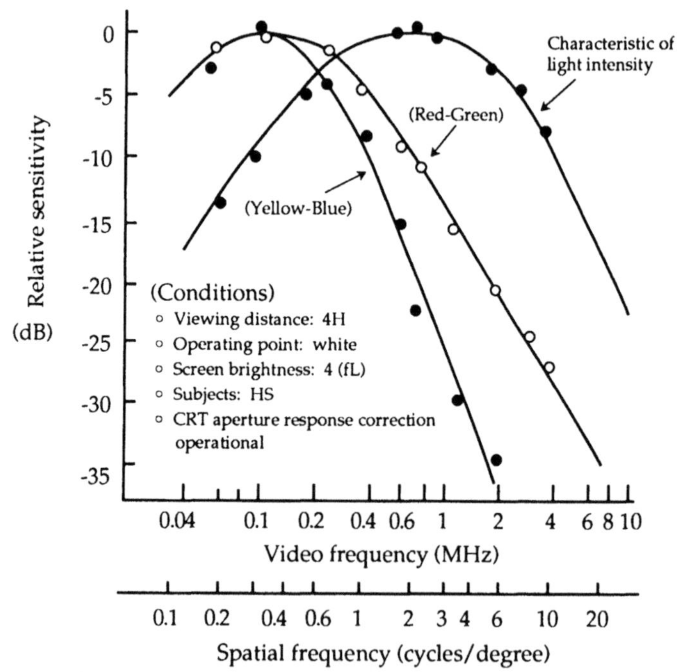

FIGURE 1.14. Example of spatial frequency characteristics for chromaticity and brightness-stripes.

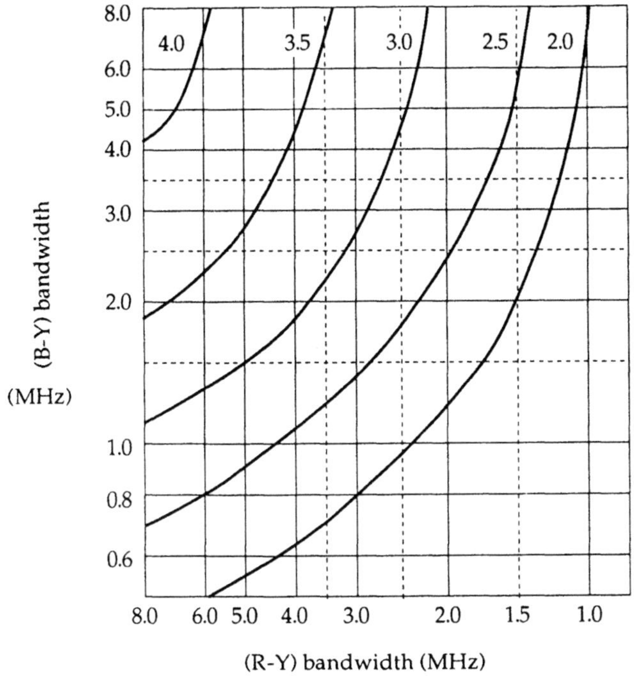

Numbers in the figure refer to the following categories:

5. Degradation cannot be confirmed.
4. Difference is negligible.
3. Difference is clear, but no problem for broadcasting .
2. Difference is significant and a problem for broadcasting.
1. Unusable for broadcasting.

FIGURE 1.15. Required chrominance signal bandwidth for 1125-line high definition television.

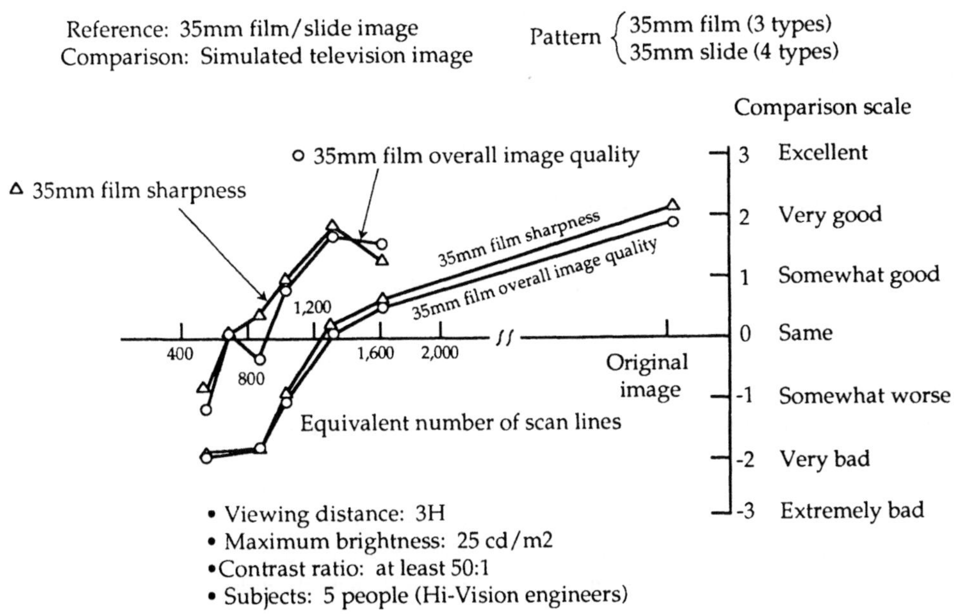

FIGURE 1.16. Image quality comparison between television and motion pictures in terms of scanning lines.

TABLE 1.3. Basic specifications of the Hi-Vision system.

Screen format	Aspect ratio	16:9
	Standard viewing distance (Field of vision)	3H/2.5 meters (30° x 20°)
Scanning format	Number of scanning lines	1125
	Frame frequency	30 Hz
	Field frequency	60 Hz
	Interlace	2:1
Video signal bandwidth format	Luminance signal	20 MHz
	Color difference signal	7 MHz
Audio format	Arrangement	3 front, 1 rear (3-1 mode)

1.2 HI-VISION STUDIO STANDARD

1.2.1 Necessary Conditions for a Studio Standard

Hi-Vision is the term coined for the 1,125-line, 60-field HDTV (High-Definition TeleVision format. In investigating HDTV studio standards, the primary consideration must be image quality. The required characteristics in HDTV image quality include not merely resolution but also telepresence and impact. The CCIR (Comité Consultatif Internationale Telegraphique et Telephonique) has decided on the following conditions for any HDTV studio standard:

1. Both the vertical and horizontal resolution must be at least twice those of current television formats.
2. The screen's width to height ratio, in other words its aspect ratio, must be wider than the 4:3 of current television formats.
3. The viewing distance must be three times the height of the screen.
4. Also, the color and motion reproduction must be at least as good as current television formats.

The BTA and SMPTE HDTV studio standard approved by both the United States and Japan fulfills these conditions.

Besides image quality, another important point in considering a studio standard is the worldwide unification of standards. Since three television standards—NTSC, PAL, and SECAM—are currently being used in the world, a format conversion is required whenever programs are exchanged or relayed between nations with different formats. Format conversions are not only inconvenient but also cause image degradation. Thus an internationally unified HDTV studio standard would greatly facilitate the exchange of programs between nations by eliminating the need for format conversion. Furthermore with the unification of studio standards we could lower the cost of studio equipment.

Another consideration for an HDTV studio standard is its downward convertability to current television formats.

1.2.2 CCIR Investigation of a Studio Standard

Based on a Question posed at the CCIR in 1974, a Study Program was begun in 1985. Investigations of an HDTV standard have continued since then. Particularly at the last meeting in 1985, a recommendation was proposed calling for the international unification of a studio standard based on the 1,125-line, 60-field parameters. However, due to opposition from several European nations at the general meeting in May 1986, the proposal did not survive. The proposal subsequently was put into the CCIR's Report 801 as Appendix II for further study. Several

European nations are currently investigating a format with 1,250 lines, or twice the 625 of their present format, and 50 fields per second.

The CCIR aims to have a recommendation on a HDTV studio standard at their general meeting in 1990. In preparation, they are planning a special meeting in May 1989 to deliberate on studio standards.*

1.2.3 The BTA and SMPTE Studio Standard

While a CCIR recommendation for a unified HDTV standard was not produced at the general meeting in 1986, an order was issued to make a detailed standards proposal for a 1,125-line, 60-field studio standard. Further, as more people began to actually use 1,125-line, 60-field format HDTV equipment in program production, the need arose for detailed standards for manufacture of the equipment. With regard to such an HDTV standard, a temporary 1125 standard had been devised to manufacture the equipment for the HDTV exhibit at the 1985 Tsukuba Science Exposition. It became necessary to revise this standard and include it as well in the proposed recommendation in Appendix II of the CCIR report mentioned above.

Based on the reasons stated above, in Japan the BTA (Broadcast Technology Association) began a detailed investigation of a studio standard in September 1986. In the United States, the SMPTE, which had been studying the 1,125-line/60-field format, joined with the BTA in conducting a detailed investigation. The SMPTE investigation had begun in August 1986. The joint investigation by the BTA and SMPTE was completed except for a portion including digital standards, and was announced as both the BTA Standard S-001[13] and SMPTE 240M.[14]

1.2.4 Content of the BTA/SMPTE Standard

The BTA/SMPTE studio standard prescribes the basic parameters for the 1125/60 high-definition television format and the video signal, synchronizing signal, and colorimetry parameters used in the studio.

In this section we will discuss the standards regarding the video signals. The synchronizing signal and colorimetry parameters will be discussed in Sections 1.3 and 1.4.

Table 1.4 shows the basic characteristics of the video signal. We will follow the table in explaining the parameters of the standard. The reader may remember that these parameters are based on the psychological experiments described in Section 1.1.

The number of scanning lines per frame is 1,125, which is a 15:7 ratio to the current 525-line format, and 9:5 to the 625-line format.

The number of effective scanning lines is 1,035, and the effective scanning ratio of 0.92 is equivalent to that of current standard television.

The interlace ratio is 2:1. The aspect ratio initially was 5:3 in Japan, but was changed after the ATSC (Advanced Television Systems Committee) in the United States investigated the compatibility to aspect ratios of movie film. As Figure 1.17 shows, by overlapping the various movie film screen sizes (the areas enclosed by boxes a and b in Figure 1.17), the composite aspect ratio was found to be 16:9. Thus if Hi-Vision were to adopt the 16:9 aspect ratio, it would have a distinct relationship to movie film screen size, and conversion going either way would be simplified. For this reason, the ATSC decided to adopt a 16:9 aspect ratio. This ratio is also used in the proposed recommendation in Appendix II of the CCIR report mentioned earlier.

The field frequency is 60.00 Hz, which is different from the 59.94 Hz of the NTSC format. A field frequency of 60 Hz causes no problems in practical use with regard to large screen flicker described in Section 1.1.

The problem of field frequency was studied from the viewpoint of converting the current standard television field frequencies. In the CCIR studio standard investigation, there were problems with converting from the 1,125/60 HDTV format to the 625/50 format, and to alleviate these problems, motion compensating HDTV-PAL format conversion equipment was developed. As a result, in practice it was shown that

*See Appendix Section A.4 for recent developments.

TABLE 1.4. Basic characteristics of the video signal and synchronizing signal for the BTA/SMPTE 1125/60 studio standard.

No.	Item		Standard
1	Scan lines per frame		1125
2	Effective scan lines per frame		1035
3	Interlace ratio		2:1
4	Aspect ratio		16:9
5	Field frequency (Hz)		60.00
6	Line frequency (Hz)		33750
7	Y, GBR signal levels		
	Blanking level (reference level)	(mV)	0
	White peak level	(mV)	700
	Synchronizing signal level	(mV)	±300
	Difference in black and blanking levels	(mV)	0
8	P_B, P_R signal levels		
	Blanking level (reference level)	(mV)	0
	Peak level	(mV)	±350
	Synchronizing level	(mV)	±300
	Difference in black and blanking levels	(mV)	0
9	Nominal video signal bandwidth (MHz)		
	Y, G		30
	P_B, B		30
	P_R, R		30
10	Synchronizing signal waveform		Bipolar ternary synchronization
11	Horizontal blanking width (μs)		3.77
12	Vertical blanking width (lines)		45
13	Digital sampling		
	Effective samples per line	Y, G	1920
		P_B/B	960 / 1920
		P_R/R	960 / 1920
	Sampling frequency (MHz)	Y, G	74.25
		P_B/B	37.125 / 74.25
		P_R/R	37.125 / 74.25

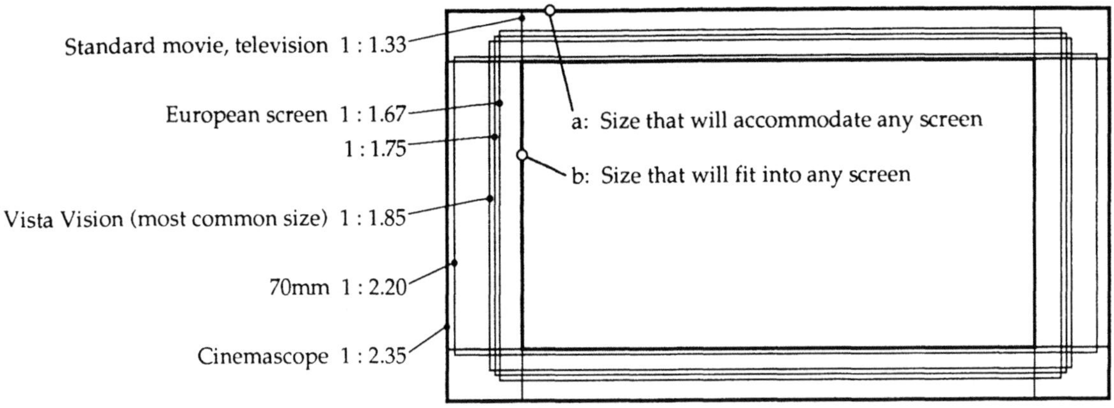

FIGURE 1.17. Comparison of screen sizes for motion picture film.

there were no problems in converting from 1125/60 HDTV to a 625/50 format.

Meanwhile, the BTA/SMPTE studio standard investigation was studying field frequency conversion with respect to format conversion from the 1125/60 HDTV format to the 525/59.94 NTSC format. If the conversion from 60 fields to 59.94 was done using the same principle as a frame synchronizer, one frame would be skipped every 33 seconds and motion continuity would be lost. Thus a motion adapting field conversion method was developed, and this will be discussed in Section 6.1. This method skips a frame when the image is fixed. HDTV programs that have been recorded on videotape can be played back in NTSC format to simplify format conversion.

Based on the results of investigations described above, both the BTA and SMPTE adopted a field frequency of 60 Hz.

(1) Video Signal Form and Bandwidth

The video signals can be either G, B, R or Y, P_B, P_R. There is no particular priority regarding which set to use, the reason being to provide flexibility in the future to accommodate technological progress and the expansion of application areas for HDTV. The Y, P_B, and P_R signals are effective in the areas of recording and transmission, while G, B, and R signals are effective in movie production, printing, and computer graphics.

The relationship between G, B, R and Y, P_B, P_R will be explained in Section 1.4. A peak-to-peak bipolar ternary synchronizing signal is added to all of the video signals.

The video signal bandwidth is 30 MHz for either G, B, R or Y, P_B, P_R. This bandwidth is prescribed as the bandwidth for signal generating sources such as cameras. The reason is that some equipment currently in use such as VTRs cannot accommodate a 30 MHz bandwidth. The 30 MHz bandwidth is also applicable to ternary parallel transmission methods for analog interfaces.

The video signals have no set-up. In other words, the difference between the signal level and the black level and blanking is zero.

(2) Digital Sampling Frequency and Horizontal Blanking Width

In Appendix II of CCIR Report 801, the HDTV digital standard is devised to have a simple relationship to CCIR Recommendation 601. For this reason, the BTA/SMPTE studio standard has a sampling frequency of 74.25 MHz, which is 5.5 times the 13.5 MHz sampling frequency in CCIR Recommendation 601. In this case, the total number of samples per line is 2,200. Further, the effective number of samples per line is specified at 1,920 in Appendix II of CCIR Report 801, and this value is also used in the BTA/SMPTE standard. The relationship between the effective number of samples of 1,920 and the 720 prescribed in CCIR Recommendation 601 for the 4:2:2 format is expressed in the following equation.

$$1{,}920 = 720 \times 2 \times (16/9) \div (4/3) \quad (1.2)$$

This equation indicates that the HDTV digital sampling frequency doubles the 4:2:2 digital sampling format and compensates for the wider aspect ratio.

The sampling frequency of 74.25 MHz is used for the G, B, R, and Y signals, while the frequency is 37.125 MHz or one-half for the P_B and P_R signals.

When the total number of samples per line and the effective number of samples are determined, the horizontal blanking width is also set. This value is 280 in terms of the number of samples, and 3.77μs in terms of time width. Thus,

$$3.77\mu s = (1 \div 33.75 \text{ kHz}) \times 280 \div 2{,}200 \quad (1.3)$$

The decision to use a horizontal blanking width of 3.77μs was based on the following factors:

1. The possibility of developing a horizontal deflecting transistor and low impedance deflection yoke to solve the problems involved in reducing the display's horizontal retrace time, such as an increased power deflection and reduced deflection linearity;
2. Amount of leeway for multiplexing command or error correction signals during the horizontal blanking interval in cameras and VTRs;
3. Compatibility with equipment manufactured to the provisional Tsukuba HDTV standard.

The 3.77μs value was adopted after determining that each of these could be accommodated.

Besides the 74.25 MHz, 3.77μs proposal for digital sampling frequency and horizontal blanking width, proposals that were considered but rejected included 77.625 MHz (13.5 MHz × 5.75) and 4.9μs, and 74.25 MHz with a digital sampling width of 3.77μs and an analog blanking width of 4.9μs. The first proposal had a sampling frequency higher than necessary and also was not compatible with the equipment made under the provisional Tsukuba standard, while the second one was rejected because the presence of two values would cause unnecessary confusion.

One proposal from SMPTE had 1,840 effective samples per line with a square lattice sampling structure. This made it convenient in applications such as computer graphics, printing, and measurement. However, the proposal was not adopted, partly because of the emphasis on the relationship with CCIR Recommendation 601 mentioned above, and because the 1,920-sample proposal, while not a square lattice, has a dislocation of 4.3% and is not a problem for ordinary video signal processing.

1.2.5 Digital Standards

As shown in Table 1.4, while the BTA/SMPTE HDTV studio standard has digital standards for sampling frequency, total number of samples per line and effective number of samples, other parameters such as the structure of the sample points and quantization bit count have not been set yet. Thus the BTA is currently investigating these digital parameters, which are shown in Table 1.5. They are looking at a quantization bit count of at least eight bits in anticipation of signal processing developments in the future.

In addition, standards are also necessary for front-end filters as well as back-end (interpolation) filters.

(2) Digital Interface

Digital interface standards are also currently being studied. As Table 1.6 shows, parallel and serial digital signal transmission methods are being investigated. The parallel transmission method performs a bit parallel transmission of digital data. In parallel transmission using digital interface standards for conventional television, the digital data for the luminance and color difference signals are transmitted at a clock frequency of 27 MHz in the sequence of Y, P_B, Y, P_R, Y.…However, the HDTV transmission methods under investigation transmit the digital data

TABLE 1.5. 1125/60 format digital parameter standards under study by the BTA.

No.	Item	Standard
1	Signal type: Y, P_B, P_R or G, B, R	All signals are obtained from gamma corrected signals.
2	Samples per line	Y 2200 / G 2200 P_B ·1100 / B 2200 P_R 1100 / R 2200
3	Sampling structure G, B, R or luminance signal Y Color difference signals P_B, P_R	Sample points are orthogonal, and repeated for each horizontal line, field, and frame. B, B, and R sample points match in relation to each other. They also overlap the sample points for the luminance signal. P_B, and P_R sample points match the odd numbered Y sample points in each line.
4	Sampling frequency	Y 74.25 MHz / G 74.25 MHz P_B 37.125 MHz / B 74.25 MHz P_R 37.125 MHz / R 74.25 MHz
5	Quantization	Linearly quantized at at least 8 bits.
6	Effective samples per line	Y 1920 / G 1920 P_B 960 / B 1920 P_R 960 / R 1920
7	Timing for analog video and digital video	The timing between the trailing edge of the digital active video and the analog horizontal reference phase is 88 clock cycles.
8	Correspondence between the video signal level and quantized level of upper 8-bits.	Scale is from 0 to 255. Y: 220 levels allotted; pedestal level is 16, peak white is 235. Signal can exceed level 235. P_B/P_R: 225 levels allotted; black level is 128. Signal can exceed level 240, 16. G, B, R: Apply correspondingly to Y.
9	Code allotment	Upper 8-bit quantized levels 0 and 255 are for synchronization only. Levels 1 to 254 are for video.

for the G, B, and R signals separately at a clock frequency of 74.25 MHz, or else transmit the digital data for the luminance and color difference signals separately and in parallel at a clock frequency of 74.25 MHz. The color difference signals are transmitted with time multiplexing in the sequence of P_B, P_R, P_B, P_R...

Although parallel transmission is done over a twisted pair line, at a clock frequency as high as 74.25 MHz even an ECL driven signal can be transmitted only 20 to 30 meters. Thus for distances longer than this, it is necessary to use a serial transmission method and optical fiber.

In a serial transmission method, the digital data of the Y, P_B, P_R or G, B, and R signals are transmitted in bit serial. The transmission bit rate in the first case is 1,188 Mb/s if the bit count is eight bits and 1,485 Mb/s for ten bits,

TABLE 1.6. 1125/60 digital interface standard under study by the BTA.

	$Y / P_B / P_R$	G / B / R
(1) Type-A bit parallel transmission (simultaneous multiple signal)	Can include ECL differential, EAV, SAV	
	Includes 2-signal 74.25 MHz clock (1) Y 74.25 Mwords/sec (2) P_B/P_R 74.25 Mwords/sec	Includes 3-signal 74.25 MHz clock (1) G 74.25 Mwords/sec (2) B 74.25 Mwords/sec (3) R 74.25 Mwords/sec
(2) Type-B bit parallel transmission (time multiplexed single signal)	Includes ECL differential, scrambled NRZ, EAV, SAV	
	$P_B/Y/P_R/Y$ 148.5 Mwords/sec	B/B/R 222.75 Mwords/sec
(3) Serial transmission (of format 2 above)	Optical transmission, scrambled NRZ, LSB	
	(1) 8 bits/word 1188 Mb/s (2) 10bits/word 1485 Mb/s	(1) 8 bits/word 1782 Mb/s (2) 10 bits/word 2227.5 Mb/s

and in the latter case, 1,782 Mb/s for eight bits and 2,227.5 Mb/s for ten bits.

Connectors are another area that need digital interface standards. As with the studio standard, the BTA and SMPTE will cooperate in investigating digital standards.

1.3 SYNCHRONIZING SIGNAL

1.3.1 Required Synchronizing Signal Characteristics

Synchronizing signals are indispensable for television scanning, and Hi-Vision and NTSC formats alike use a complex synchronizing signal that combines horizontal and vertical synchronizing signals. The vital role of the synchronizing signal is to transmit the reference phase for scanning. However, due to the influence of transmission path characteristics, there is normally some error in the reference phase reproduced from the transmitted synchronizing signal. The synchronizing signal thus should have a signal format that is not easily affected by characteristics of the transmission path or by the synchronization reproduction circuit.

The Hi-Vision studio transmits the G, B, and R or Y, P_B, P_R component signals in parallel over three channels using three coaxial cables. If transmission delay time differences or errors in the reproduced synchronizing phase occur among the channels, the result is misregistration or degradation of the resolution. Thus the delay time difference between each channel must be minimized during the parallel transmission of the component signals.

Since the limit of detection for delay time differences among the G, B, and R channels of the Hi-Vision signal is 3.5ns, the delay error between each of the channels during parallel transmission must be kept below this level. Coaxial cable, which is widely used in transmission paths, has a delay time of 5ns per meter and a deviation of $\pm 2\%$.* This means that a 100 meter coaxial cable can have a delay time difference of $\pm$ 10ns. A bundled cable consisting of several coaxial cables with this characteristic has a cable length error of 0.1% before any special measures are implemented. For this reason a Hi-Vision studio requires the interpolation of a signal with an accurate phase reference to manage the phases between each channel. The synchronizing signal should be usable as the reference signal for this phase management.

If several transmission stages are being connected, the inter-channel delay time difference across the entire system will be the sum of delay time differences that occur in each transmission

*JIS C3501

path. The required Hi-Vision studio inter-channel delay time difference of below 3.5ns is for a transmission through the entire system, and so the allowable inter-channel delay time difference in each path must be smaller than this value. The inter-channel delay time difference that occurs in each path is often caused by minor deviations in the cable length or frequency characteristics between the channels. These deviations can be considered to be randomly distributed occurrences in the paths. Thus between the inter-channel delay errors of each path and of the system as a whole there exists an rms additive rule.** For an ordinary Hi-Vision studio, it is sufficient to assume 10 stages serial transmission connections, in which case the allowable inter-channel delay error for each path is about 1.2ns.

For transmission of component signals with Time Compressed Integration (TCI) or subsampling transmission, the phase error of the clock that is reproduced by the decoder must be less than one-tenth of the clock cycle. The 74.25 MHz studio standard clock thus has an allowable reproduced clock phase error of 1.3ns.

In separating the synchronizing signal that has been multiplexed into the composite video signal (the Hi-Vision studio signal), the ability to detect the synchronizing signal using an amplitude separated signal or to detect the vertical synchronizing signal with integral separation are important characteristics for the purpose of simplifying the hardware configuration.

The discussion above on the characteristics required for a Hi-Vision synchronizing signal can be summarized as follows:

- Inter-channel delay time error: less than 3.5ms (for system as a whole); less than 1.2ns (for each component).
- Reproduced clock pulse phase error of less than 1.3ns.
- Unaffected by band limitation or nonlinear distortion in transmission system.
- Stiff against noise.
- Ease of separation and reproduction from composite video signal.
- Simple waveform and ease of signal generation.

**If the delay error is d_k for the individual transmission paths and d_t for the system as a whole, their relationship is $d_t = \sqrt{\Sigma d_k^2}$.

1.3.2 Synchronizing Signal Waveform

To simplify the separation and detection of the video signal from the synchronizing signal, the best method is to add a negative pulse to the video signal's blanking period. There are several possibilities: (1) a method resembling the NTSC synchronizing signal (binary waveform), (2) black burst, and (3) positive bipolar pulse (ternary waveform). These waveforms are shown in Figure 1.18. While any of these waveforms can be used for the Hi-Vision synchronizing signal, the waveform varies depending on the tolerable characteristic fluctuation of the transmission system and the inter-channel deviation.

(1) Binary Waveform

As Figure 1.18(a) shows, if the waveform is corrupted by variations in the signal amplitude or the band limitation, the reproduced reference phase will fluctuate. If the rise and fall time of the pulse is 50ns (which corresponds to band-limiting the rectangular pulse to about 10 MHz), then the variation in the synchronization discrimination level (slice level) and the inter-channel deviation must be below 2.4% (1.2ns/50ns) of the synchronizing pulse amplitude. If the synchronization discrimination level is set at one-half of the normalized pulse amplitude, the allowable variation in pulse amplitude and the inter-channel deviation are about 5%. If both the pulse amplitude and discrimination level are normalized, the fluctuations in the pulse rise and fall times that can be allowed and the deviations are 2.4ns. In practice, because these fluctuations and deviations occur simultaneously, the tolerances are more severe than described above. In particular, it is quite difficult to manage the pulse rise and fall times with an accuracy of within 2.4ns. Considering the extremely severe conditions in maintaining characteristics for the transmission paths and equipment, the use of a binary waveform for the Hi-Vision synchronizing signal is not practical.

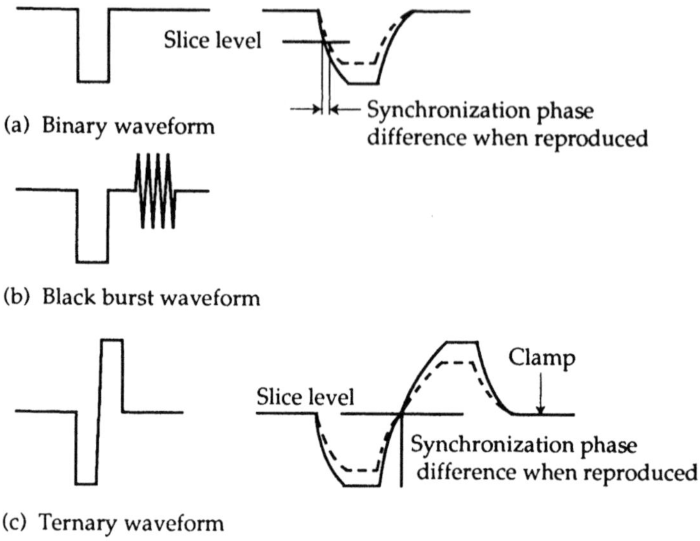

FIGURE 1.18. Synchronization signal waveforms.

(2) Black Burst

Equipment handling video signals ordinarily reproduce the pedestal level with ample accuracy. When a burst signal is applied to the equipment with the standard phase set at a zero-cross point (the point where the burst wave crosses over the pedestal level), the reproduced phase standard will be unaffected by fluctuations and deviations in the signal amplitude and discrimination level (pedestal level). However, because the burst signal spectrum is concentrated in a relatively high frequency region (Figure 1.19), it is easily influenced by the band limitation of the transmission. Many low-pass filters used in Hi-Vision have a group delay ripple of several nanoseconds at frequencies above several MHz. For this reason, even if the inter-channel burst phases are strictly matched, the phases of the video signal may not necessarily match. Although this problem can be reduced by choosing a lower burst frequency, doing so would require a long synchronizing signal interval and make TCI transmission disadvantageous.

(3) Ternary Waveform

Because the ternary waveform in Figure 1.18(c) uses the pedestal level as the discrimination level, it is similar to the black burst waveform in that the reproduced reference phase does not vary with fluctuations in the discrimination level or pulse amplitude. Furthermore, whenever the discrimination level accurately matches the pedestal level, the waveform is unaffected by variations in the pulse rise time. Since this waveform is composed of relatively low frequency components (Figure 1.19), it is resistant to the effects of the high band characteristics of the transmission system. Thus the ternary waveform is strong against fluctuations in characteristics (amplitude fluctuation and high band phase fluctuation) in the transmission system (video distribution amplifier–coaxial cable–equalization amplifier).

One factor that does affect the reproduced reference phase with this waveform is the vari-

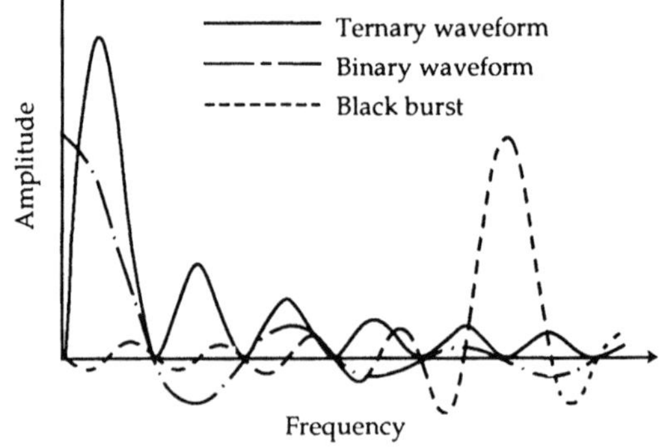

FIGURE 1.19. Spectra of synchronization signals.

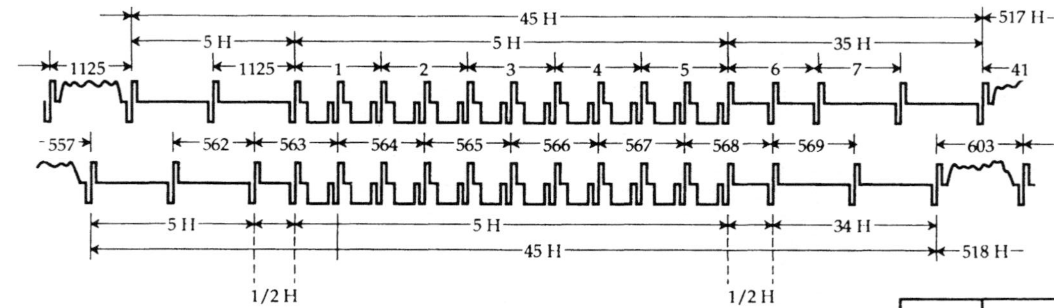

Vertical reference phase

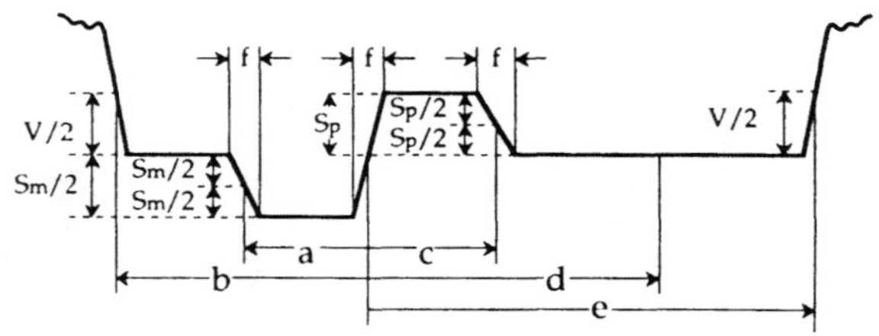

Horizontal reference phase

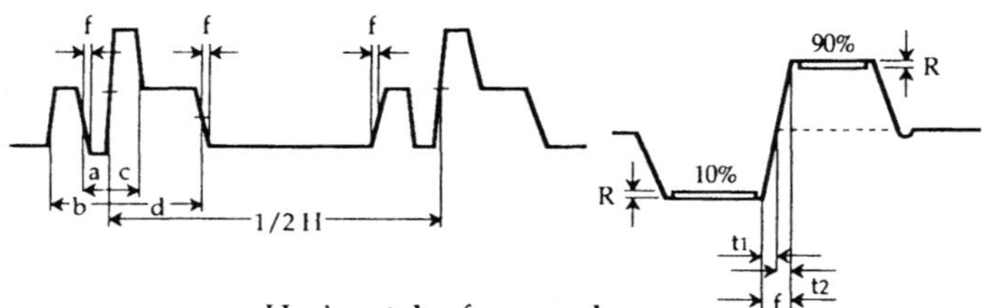

Horizontal reference phase

Item	Name	Reference value	Tolerable deviation
a	Negative pulse starting point	0.593 µs	±0.040 µs
b	Video signal end point	1.185 µs	±0.040 µs
c	Positive pulse end point	0.593 µs	±0.040 µs
d	Clamp end point	1.778 µs	±0.040 µs
e	Video signal starting point	2.586 µs	±0.040 µs
f	Pulse rise time	0.054 µs	±0.020 µs
$t_2 - t_1$	Symmetry of pulse rise	--	±0.002 µs
S_m	Negative pulse amplitude	300 mV	±6 mV
S_p	Positive pulse amplitude	300 mV	±6 mV
$S_m - S_p$	Pulse pp value deviation	--	±6 mV
V	Video signal amplitude	700 mV	
R	Synchronization overshoot	--	< 30 mV
	Line frequency	33750.00 Hz	±10 ppm

FIGURE 1.20. Hi-Vision studio synchronization signal waveform.

TABLE 1.7. Relationship between pulse timing and clock frequency.

	a	b	c	d	e	f
Reference value	0.593 μs	1.185 μs	0.593 μs	1.778 μs	2.586 μs	0.054 μs
74.25 MHz	44	88	44	132	192	4
74.25 MHz/4	11	22	11	33	48	1
13.5 MHz	8	16	8	24	approx. 35	–

ation in the discrimination level, for which the allowable value is about 14mV (0.6V × 1.2ns/50ns). This corresponds to 2% of the video signal, and in practice fluctuations and deviations in the discrimination level are not a problem because the pedestal fluctuation is ordinarily below this level in the equipment handling the video signal. In the equipment for handling the synchronizing signal alone, the zero potential (earth level) can be used as the discrimination level because the APL (Average Picture Level) of the ternary waveform is zero.

1.3.3 Hi-Vision Studio Synchronizing Signal

As the discussion above indicates, the best suited waveform for the Hi-Vision synchronizing signal is the ternary waveform. In practice, the Hi-Vision sychronizing signal is a combination of a ternary waveform horizontal synchronizing signal, a binary waveform vertical synchronizing signal, and a ternary waveform equalization pulse. Details of these waveforms are shown in Figure 1.20.[13] As made clear by the study of waveforms, the timing in which the ternary pulse intersects the pedestal level is used as the horizontal synchronizing reference phase. Furthermore, the first ternary timing after the vertical synchronizing pulse (the timing that intersects the pedestal level) is used as the vertical reference phase. The phase relationships of the various pulses is shown in Table 1.7. These values were selected so as to be obtained by counting the Hi-Vision studio clock pulse frequency (74.25 MHz) or one-fourth this fre-

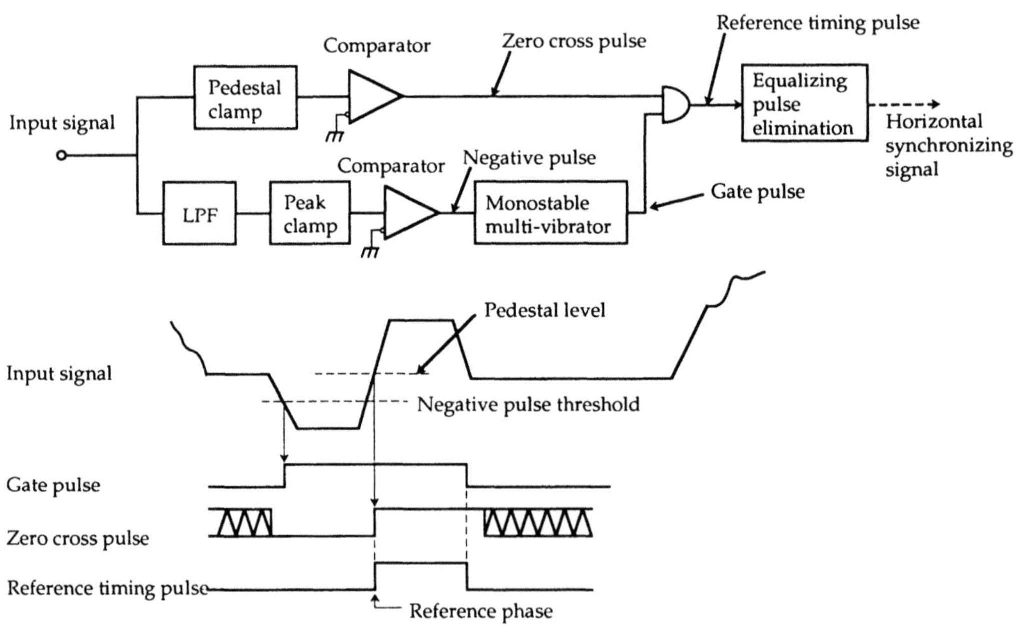

FIGURE 1.21. The synchronization separation circuit and its operation.

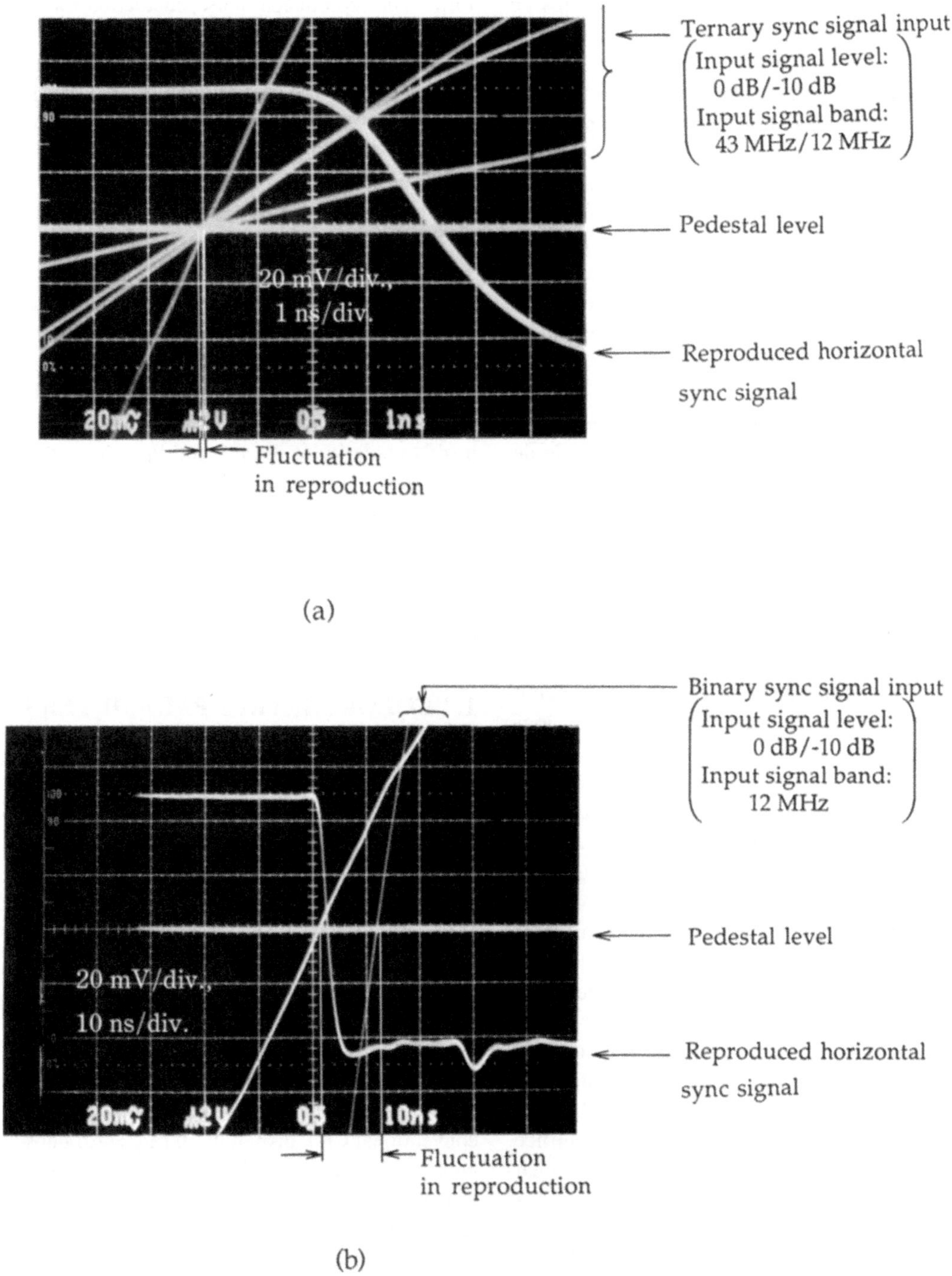

FIGURE 1.22. Fluctuations in the reproduced horizontal synchronizing signal due to fluctuations in the input signal.

quency (18.5625 MHz). The relationships between the pulse timings and the clock frequency are shown in Table 1.7.

Figure 1.21 is an example of the synchronizing signal separation circuit. The input signal is divided in two. One part of the signal undergoes peak clamping of the synchronization signal (negative peak clamp), after which the negative pulse is detected in the compiler. The other part of the signal undergoes a pedestal clamp, after which its amplification is compared to that of the pedestal level (noted as earth level in the figure). Since output of the compiler (zero cross pulse) has a logic value of 0 for the negative pulse interval and 1 for the positive interval, the timing of the changes between 1 and 0 are detected as the horizontal reference phase. However, because the zero cross pulse sometimes switches from 0 to 1 in the video signal interval or the front porch and back porch intervals, it is eliminated by using the gate pulse made from the negative pulse. Following this, the equalization pulse of the vertical synchronizing signal interval is eliminated and the vertical synchronizing signal is separated using ordinary logic circuits.

In applications requiring a precise synchronizing reproduction of the phase, the circuit configuration described above is used. However, if the reproduced phase need not be all that precise, it is possible to ignore the positive pulse and use only the negative pulse. In this case, the synchronizing separation circuit can have the same configuration as that used for the NTSC format. For the vertical synchronizing separation as well, a ternary pulse is used to detect the vertical reference phase for applications requiring precision, while for simpler applications separation by an integration circuit will suffice.

Figure 1.22(a) shows the fluctuation characteristics of the reproduced reference phase for an actual Hi-Vision signal. The reproduced horizontal synchronizing signal phase fluctuation was less than 0.2ns for an input signal band of 43 MHz (pulse rise time of 15ns) and 12 MHz (50ns), a normal signal amplitude (0.6 V_{pp}) and one-third amplitude (0.2 V_{pp}). Figure 1.22(b) shows the situation when the positive pulse is ignored and the signal input has a binary waveform. The reproduced phase fluctuation for the same input signal fluctuation range was 17ns. These results confirm that the reference phase for the Hi-Vision synchronizing signal can be reproduced with adequate precision despite characteristic fluctuations in the transmission path.[15]

The ternary synchronizing signal is added to each of the component signals (G, B, and R, or Y, P_B, and P_R) in the Hi-Vision studio. For G, B, and R transmission, because the video signal is positive, the negative pulse can be detected with amplitude separation and synchronizing separation is possible for any of the channels. However, for Y, P_B, and P_R transmission, the positive polarity of the P_B, and P_R color difference signals prevents a direct synchronizing separation. Thus it is necessary to make some modification, such as using the synchronizing signal separated from the Y signal (or the gate pulse) to detect the synchronizing signal in the color difference signals.

1.4 COLORIMETRIC PARAMETERS

The colorimetric parameters for the BTA and SMPTE studio standard are shown in Table 1.8.

1.4.1 Chromaticity Points for the Three Primary Colors

We adopted the SMPTE C primary colors shown in Figure 1.23 as the three primary chromaticity points. To reproduce the color range as much as possible, one proposal called for the use of the G primary color from the NTSC format and R and B from EBU. However, as the figure shows, since the three chromaticity points for a conventional CRT display with rare earth phosphors differ from the NTSC chromaticity points—in particular the G color—errors occur in color reproduction. Although this color reproduction error can be corrected with the linear matrix located in the display, to do so, the signal, which has undergone gamma correction in the reverse gamma circuit of the display, must be reconverted into a linear signal with nonlinear processing. The problem with this processing, however, is that analog circuits have not proven their

TABLE 1.8. Colorimetric parameters for the BTA and SMPTE 1125/60 format studio standard.

No.	Item	Standard
1	Chromaticity of three primary colors	1. CIE chromaticity index is as follows: x y G 0.310 0.595 B 0.155 0.070 R 0.630 0.340
2	Reference white	1. Set at D_{65} CIE chromaticity index is as follows: $x = 0.3127$ $y = 0.3291$
3	Transmission of primary colors	1. There are two sets: luminance signal and color difference signals, and G, B, R 2. Luminance signal and color difference signals $Y = 0.701G + 0.087B + 0.212R$ $B\text{-}Y = -0.701G + 0.913B - 0.212R$ $R\text{-}Y = -0.701G - 0.087B + 0.788R$ G, B, and R signals will have undergone gamma correction. 3. Y, P_B, P_R $P_B = (B\text{-}Y) / 1.826$ $P_R = (R\text{-}Y) / 1.576$ However, only if Y, P_B, and P_R have same p-p value.
4	Gamma correction	1. Gamma correction is done on transmitting side. 2. The conversion equation between the luminance and video signals will follow this theoretical guideline: $L = \{(V + 0.1115) / 1.1115\}^{(1/0.45)}$ $V \geq 0.0913$ $L = V / 4.0$ $V < 0.0913$

reliability, while digital circuits at present are high in cost. For this reason, it was decided not to use a linear matrix in the display, and instead to prevent color reproduction error by matching the chromaticity points for the three primary colors to those of the phosphors on a conventional CRT.

If linear matrix processing in the receiver becomes possible in the future, the SMPTE will reexamine the three primary colors and look into expanding the color reproduction range.

Table 1.9 shows the luminance signal equations for various primary color and reference white values.

1.4.2 Reference White

Although Japan currently uses C illuminant as the reference white for conventional television, we adopted CIE illuminant D_{65} to conform with the United States and Europe.

1.4.3 Transmission of the Primary Colors

The primary colors are transmitted by using either the three primary color signals of G, B, and R or the luminance and color difference signals. The luminance and color difference signals are obtained from gamma corrected G, B, and R

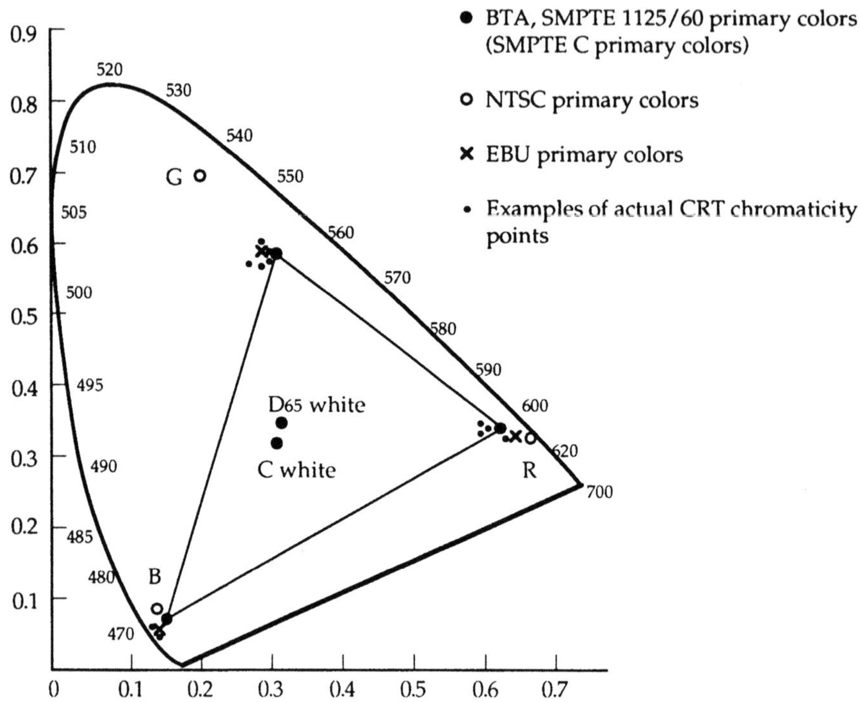

FIGURE 1.23. Chromaticity points of the three primary colors in the 1125/60 format.

signals using the equations in Table 1.9 The P_B and P_R signals are made by multiplying the B-Y and R-Y color difference signals by 1/1.826 and 1/1.576 respectively.

Figure 1.24 shows the waveforms of color bar signals for Y, P_B, and P_R.

1.4.4 Gamma Correction

We investigated gamma correction performed at the image transmitting end as done in conventional television, as well as on the receiving end.

While gamma correction at the transmission

TABLE 1.9. Luminance signal equations for various primary colors and reference white values.

	BTA SMPTE 1125/60 reference primaries		NTSC reference primaries		NTSC reference primaries		EBU reference primaries	
	x	*y*	*x*	*y*	*x*	*y*	*x*	*y*
Colorimetry R	0.630	0.340	0.67	0.33	0.67	0.33	0.64	0.33
of reference G	0.310	0.595	0.21	0.71	0.21	0.71	0.29	0.60
primaries B	0.155	0.070	0.14	0.08	0.14	0.08	0.15	0.06
Reference	D_{65}		C illuminant		D_{65}		D_{65}	
white	0.3127	0.3291	0.310	0.316	0.313	0.329	0.313	0.329
Luminance signal equation Y	Y = 0.212R + 0.701G + 0.087B		Y = 0.30R + 0.59G + 0.11B		Y = 0.29R + 0.61G + 0.10B		Y = 0.22R + 0.71G + 0.07B	

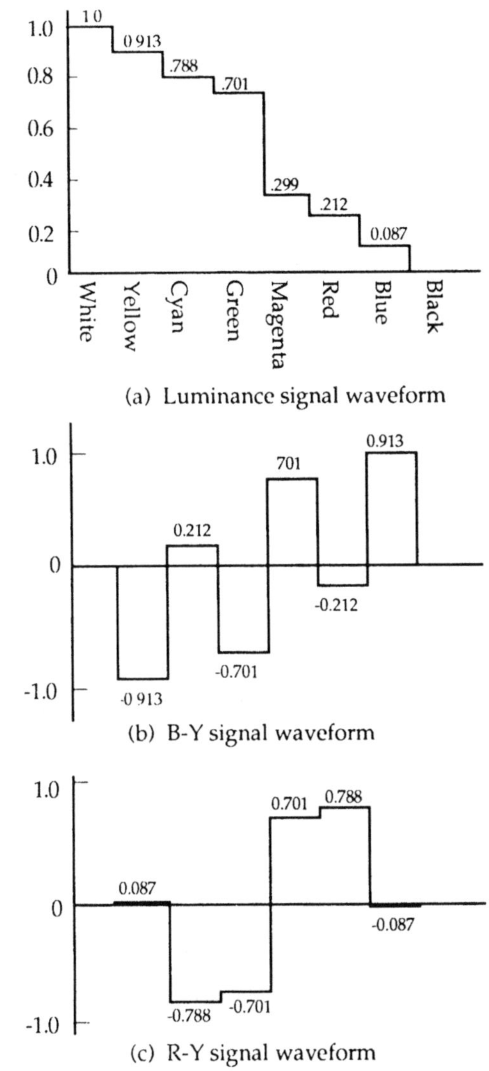

FIGURE 1.24. Color bar signal of the 1125/60 format.

end has the advantage that correction is not required at the display, it has the disadvantage that the constant luminance principle is not fully established, thereby causing the resolution of the luminance signal's high band to deteriorate in highly saturated images. On the other hand, when gamma correction is performed at the receiving end, this disadvantage is eliminated because the fixed luminance principle is established, and the color signal bandwidth can be reduced further without causing image degradation.

However, even though linear gamma dis-

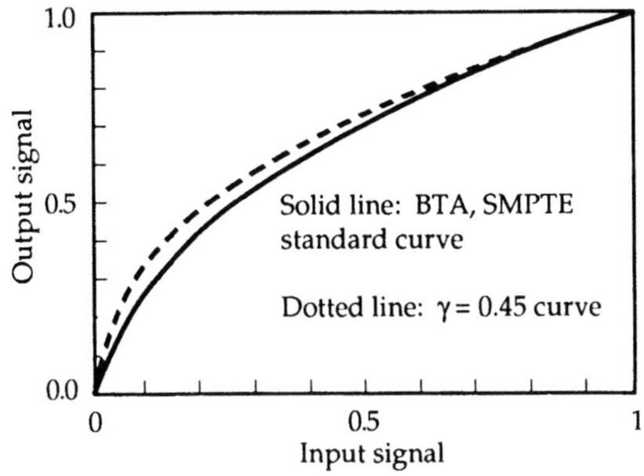

FIGURE 1.25. Gamma correction curve.

plays that differ from existing CRTs may be developed in the future, CRTs will continue to be widely used. Thus gamma correction will need to be done on the display end, which raises problems as explained in Section 1.4.1.

After deliberating on the matter, both the BTA and SMPTE have decided that gamma correction would continue to be done at the transmitting end.

The gamma correction curve approximates that of the $\gamma = 0.45$ curve used in standard television, and the specification of this characteristic has not been completely unified. However, because the gamma characteristic must be adjusted to film when recording HDTV signals onto film, it is necessary to accurately know the gamma characteristic of the original video signal. Furthermore, recent advances in post-production signal processing are enabling the video signal to be linearized and filtered. This requires reverse gamma processing, which in turn requires that the original gamma characteristic be accurately known. Thus in the SMPTE HDTV studio standard, the gamma characteristic has been set with the equations shown in Table 1.8. The BTA also uses this gamma characteristic as a guideline. The gamma characteristic is shown in Figure 1.25.

1.5 THE HI-VISION STEREO AUDIO SYSTEM

By using the visual capabilities of a large display with high resolution, Hi-Vision is able to far exceed conventional television in delivering a sense of telepresence. A suitable stereo audio system must consider the interdependence of visual and auditory senses and strike a balance between the maximum impact that can be delivered and the technological and economic considerations. The Hi-Vision stereo system meets these considerations by incorporating the following features:

1. Four independent channels produce a sense of telepresence that far exceeds that of conventional television.
2. So that as many people as possible experience the stereo effect whether in a living room or video theater, there is a central speaker in front with an independent channel. Thus three of the four channels are placed in front.
3. An independent audio channel in the back of the room surrounds the viewer with sound and reproduces sounds coming from behind the viewer. The remaining channel (using several speakers) is used for this purpose.

Thus the stereo audio system reproduces four audio channels, with three in front and one in back. Called the 3–1 mode, this arrangement is shown for a household Hi-Vision system in Figure 1.26. Following is a discussion of the basis for choosing this mode.

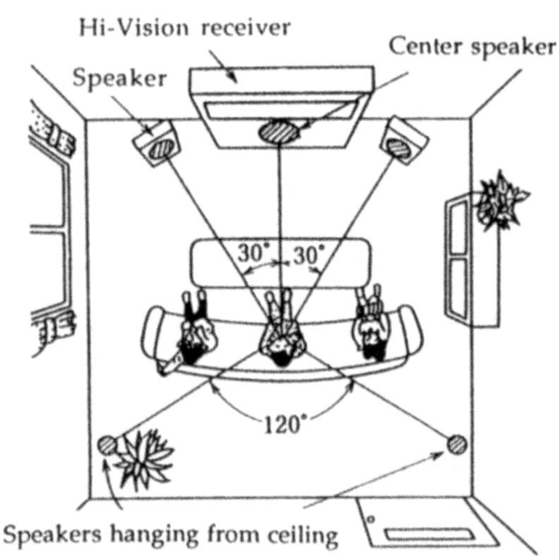

FIGURE 1.26. Basic speaker arrangement for the 3-1 mode 4-channel stereo system.

1.5.1 The Number of Independent Channels and Speaker Arrangement

(1) Purpose of Experiment

Although more independent audio channels in general increase the sense of realism, considerations involving program production as well as technological and economic issues make it impractical to transmit more than five audio channels. Figure 1.27 shows seven speaker arrangements, including one two-channel, one three-channel, and five four-channel speaker arrangements that were tested in a psychological evaluation experiment to determine the optimal number of independent audio channels and speaker arrangement. It should be noted that rear speakers L_B and R_B in arrangement 3–1(B) reproduced the same auditory signal in synchronous phase.

(2) Experimental Method

The program source used for evaluation was a variety show recorded on a Hi-Vision VTR and the audio was recorded on 24 channels. To ensure that the material would be mixed to optimal effect for each speaker arrangement, seven experimental programs were produced. The video image consisted mostly of a head-on shot of the singer on center stage. The sound mixing basically put the singer at front and center, with the accompaniment, auditorium noises and echoes on either side and in the back.

The viewing conditions consisted of an audition studio with a relatively short reverberation time, and a 5:3 aspect ratio, and a 67-inch projection display placed 2.8 meters in front of the subject. To be able to evaluate all seven speaker arrangements from Figure 1.27, nine speakers were placed around the subject as shown in Figure 1.28. The subject was placed directly in front of the display in the first experiment as shown in Figure 1.28(a), and 70cm to one side of the center in the second experiment as in Figure 1.28(b).

The first experiment involved twenty men and women 20 to 40 years in age, including mixers and sound engineers, and the second experiment involved sixteen people.

In the first experiment, the subject was ex-

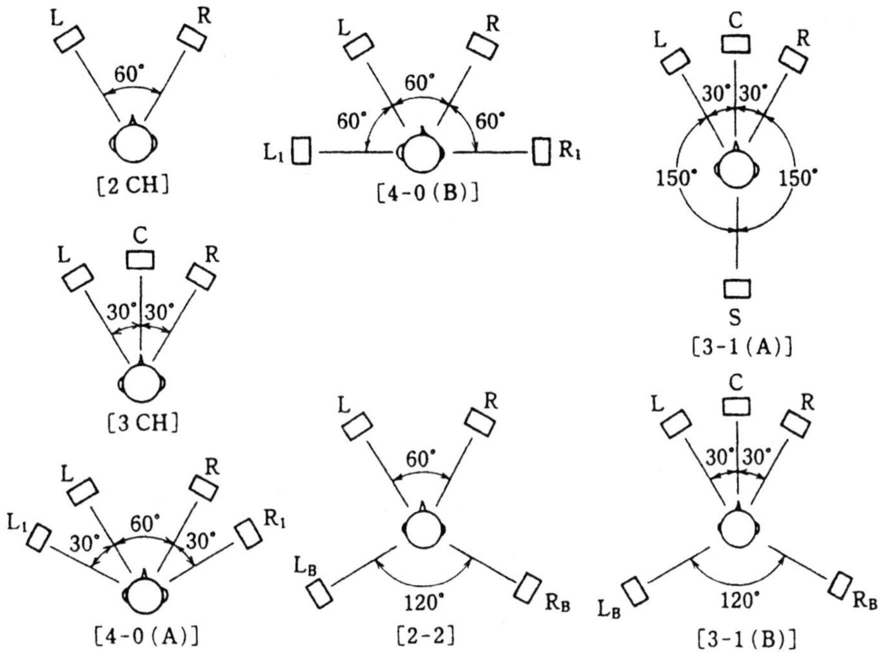

FIGURE 1.27. The seven types of speaker arrangements used in psychological experiments.

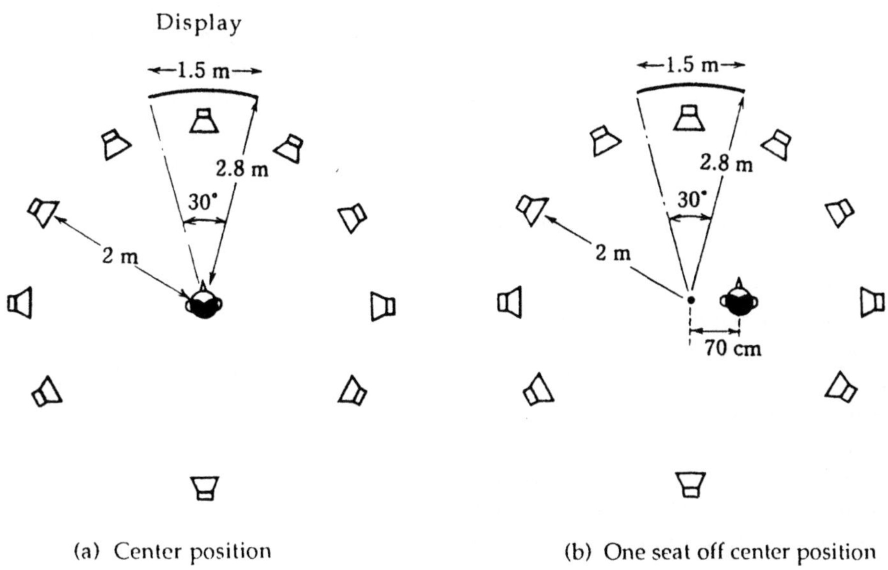

FIGURE 1.28. Positioning of subjects in the psychological experiments.

TABLE 1.10. Evaluation areas for psychological experiments.

Spatial impression of sound	1. The applause and audience noise is all around you.
	2. The sound stage (area over which sound of accompaniment is spread) is wide.
	3. You are surrounded by the sound reverberation.
	4. You feel like you are in a large auditorium.
	5. The vocal sound image is sharp.
	6. The vocal sound image is rising.
Overall stereo effect	7. It does not sound unnatural.
	8. It sounds real.
	9. You like the experience.
Fusion of audio and video	10. When the singer appears to be at a distance, there is no discrepancy in the orientation of the image and the sound field.
	11. In a close-up of the singer, the sense of distance from the sound matches that of the image.
	12. The sound of the vocal appears to be coming from above the position of the mouth.
	13. Overall, the video and audio are in harmony.

posed to a one minute forty second program of Hi-Vision video and audio for each speaker arrangement, after which we noted their psychological impressions. The subject evaluated the programs, which were presented in a different sequence for each subject, in the thirteen areas in Table 1.10 on a scale of seven, with +3 being strongly affirmative and −3 being strongly negative.

In the second experiment, the subject used the same scale to evaluate the three items 7, 8 and 9 in Table 1.10, as well as a fourth item combining items 10, 11, and 12 in the table as to how well the image was oriented to the sound.

(3) Experimental and Analytical Results

To obtain the interrelationships of the overall psychological impacts with regard to the speaker arrangements from the data obtained in the first experiment, as well as to gain a better understanding of the evaluations in Table 1.10, we analyzed the results with both the multidimensional scaling method (MDS) and multiple regression analysis. To do this, we first calculated the total scores for each of the items in Table 1.10. These results were used to calculate the Euclidean distance between the various arrangements, and a distance matrix was made showing the magnitude of the differences in psychological effects among the various arrangements, and this distance matrix was analyzed with classical nonmetric MDS. The results are shown in Figure 1.29. The dots show how the seven speaker arrangements are distributed over the two-dimensional psychological space of the subjects. The horizontal axis corresponds closely to the number of audio channels. The top half of the vertical axis corresponds to speaker arrangements having a central front speaker and

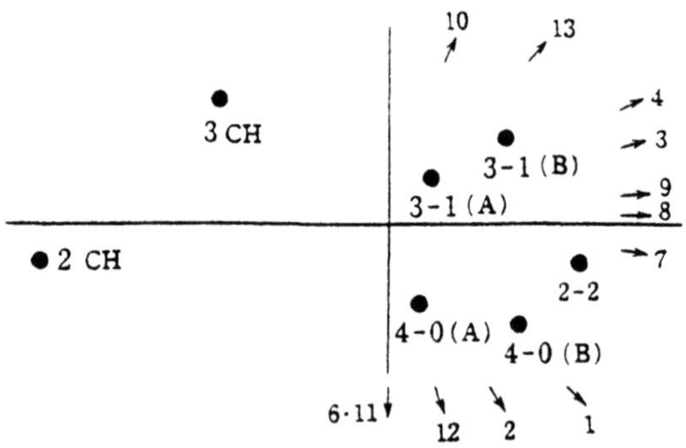

FIGURE 1.29. Interrelationship of psychological impressions for the various speaker arrangements.

the bottom half to arrangements without a central front speaker. The arrows indicate the direction closest to the affirmative direction in the two-dimensional psychological space for the items in Table 1.10. However, item 5 has been excluded due to ambiguity.

If we look at how the psychological impressions correspond to the axes, the horizontal axis corresponds to the sense of realism (item 8), and as the number of channels increases so too does the sense of realism. It has been known that in general the phantom sound image at front and center (made by the left and right speakers) tends to rise above ear level. On the vertical axis, there is no increase in the positive direction because the central sound image is the real sound image, while there is an increase in the negative direction due to the phantom sound image (items 6 and 12).

In the results of the second experiment, the evaluation of realism was almost identical to the results of the first experiment. The results for naturalness and attractiveness were one rank lower for arrangements 4–0(A), 4–0(B), and 2–2, but the other results were quite similar. The results on whether the video and sound images match are shown in Figure 1.30. These results show that the arrangements with a speaker in front and center did considerably better than the other arrangements.

The experimental and analytical results discussed above indicate that the optimal number of channels is four, and that the preferred speaker arrangements for a viewer positioned directly in front of the display are either 2–2 or 3–1(B), while if the viewer is off to one side then either 3–1(A) or 3–1(B) is preferred.

Since Hi-Vision is intended to be viewed by several people at any given time, we determined that the best speaker arrangement of the seven is 3–1(B).

1.5.2 Comparison of 3–1 Mode 4-Channel Stereo and 2-Channel Stereo

(1) Purpose of Experiment

The best position from which to view a Hi-Vision screen or listen to stereo is along the line centered in front of the display or between the left and right speakers. Particularly with 2-channel stereo, listening from any place off this central axis moves the virtual sound image away from the center and toward the speakers. However, in a household or video theater, only one or at most a few people can sit in the optimal viewing position, which inevitably situates others away from the central axis. Thus we investigated how the overall viewing quality would differ for a group viewing Hi-Vision if the stereo audio system were a 2-channel system or a 3–1 mode 4-channel system.

(2) Experimental Conditions

The audition studio was arranged as shown in Figure 1.31, with a 50-inch rear projection Hi-Vision display, and three speakers in front and

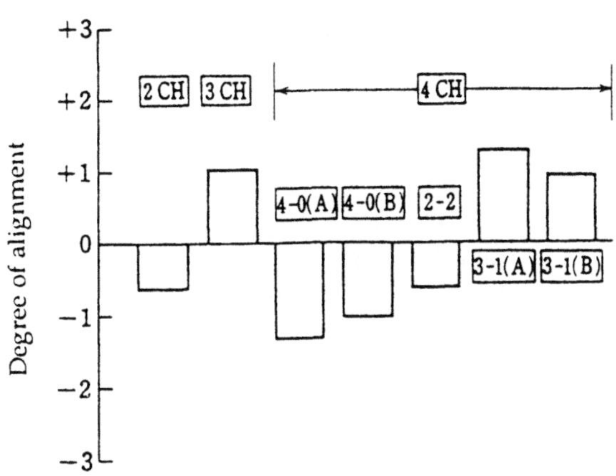

FIGURE 1.30. Alignment between audio and video images when the subject is off center.

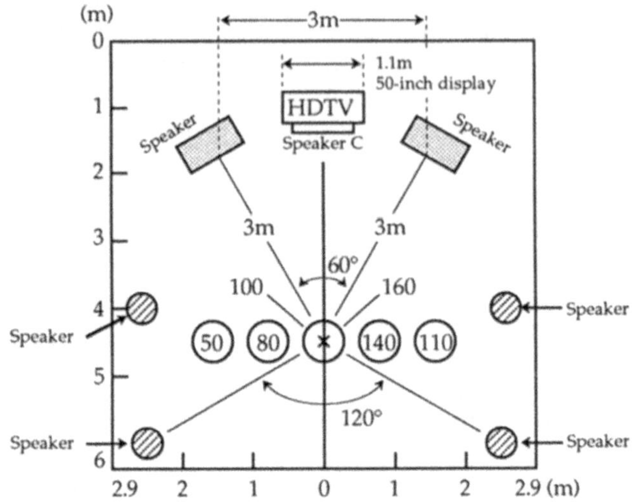

FIGURE 1.31. Comparison of overall Hi-Vision evaluations for 2-channel (circled number on left) and 4-channel arrangements (circled numbers on right); speakers are shaded.

four in the rear. The central front speaker was located directly under the display, while the four rear speakers were placed 2.5 meters above the floor. The subjects were seated at the locations marked by the circles, 3 meters in front of the screen and 75cm and 150cm off the central axis. For 2-channel stereo the front left and right speakers were used, while the 4-channel stereo used all seven speakers, with a delay of 0 to 14ms for the rear speakers.

The program sources used for evaluation were a pipe organ recital at NHK Hall and a variety show, both recorded with a Hi-Vision VTR and a 24-channel PCM recorder. The evaluation programs were three minutes long and produced to achieve the best possible 2-channel and 4-channel effects from the recordings. Both programs used the rear speakers to reproduce auditorium noises and echoes. Including sound technicians, there were 25 subjects ranging from 20 to 40 years in age.

(3) Evaluation Test Method

The subjects evaluated both programs using a magnitude inference method as follows. First sitting in the middle seat, each subject watched the Hi-Vision images while listening to 2-channel stereo, and evaluated the overall quality of the video and audio (including the audio visual synergistic effect). This subjective experience was given a reference value of 100. The subject then sat 75cm and then 150cm off center and compared the quality to their reference value. The same reference value was also used to evaluate the 4-channel stereo from the three seating positions.

(4) Experimental Results

The average results obtained from both programs are summarized as follows:

1. The ratio of the subjective score for the center seat to that of the seat 150cm off center was 2.0 for 2-channel stereo and 1.45 for 4-channel stereo. Thus the difference in quality due to the seating position was found to be smaller for 4-channel stereo than for 2-channel stereo.
2. The ratios of 4-channel to 2-channel scores for the three seating positions were:

Center	1.6
75cm off center	1.75
150cm off center	2.2

TABLE 1.11. Basic transmission standard for in-studio use.

No.	Item	Standard
1	Audio signal band	20 kHz
2	Sampling frequency	48 kHz
3	Quantization bit count	16 (uniform)
4	Sampling time	Same as for stereo
5	No. of channels	1 to 4

In other words, the disparity between 4-channel and 2-channel stereo increases as the distance from the central axis increases.

3. A 4-channel system heard at 150cm off center has a higher quality than a 2-channel system at 75cm off center.

These results indicate that the 3–1 mode 4-channel stereo system is the preferred system when several people are viewing Hi-Vision at the same time.

1.5.3 Downward Compatibility with Conventional 2-Channel Stereo

Since some households receiving broadcasts of the 3–1 mode 4-channel stereo signal will have a 2-channel stereo system, we conducted the following test to determine how much the quality would differ from a program originally produced in 2-channel stereo. In a psychological experiment, we compared the stereoscopic quality of a program originally recorded in 2-channel stereo to a program recorded in 4-channel stereo and subsequently converted to 2-channel stereo (L',R') using the following conversion matrix:

$$L = +6\ 0.7\,C + 0.7S \qquad (1.4)$$
$$R' = R + 0.7\,C + 0.7\,S$$

where L: Left channel
R: Right channel
C: Center channel
S: Rear channel

The result was that on a scale of seven, the converted program had a rank of 0 or −0.5 in relation to the original program, which is within the tolerance range. However, it should be noted that programs that have sounds emanating from behind the viewer can be produced with 4-channel stereo but not with 2-channel stereo.

1.5.4 Compatibility with Motion Picture Sound

Since movies will be broadcast in Hi-Vision, and Hi-Vision programs will be viewed in movie theaters, compatibility between motion picture and Hi-Vision stereo formats must be assured. Presently the main method for recording motion picture sound is the Dolby method, which op-

TABLE 1.12. Channel location.

Audio mode	Location
4-channel stereo	Left front, right front, front and center, left and right rear*
3-channel stereo	Left front, right front, front and center
2-channel stereo	Left front, right front
Monophonic	Front and center

* The rear two speakers carry the same channel.

tically records on two channels. However, it is premised on being replayed in a 3–1 stereo mode with multiple speakers, and incorporates some modifications in the recording method. The output of the two independent channels undergo additive, subtractive, and other operations, and the decoder output has the appearance of a 3–1 stereo mode. Thus along with the true 3–1 stereo mode of Cinemascope, motion picture sound systems can be said to be compatible with the Hi-Vision stereo format.

1.5.5 Audio Signal Standard

Standardization of the Hi-Vision audio signal is currently underway, with the focus of the effort being the 3–1 stereo mode. Proposals for the basic transmission standard for in-studio use and channel locations for playback are presented in Tables 1.11 and 1.12.

REFERENCES

1. Hiwatashi, "The Future of the Information Industry and Electronic Imaging Systems," *Journal of the Institute of Television Engineers of Japan,* (ITEJ) Vol. 26, No. 11, pp. 885–891 (1972).
2. Otani, Mitsuhashi, "Image Quality Evaluation and Psychological Factors," *NHK Technical Report,* Vol. 18, No. 12, pp. 462–469 (1975).
3. Hatada, Sakata, Kusaka "Directional Inducement Effect and Screen Size—Basic Experiments with Telepresence and Large Screens," *ITEJ Journal,* Vol. 33, No. 5, pp. 407–413 (1979).
4. Otani, Fujio, Hamazaki, "Screen Mode and Subjective Image Quality for High Definition Television," *NHK Technical Journal,* Vol. 28, No. 4, pp.1–19 (1976).
5. Mitsuhashi, Hatada, "Image Quality of High Definition Television," *ITEJ Journal,* Vol. 36, No. 10, pp. 873–881 (1982).
6. Nishizawa, "Visual Effect of Interlace," *ITEJ Visual Information Research Report,* Vol. 1961-1 (1969).
7. Fukuda, "Differences in Sensitivity Between Central and Peripheral Vision as Indicated by CFF," *ITEJ Journal,* Vol. 22, No. 3, pp. 210–216 (1968).
8. Miyahara, "The Relationship Between Visual Characteristics and Image Quality in Moving Images and its Application to Television Signal Bandwidth Compression," *NHK Technical Journal,* Vol. 22, No. 3, pp. 141–171 (1975).
9. Watanabe, Mori, "Spatial Frequency Characteristics of Vision," *NHK Technical Report,* Vol. 8, No. 11, pp. 661–668 (1965).
10. Sato, Kubo, "Image Quality of High Resolution Wide Television," *NHK Technical Report,* Vol. 22, No. 4, pp. 124–131 (1979).
11. Sakata, Isono, "Spatial Frequency Characteristics of Chromaticity in Vision," *Journal of the Institute of Television Engineers of Japan,* Vol. 31, No. 1, pp. 29–35 (1977).
12. Ishida, "Required Bandwidth for the Chrominance Signal in 1,125-Line High Definition Television," *ITEJ Technical Report,* VVI 17-4, pp. 1–7 (1976).
13. Broadcast Technology Association, "BTA S-001 1125/60 High Definition Television Format Studio Standard," (August 1987).
14. Proposed American National Standard, 1125/60 High-Definition Production System, SMPTE 240M, *SMPTE Journal,* Vol. 96, No. 11, pp. 1150–1152 (November 1987).
15. Kumada, "The Hi-Vision Synchronizing Signal Standard," *ITEJ Technical Report,* Vol. 11, No. 9, pp. 13–18, PPOE '87–17 (June 1987).
16. Ohgushi, Komiyama, Kurozumi, Morita, Ujihara, Tsujimoto, "Subjective Evaluation of Multi-Channel Stereophony for HDTV," *IEEE Trans. Broadcasting,* BC-33, No. 4, pp. 197–202 (December 1987).
17. Ohgushi, "The Hi-Vision 3–1 Mode 4-Channel Stereo System," *JAS Journal,* Vol. 27, No. 11, pp. 5–12 (1987).

2

Imaging Technology

Masahide Abe, Junji Kumada, Keiichi Shidara, Hiroshi Hirabayashi

2.1 IMAGING DEVICES

2.1.1 Saticon

The foremost requirement for a Hi-Vision camera tube is high resolution. To meet this requirement, 1.5-inch vidicons, 2-inch return beam vidicons, and 2-inch return beam saticons (RBS tubes) were employed in the early research on Hi-Vision.[1] While those camera tubes contributed significantly to the success of system experiments as the signal sources for the image evaluations and instrument testing, their performance was not sufficient for commercial cameras. To meet this new requirement, a 1-inch MM type (Magnetic field-converging, Magnetic field-deflecting) camera tube called DIS (Diode-gun Impregnated-cathode Saticon) was developed.

Thus far, photoconductive camera tubes with blocking targets have been used in color cameras for broadcasting so that the low dark current, low after-image and other excellent characteristics of the targets can contribute to the high image quality, an important requirement for the broadcasting color cameras. Because a saticon target has more than sufficient resolution by itself, the resolution of a camera tube using the target depends on the diameter of the scanning beam. The DIS electron gun was developed to take full advantage of the feature of the saticon target in the manufacture of a camera tube.

Figure 2.1(a) shows the structure of the gun for the DIS. This electron gun differs from a crossover three-electrode electron gun, whose structure is shown in Figure 2.1(b), in the application of a positive voltage to the first grid and the absence of the beam crossover point. A diode gun[1] is essentially an electron gun with two electrodes–a cathode and a beam-extracting electrode. The electron gun used in the DIS is also called a diode gun because the voltage is applied in a way that is similar to the way the voltage is applied to a regular diode gun.

The design goals for the DIS electron gun are to secure the required beam quantity and thereby obtain a high resolution, to minimize resolution loss in the fringe area, and to minimize capacitive image retention. To achieve these goals, the gun of this type has an electron beam-restricting hole with a diameter of about 12 μm in the first grid, which is next to the cathode. Potentials are applied to the electrodes, at the levels shown in the figure, in such a way that electrons are drawn out as parallel as possible to the tube axis to minimize the angle of divergence. In this manner, the loss of the fringe resolution by the cross term with deflection (i.e., deflecting defocusing)[1,2] can be minimized.

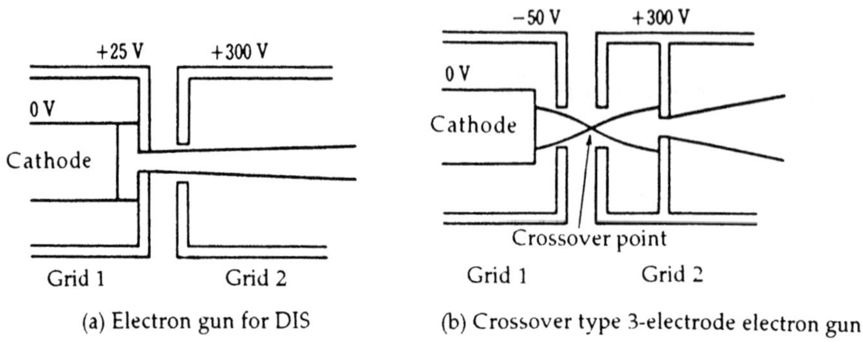

FIGURE 2.1. Comparison of electron gun structures.

With the DIS electron gun, no crossover is formed. Since a crossover dramatically increases local current density, the interaction among electrons is said to expand the velocity dispersion (increase the equivalent beam temperature) in the tube axis direction.[2] The increase in velocity dispersion means an increase of the percentage of electrons that can adhere to the target even when the scanned surface voltage has become negative, and thus a less steep beam landing characteristic curve. In other words, the efficiency in picking up low level signals falls, and the capacitive image retention increases. Accordingly, the DIS electron gun, which does not have a crossover point, promises a low image retention.

Figure 2.2 shows the observed beam-landing characteristics of various electron guns. Compared with crossover type electron guns, the DIS electron gun shows a considerably steeper characteristic curve. The equivalent beam temperature of the DIS electron gun calculated from these results is about 1700 K, compared with about 3500 K for a crossover type electron gun. However, since the beam quantity required for a Hi-Vision camera tube must be obtained on the very small cathode surface of the diode gun, the load on the cathode is increased. To ensure

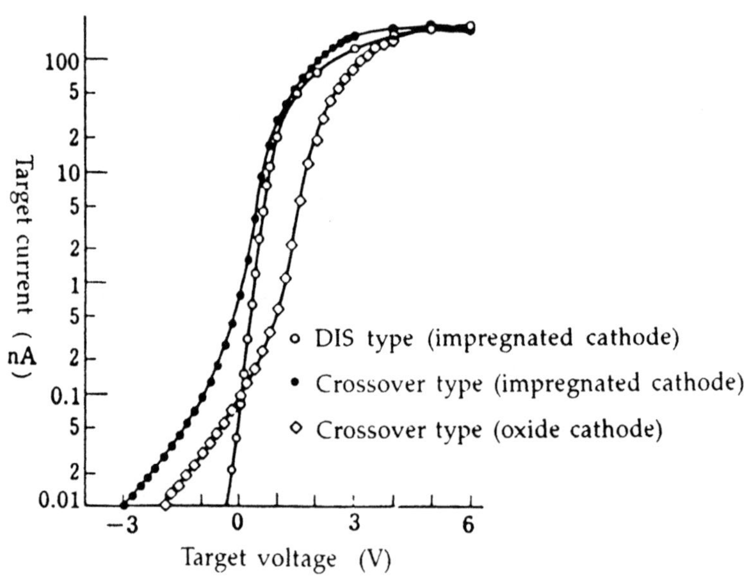

FIGURE 2.2. Comparison of beam landing characteristics.

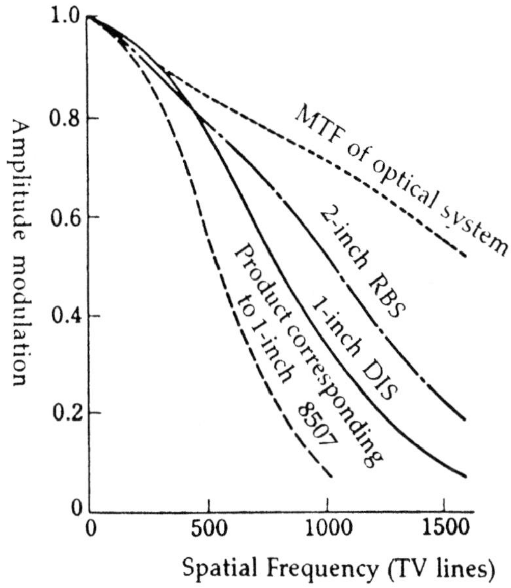

FIGURE 2.3 Resolution characteristics (including that of the camera lens).

safety under this high current load, a barium-impregnated cathode is adopted in place of the generally used oxide cathode.

The target structure of the DIS is the same as that of the saticon for the standard 525-line system, and is capable of the kind of high resolution which is shown in Figure 2.3. With a target film thickness of 4 μm, the storage capacitance of the target increases to about 1600 pF. However, because of the good electron beam characteristic, its after-image of less than 1% (after 3 fields) is sufficient for practical use.

A 1-inch and a 2/3-inch MS tube (magnetic field convergence, electrostatic field deflection type) have been developed for Hi-Vision. Both of them use DIS electron tubes.

2.1.2 HARP

The fact that Hi-Vision's high SN ratio requires a larger standard signal current than does the present television format, together with the need to increase the depth of field to obtain sharp images, makes it necessary to use a high sensitivity camera tube. However, because the structure of a blocking type target, in which the charge injection is blocked both at the signal electrode side and at the electron beam scanning side of the target film, the number of signal charges cannot exceed the number of incident photons (i.e., the state of a quantum yield (η 1). This is the theoretical limit of sensitivity, beyond which improvement was thought to be impossible. HARP (High-Gain Avalanche Rushing Amorphous Photoconductor) is able to realize a sensitivity beyond the limit of the quantum yield of 1 by applying a high voltage to a blocking type target to multiply the signal charges.[3]

Figure 2.4 shows a side view of the structure of a HARP target. On the glass plate is a signal electrode transparent to visible light (Nesa), over which is a supplemental blocking layer consisting of a CeO_2 layer about 0.02 μm thick. The photoconductive layer consists of high vacuum-deposited Se film, that is, an amorphous Se film. An Sb_2S_3 porous film on the scan surface of the photoconductive layer blocks electron injection. Figure 2.4 shows a 2/3-inch MS camera tube for Hi-Vision in which a HARP target is used. As the diagram in Figure 2.5 indicates, the camera tube has the same shape as a regular camera tube. The difference is the above-described target.

The signal current caused by blue light ir-

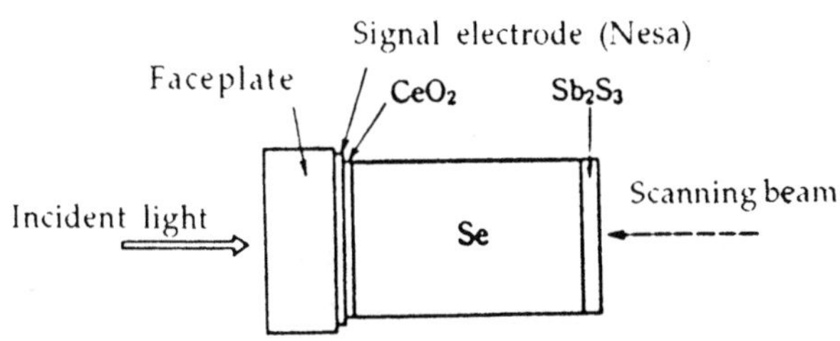

FIGURE 2.4. Target structure.

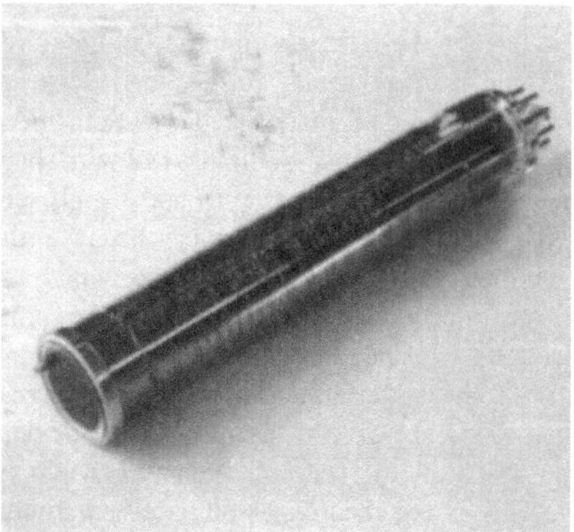

FIGURE 2.5. MS type 2/3-inch tube for HDTV.

radiation changes with increases in the target voltage as shown in Figure 2.6. As the target voltage is increased from zero, there is a region in which the signal current increases sharply. This is because hole-electron pairs are separated by the incident light and the signal current is starting to flow. Then the signal current shows a tendency toward saturation, reflecting a state in which most of the excited hole-election pairs are being emitted as signal current. Conventional blocking targets are used in this state. When the target voltage increases further, the signal current starts another sharp increase, surpassing the state of quantum efficiency 1. In this state the signal current flows as a result of avalanche multiplication. Dark current also in-

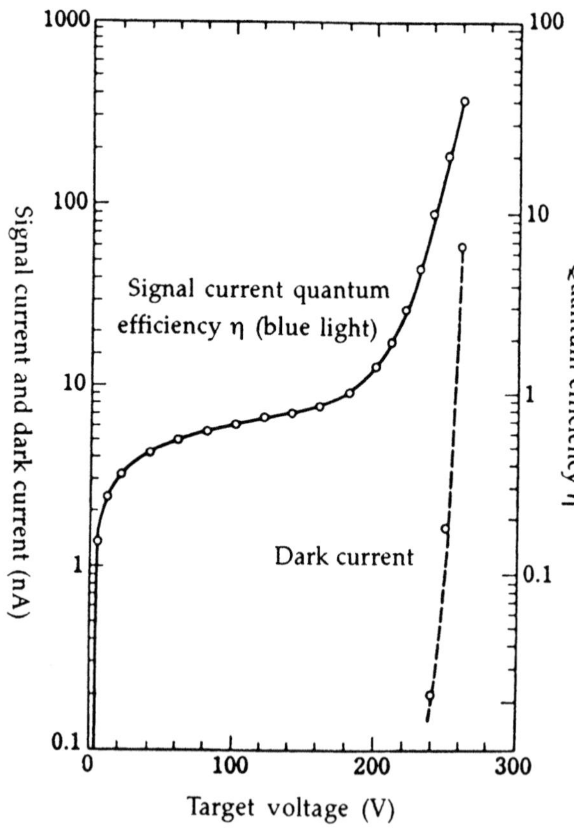

FIGURE 2.6. Voltage-current characteristics.

creases with avalanche multiplication. However, as shown in the same figure, it remains at an extremely low level, below 0.2 nA if target voltage does not exceed 240 V (η = 10). Thus we can expect a camera tube with this target, when used in a color camera, to produce the same good image quality as a photoconductive camera tube with the conventional blocking target.

The avalanche multiplication phenomenon is caused by both holes and electrons. However, in amorphous Se, the avalanche multiplication caused by holes starts at a lower electric field than one caused by electrons. In the HARP system, this difference is utilized to obtain a state in which only holes undergo avalanche multiplication, thus producing a stable and high quality image.

Figure 2.7 shows that the photoelectric conversion characteristic of the HARP tube is about 10 times as sensitive as a saticon. Because the HARP also operates as a blocking target, its gamma value at low light levels is the same as a saticon, about 0.95. In the high illuminance region, however, the value is slightly lower because signal current (that is, the plane potential changes on the scanning surface) increases in the high illuminance region, thus decreasing the

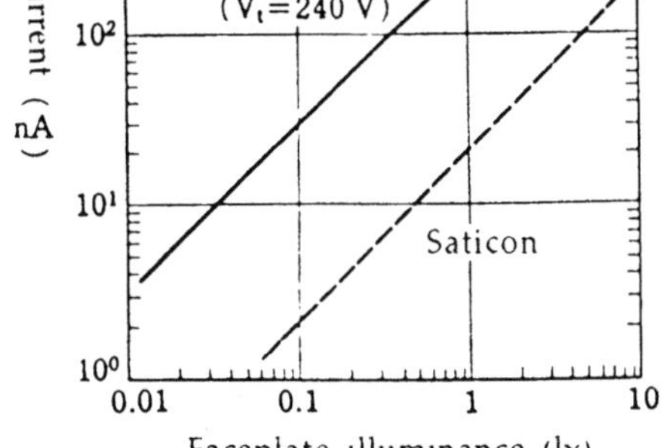

FIGURE 2.7. Photoelectric conversion characteristics.

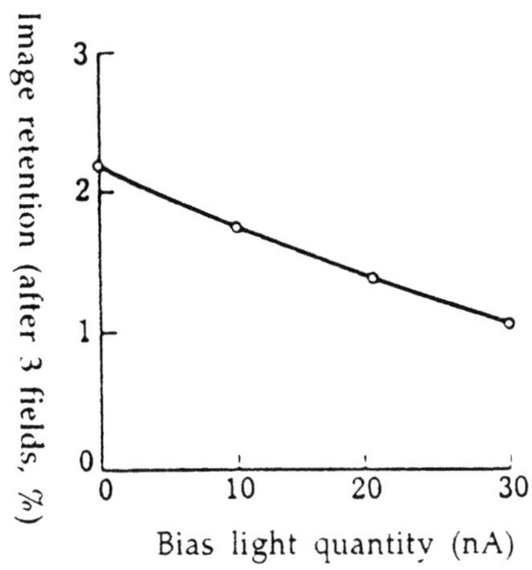

FIGURE 2.8. Bias light dependence of after-image.

voltage actually applied to the photoconductive film and inhibiting the multiplication.

Avalanche multiplication is a phenomenon within the target film. With regard to the scanning electron beam, the HARP target is the same as a regular blocking target. Because it does not show an increase in photoconductive image retention caused by multiplication, its image retention characteristic depends on the storage capacitance and electron beam characteristic of the target just as the image retention of the conventional photoconductive camera tube does. The Hi-Vision 2/3-inch tube shown in Figure 2.5 has a DIS electron gun and a target with a film thickness of 2 μm. A value of 2.2% (after 3 fields) has been obtained without bias light. This image retention characteristic can be improved with bias light in a manner shown in Figure 2.8.

Resolution depends on the properties of the target and the performance of the scanning electron beam. The dark resistivity of the material of the target, amorphous Se or Sb_2S_3, is higher than 10^{12} Ωcm, sufficient for having the target perform a storing operation. Measurement under varied target voltage shows that no deterioration in resolution has resulted from the avalanche operation state. Shown in Figure 2.9 is an example of the resolution characteristic of a 2/3-inch Hi-Vision tube.

In the past, targets with amorphous Se were unable to obtain sensitivity to long wavelength

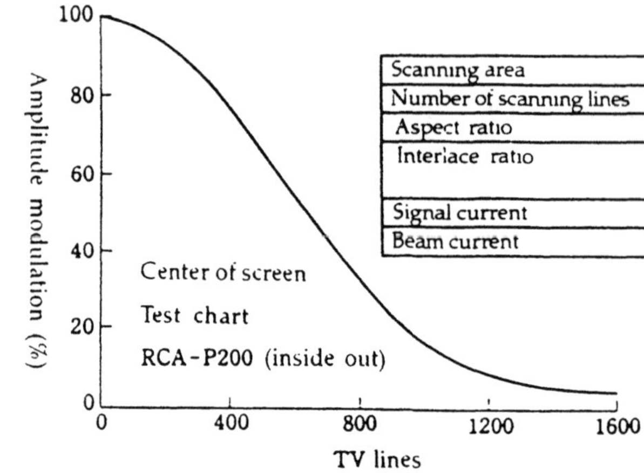

Scanning area	8.8x6.6 mm	Heater voltage	6.3 V
Number of scanning lines	1125	G1 voltage	140 V
Aspect ratio	4 : 3	G3 voltage	500 V
Interlace ratio	2 : 1	Horizontal deflection voltage	109 Vpp
Signal current	0.3 μApp	G4 (mesh) voltage	1000 V
Beam current	0.6 μA	Target voltage	237 V

FIGURE 2.9. Resolution characteristics.

light. The reason for this shortcoming has been attributed to the inability of amorphous Se to effectively pick up a signal current out of the hole-electron pairs because the hole-electron pairs produced by light absorption disappear by recombination. This tendency shown by amorphous Se is stronger toward the lower electric field and longer wavelength light.[4] However, the HARP system, because of the presence of a strong electric field, can obtain a sufficiently large signal current to respond to light with a wavelength of up to 620 mm, a limit imposed by the amorphous Se band gap of about 2.0 eV. Good color reproduction with a color camera requires sensitivity to light with wavelength longer than 620 mm. An improvement that involves the addition of Te has been proposed.[5]

The camera tubes mentioned thus far have a target film thickness of 2 μm. However, as shown in Figure 2.10, under the same electric field, the multiplication effect increases with the target film thickness. Thus a further increase of the sensitivity of the camera tubes of this type is theoretically possible.

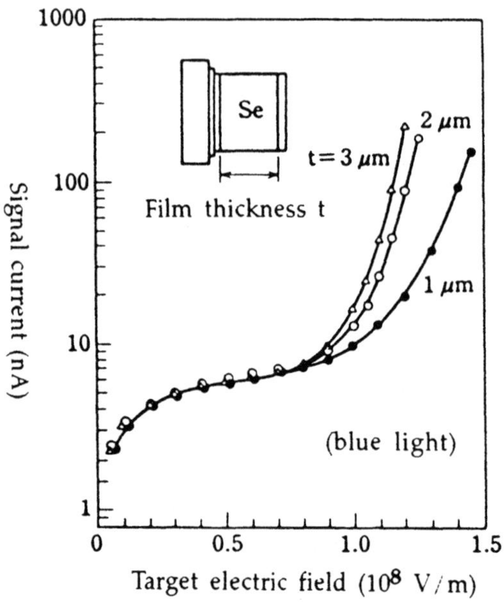

FIGURE 2.10. Electric field dependence of signal current in the target.

2.1.3 CCD (Charge Coupled Device) Image Sensor

Although solid state image sensors have been proposed for a long time, a particularly important one is the CCD announced in 1970 by Woyle and others at Bell Laboratories. This CCD, supported by the rapid advances in LSI technology that followed, developed into the image sensor. Today, the use of solid state image sensors is not restricted to consumer video cameras. The adoption of them for ENG (Electronic News Gathering) cameras has already started. Even in studio cameras, in which high image quality is of particular importance, the CCD image sensors started being employed around 1988.

The adoption of CCD image sensors for current TV cameras was prompted by the following reasons. First, the CCD image sensor is compact and lightweight, and also has mechanical strength and stability which the conventional camera tube does not have. Second, production costs of CCD image sensors fell drastically due to advances in semiconductor technologies. Third, the remarkable progress made in the basic performance (in such areas as sensitivity and resolution) of the CCD image sensor has reached such a level that the CCD sensors can even surpass camera tubes in some performance areas. Because of these advantages of CCD sensors, many reports have been published on CCD research and prototypes, and these will be described later in this chapter. The performance of the CCD sensors in these reports has already reached a significantly high level. However, for the adoption of CCD image sensors in commercial Hi-Vision products, some problems (not simply limited to the number of pixels) need to be solved.

(1) Present State and Characteristics of CCD Image Sensors

The characteristics that determines the performance of an image sensor, like those that determine the performance of a camera tube, are numerous. With regard to these performance-determining characteristics, the image sensor and camera tube share such characteristics as sensitivity, noise, image retention, resolution, and blooming. However, some characteristics such as smear, moire, fixed-pattern noise, dark current irregularities, and spots are specific to the CCD image sensor. Table 2.1 compares the main characteristics of the camera tube and CCD image sensor.

Blooming and smear are phenomena that occur when a strong light enters a part of the light-receiving surface of a CCD image sensor and generates an overflow of signals from the pixels. Blooming shows up in the fringe area of a highlighted section in an area several times larger than the highlighted section. Smear appears as spurious signals that form vertical streaks. Various methods have been proposed to prevent blooming and smear from occurring. One of them provides a kind of bypass so that the excess charge will not overflow and enter the other pixels and charge transfer routes and that the charge then will be absorbed by the silicon substrate.

TABLE 2.1. Comparison between CCD image sensor and camera tube.

Item	CCD image sensor	Camera tube
Sensitivity and noise	• Low light utilization—sensitivity improvement is needed. • Random noise has been decreased by the improvements in detecting amplifier and circuit technology • Fixed pattern noise has been decreased by advances in device-manufacturing techniques.	• Because the initial step FET amplifier characteristics determine the level of noise, a marked further noise reduction is difficult to realize. • Fixed pattern noise has been decreased by advances in device-manufacturing techniques.
Image quality	• Blooming does not pose problems in actual viewing. • Smear suppression exceeds 80 dB. • To reduce moire, the use of optical filter and other means are necessary.	• Blooming is not a threat in practical use. • In principle, the camera tube is smear-free.
After-image and burn-in	• Almost no after-image and burn-in occur • In a photoconductive film-laminated type, after-image and burn-in are problems.	• After-image is about 2% or less (after 3 fields). • Burn-in may occur.

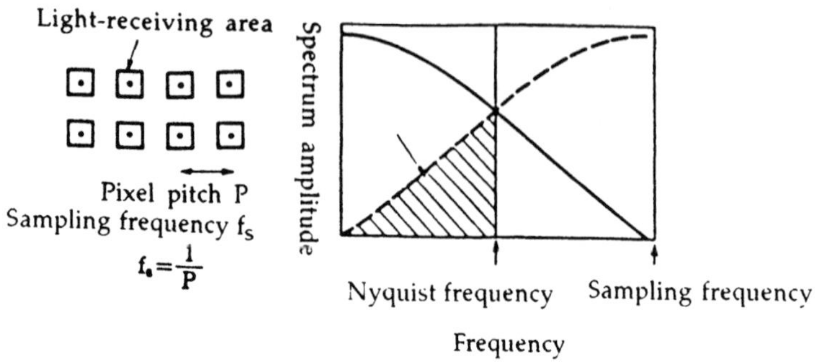

FIGURE 2.11. Sampling and response in CCD.

With these efforts described above, the CCD image sensors for the current standard television have been improved to a level that will not cause problems in actual use. However, these improvements, which are satisfactory for the current standard TV may not be sufficient for Hi-Vision CCD sensors because their light-receiving areas and transfer route size are smaller. Still, with further improvements in the above-described techniques, blooming and smear will cease to be major problems.

In the case of CCDs, image retention is partially nil. However, with a type of CCD in which photoconductive films are laminated, the film itself may have a characteristic problem of image retention similar to that shown by camera tubes.

(2) Issues for Hi-Vision CCD Image Sensors

(a) Resolution and Moire. Moire is a problem specific to CCDs and other solid-state image sensors. Moire is closely related to the resolution of the image sensor. In CCD image sensors, in which the light receiving area of each pixel is separate from other pixels, as shown in Figure 2.11, the optical image of the object is spatially sampled. For this reason, the input of an optical image having a frequency component higher than the 1/2 of the sampling frequency (Nyquist frequency) generates a spurious signal called a folded-back distortion (the hatched area in the figure). This shows up on the screen as a glittering image called moire. In general moire, due to the visual disturbances which it causes to human vision, is perceived as a deterioration in resolution.

For a Hi-Vision image sensor, which requires a high resolution, (1) a large number of pixels is necessary to increase the sampling frequency, and (2) the limitation of the light input signal bandwidth is necessary to suppress the folded-back distortion. To suppress the unwanted frequencies, usually an optical low pass filter which utilizes the quartz crystal birefringence, is employed. With today's advanced semiconductor processing technologies capable of fine processing, prototypes of CCD image sensors having about two million pixels are being produced. In this situation, it is not so difficult to increase resolution. However, with an increasingly large number of pixels, it becomes more and more difficult to maintain high levels of sensitivity and SN ratio.

(b) Signal Charge Quantity and Noise. To obtain a high sensitivity and a broad dynamic range, an increase in signal charge and a noise-reducing device are necessary. To increase the signal charge, it is necessary to increase the photoelectric conversion, which is determined by the aperture and quantum efficiency* of the light-receiving section. The aperture is defined as the ratio of the light-receiving area (of a photodiode or the like) in a pixel. Although efforts are being made to increase this ratio, usually the aperture value ranges from 0.3 to 0.5.

Noises can be either Fixed Pattern Noises (FPN) which are immobile on the screen, or random noises such as thermal noises. The main

*This is the number of electrical charges generated by one photon, and depends on the light's wavelength and the material.

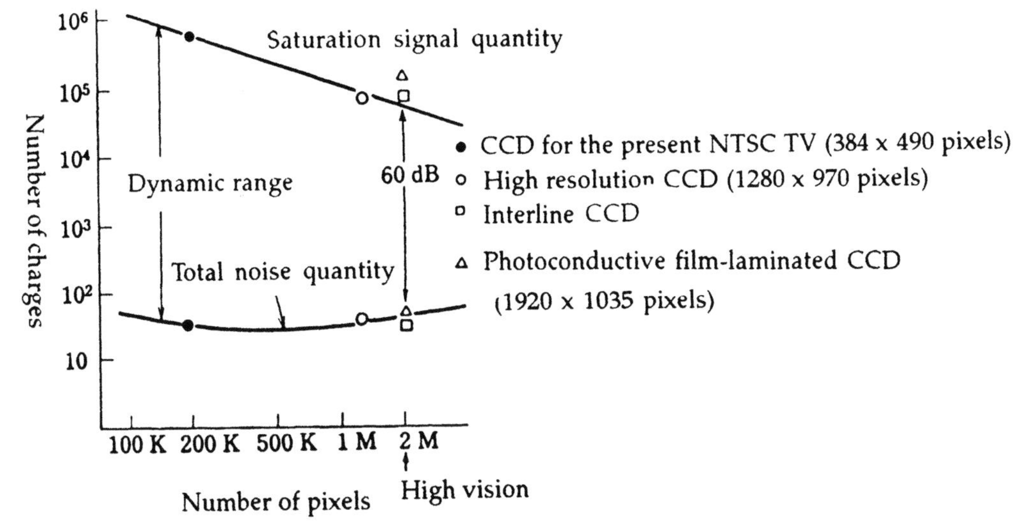

FIGURE 2.12. Relationship between number of pixels and dynamic range in 1-inch CCD image sensor.

source of the FPN in CCD is the scattering of dark current levels among pixels. However, with advances in manufacturing techniques, the FPN of this type have been decreased to a level that does not cause problems. The random noises are mainly caused in the signal charge detecting amplifier and in the transistors nearby. Because on-chip installation of the detecting amplifier on the CCD device is possible, noise reduction is easier with a CCD than with a camera tube. However, noise reduction needs to go much further in Hi-Vision than in standard television.

(c) Dynamic Range. Let us shift our viewpoint in examining the relationship between the signal charge quantity and noise in a Hi-Vision CCD Image sensor. Figure 2.12 shows how the number of pixels in a solid state image sensor is related to noise, saturation charge, and dynamic range.[6] What is shown here is the calculated performance of a 1-inch image sensor as the number of pixels increases. It is based on data from a commercial 2/3-inch CCD image sensor. The results in the figure indicate that an attempt to obtain a dynamic range of 80 dB (which is the dynamic range of the current standard TV) with a 2-million pixel Hi-Vision CCD will end up short of the target by about 20 dB. To attain the goal, more technological advances are needed.

(3) Actual Examples of Hi-Vision CCD Image Sensors

Several prototypes and proposals have been presented for image sensors for Hi-Vision cameras. Shown in Figure 2.13 is a planar view of a CCD structure with two million pixels (1920 (H) × 1035 (V).[7] It has a detecting amplifier whose required bandwidth has been reduced to 1/2 by the use of two parallel horizontal transfer CCDs. Further, the detecting amplifier itself has been improved to reduce noise.

Figure 2.14 shows a cross-sectional view of a CCD image sensor revealing a 2-story structure formed by the lamination of photoconductive films.[8] The planar view of this image sensor is about the same as that of the sensor shown in Figure 2.13. In this sensor, an ideal aperture of 100% has been achieved by the lamination of amorphous silicon (a-Si) layers. The sensitivity and dynamic range have also been improved in this sensor. Both of these two sensors have achieved the limiting horizontal resolution of 1000 TV lines. In addition to these image sensors, a preprocessing type CCD in which spatial filter function has been given to pixels (shown in Figure 2.15) has been proposed.[6] This type of CCD has been developed to suppress the folded-back distortion and to obtain a broad dynamic range.[6] In this CCD sensor, the signals

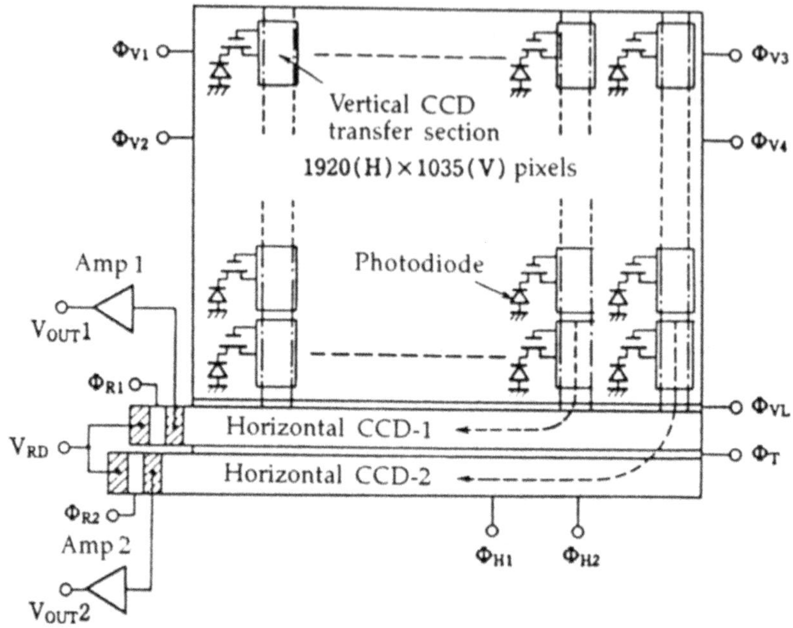

FIGURE 2.13. A 2–million pixel CCD image sensor.

of four light-receiving areas are added up within a CCD register by varying the combination for each field before the signal is output. This sensor, which increases the saturation charge level and decreases the driving frequency, has an advantage of low noise.

The main characteristics of Hi-Vision CCD image sensors that have been proposed or built as prototypes are shown in Table 2.2 and Figure 2.12. While the characteristics of these CCD image sensors have reached certain acceptable levels, there is much room for improvement. In

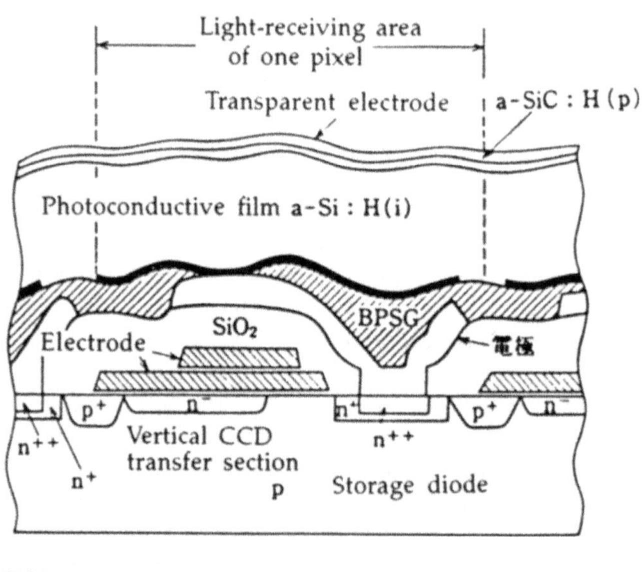

FIGURE 2.14. Cross sectional structure of a photoconductive film-laminated CCD image sensor (2 million pixels).

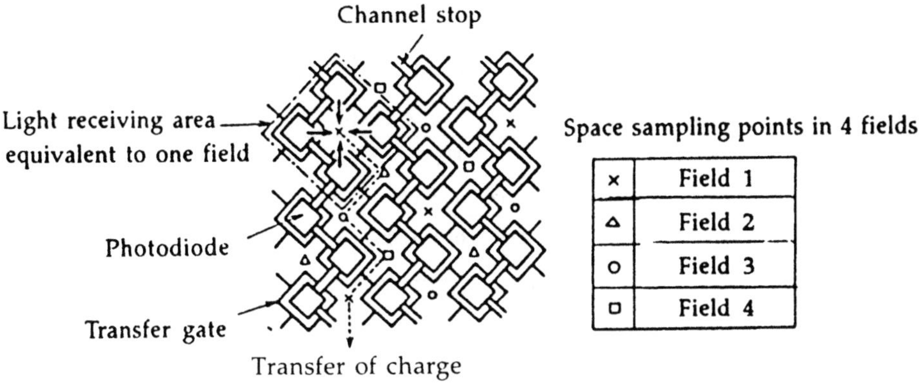

FIGURE 2.15. Preprocessing CCD image sensor (4–pixel addition type).

addition to the improvements in the sensor performances, efforts to reduce manufacturing costs are also needed. Considering the remarkable advances made in the field of semiconductor technologies, we can safely wager than Hi-Vision CCD image sensors will be realized.

2.2 CAMERAS

Hi-Vision cameras based on the 1125 system were first developed by NHK science and technical Research Laboratories in 1973. Since then astounding improvements in performance have been achieved. Today, cameras ranging from studio cameras to portable cameras have been developed and used to produce many Hi-Vision programs. Table 2.3 shows the characteristics of various cameras that have been used to produce Hi-Vision broadcast programs. In addition to the cameras shown in this table, many prototypes have been made by manufacturers. Research and development related to Hi-Vision cameras are continuously advancing, as evidenced by the development of high sensitivity camera tubes and prototypes of solid state image sensors (although still experimental) that meet Hi-Vision standards.

2.2.1 Basic Camera Performance

Although there are many characteristics used to evaluate camera performance, the major ones are sensitivity, resolution, image retention, SN ratio, dynamic range, and registration. As many

TABLE 2.2. Comparison of Hi-Vision CCD image sensors.

Item	Interline CCD	Laminated CCD	Pre-process CCD (estimated)
Max. no. of signals	80,000	200,000	260,000
Noise (no.)	22	52	26
Dynamic range (dB)	71	72	80
SN ratio (dB)	40	58	60
At 2,000 lx, F-8			
Max. horizontal resolution (TV lines)	1,000	1,000	1,000
After-image (%)	None	1.3	None

TABLE 2.3. Characteristics of Hi-Vision cameras.

Camera	NHK HDCC-3, 4	Ikegami HDS-71	Sony HDM-71	NHK HDCC-5
Year developed	1980	1984	1983	1984-6
Configuration	Standard studio type	Standard studio type	Portable	Handy camera
Camera tube	1-inch MM Saticon (DIS)	1-inch Saticon (DIS)	1-inch MS Saticon	2/3-inch MS Saticon (DIS)
Resolution (800 TV lines)	50%	35%	30%	30%
SN ratio	44 dB	44 dB	44 dB	41 dB
Sensitivity (2,000 lx)	F 2.8	F 3.3	F 3.3	F 2.8
After-image (after 3 fields)	Less than 1%	1%	3%	1%
Registration (full screen)	Less than 0.025%	Less than 0.05%	Less than 0.05%	Less than 0.02%
Max. camera cable length	200 meters	1 km optical fiber cable	1 km optical fiber cable	200 meters
Camera head weight	48 kg	39 kg	10.5 kg	6 kg
Other	DRC (15 X 25); Dynamic range correction; VF deflection expansion	DRC (16 X 27); Dynamic range correction; VF deflection expansion; Focus indicator	DRC (13 x 13); Dynamic range correction; VF deflection expansion; Focus indicator	Battery operation; Full auto registration; Outline flicker

of these characteristics are trade-offs, it is important when designing a Hi-Vision camera to maintain a balance among these characteristics. The following sections explain these performance characteristics.

(1) Resolution

The resolution of a camera depends on characteristics such as the camera's optical lines, three-color separation prism, camera tube, and edge compensation circuit. The most important factor for resolution of the camera is the camera tube, followed by the optical lens. The color separation prism does not pose any problem if its size is 1 inch (16 mm in the diagonal measurement of the imagining surface) or larger. However, if its size is 2/3 inch (11 mm in the diagonal length of the imaging surface) or smaller, its characteristics need to be carefully considered. The edge compensation circuit is employed to correct the deterioration in resolution caused by factors such as the camera tube.

(a) Resolution of the Camera Tube. Figure 2.16 shows the resolution characteristics of camera tubes used in Hi-Vision cameras developed by NHK Science and Technical Research Laboratories. Although the industry convention is that the resolution of a camera tube be expressed by its responses at 800 TV lines, because this expression is difficult to use in mathematical calculations, we will use a different expression for camera tube resolution. The characteristic curves of various camera tubes shown in Figure 2.16 indicate that they can be approximated by Gaussean curves. Accordingly, let us assume that the resolution response to an x number of TV lines, $A_r(x)$, can be expressed by the following equation:

$$A_r(x) = \exp\{-(x/a)^2\} \qquad (2.1)$$

where a is a resolution of TV lines that produces $A_r(a) = 1/e\ (-0.37)$.

In investigating numerous DIS tubes, Equa-

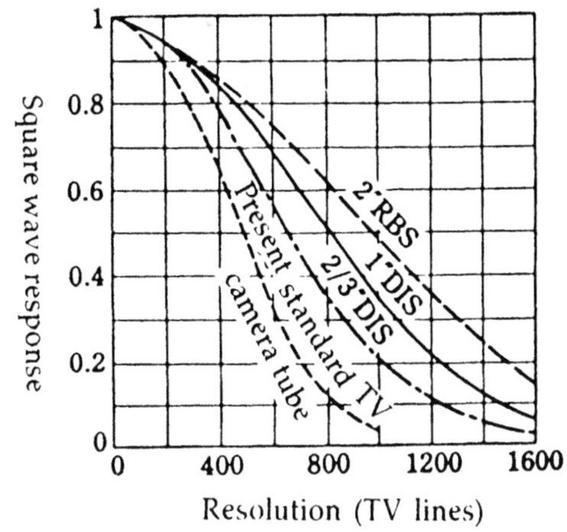

FIGURE 2.16. Resolution characteristics of various camera tubes.

tion 2.1 has been found to hold within an error of a few percent. According to Equation 2.1, resolution can be expressed in terms of the value of a in place of the response at 800 TV lines. For example, the resolution of a 1-inch DIS tube would be 950 TV lines. Incidentally, the value a does not give the value of limiting resolution. It is empirically known that the limiting resolution is about twice the value of a.

(b) Edge Compensation circuit. The deterioration in resolution described above is compensated for by an edge compensating circuit. The appropriate type of compensation depends on the characteristics of the display and other components. Taking an actual example of a camera, let us assume that an edge compensation should be made so that a 100% response is obtained for signals from the camera within the ranges of 870 horizontal TV lines (equivalent to 30 MHz) and 1035/2 vertical TV lines (half of the effective scan lines).

While edge compensation improves resolution, it also deteriorates the SN ratio of the output signal. This is because the noise component in the output signal from the camera preamplifier is enhanced by the edge compensation. For this reason, a large amount of edge compensation will produce an image full of noise. The results of the calculation of the noise increase are shown in Figure 2.17. In this calculation, triangular noise (a noise whose amplitude increases in proportion with frequency) is assumed to be the noise existing before the edge compensation, and the resolutions of the camera tube in the horizontal and vertical directions are assumed to be equal to each other.

As indicated in Figure 2.17, noise increases sharply when the resolution a is below 1000 TV lines, and the SN ratio deteriorates by the same proportion. In the actual Hi-Vision camera, the SN ratio of the output signal of the preamplifier is about 45 dB, while the noise detection limit for triangular noise is about 35 dB in terms of SN ratio. It follows that a Hi-Vision camera tube needs to have a resolution in which the noise increase caused by edge compensation is less than 10 dB, that is, a resolution a of 800 or more TV lines. This finding agrees with an empirically known value necessary for a Hi-Vision camera tube: 800 TV lines with 40% or more response.

(c) Maximum Signal Current and Resolution. In general, the maximum value of the signal current that a camera tube can handle and the resolution of the camera tube have a reciprocal relationship. In other words, a camera tube with a large maximum signal current i_s has a low resolution, while a camera tube with a high resolution tends to have a small maximum signal current. This relationship can be easily explained by assuming that the scanning beam cur-

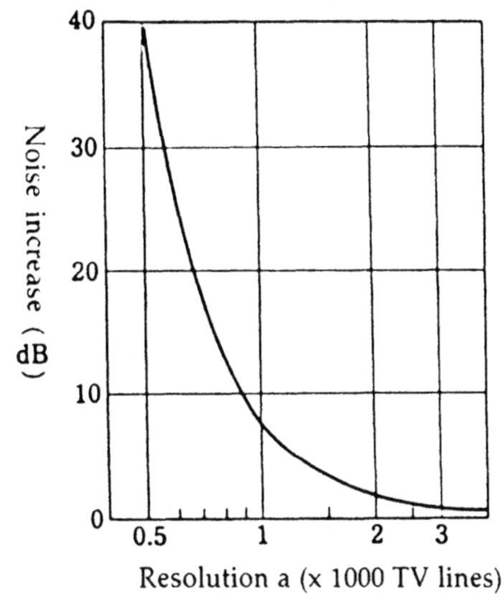

FIGURE 2.17. Increase in noise caused by edge compensation.

rent density of the camera tube is constant, and that the resolution depends on the beam diameter ϕ, as follows:

$$a = k_1/\phi$$

$$i_s = k_2/\phi^2$$

From the above, the following equation can be derived.

$$i_s \propto 1/a^2 \tag{2.2}$$

where k_1 and k_2 are proportional constants. Figure 2.18 shows the results of the measurement of this relationship for a DIS tube. Here the maximum signal current, more or less inversely proportional to the square of resolution a, proves that the relationship in Equation 2.2 holds.

A camera tube with large maximum signal current, which is able to set the reference signal current at a high level, allows a high SN ratio to be set for the preamplifier output signal. However, with a low resolution, this camera has a large noise increase in the process of edge compensation. As a result, the SN ratio of the final camera output signal obtained with this camera tube is not necessarily good. On the other hand, because a camera tube with a low maximum signal current has a high resolution, the SN ratio of the preamplifier output signal is low even though the noise increase due to edge compensation is small. Accordingly, the SN ratio obtained after edge compensation has a maximum at a certain resolution value. The results of the calculation of the relationship are shown in Figure 2.19. The figure shows that the maximum SN ratio is obtained at a = 900 TV lines. The resolution of the currently available 1-inch DIS tube, for which a = 950 TV lines, is about the optimum resolution for a Hi-Vision tube.

(d) Optical Lens. An example of the resolution characteristic of an optical lens is shown in Figure 2.20. The characteristic varies depending on the f-stop and zoom ratio. However,

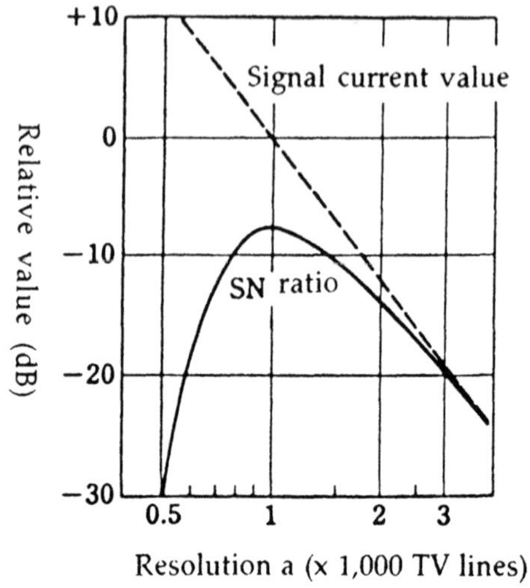

FIGURE 2.19. Optimal resolution giving maximum SN ratio.

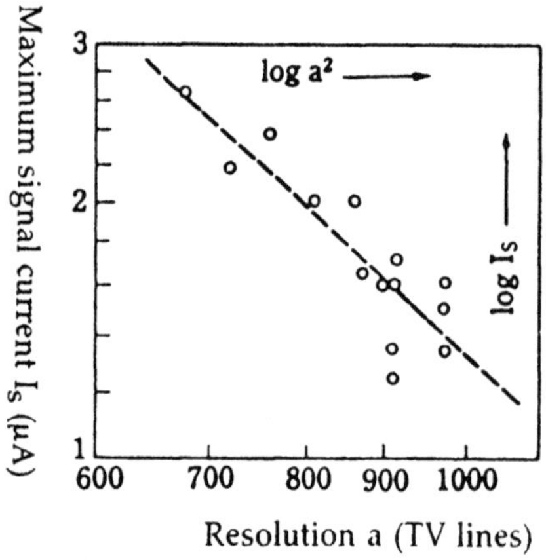

FIGURE 2.18. Relationship between maximum signal current value and resolution in a DIS tube.

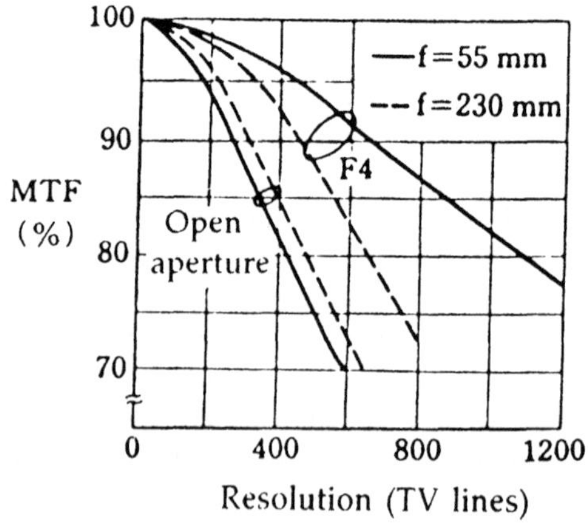

FIGURE 2.20. Example of zoom lens resolution (G channel).

when the aperture is open, the resolution characteristic of the lens becomes about equal to or lower than the resolution of the camera tube, thus becoming a camera resolution-limiting factor. The lens resolution generally starts to deteriorate at an f-stop of 2.8, although this value may vary with the type of lens. At present, a lens at an f-stop lower than 2 cannot produce a satisfactory resolution.

(2) After-images

After-images can be caused by many factors. However, the main cause of after-images in currently used camera tubes such as saticon and Plumbicon is usually capacitive after-images. Capacitive after-images occur when a signal charge stored in a photoelectric conversion film cannot be discharged completely by the first scanning beam. Assuming that the beam resistance is constant regardless of the signal discharge (potential of the scanning surface of the target), the capacitive after-image can be analyzed in a manner similar to the analysis of the discharge current (signal current) of the primary capacitance-resistance (CR) circuit, which discharges the charge stored in capacitance C through resistance R.

Assume that the ratio of electrical charge left undischarged by the first scan to be b, capacitive after-image can be expressed by the following equation.

$$y = (1 - b(x_n + \Sigma\, b^k x_{n-k}) \qquad (2.3)$$

y_n: output signal from the camera tube at the nth field

x_n: incident light at the nth field

After-image is usually expressed as the value observed 3 fields after. The relationship between this (Lag) value and b in Equation 2.3 is given by (Lag) $= b^3$.

Compensating for the after-image requires that we calculate the correct incident light x_n. By solving Equation 2.3 with respect to x_n, the following equation is obtained.

$$x_n = (y_n - by_{n-1})/(1 - b) \qquad (2.4)$$

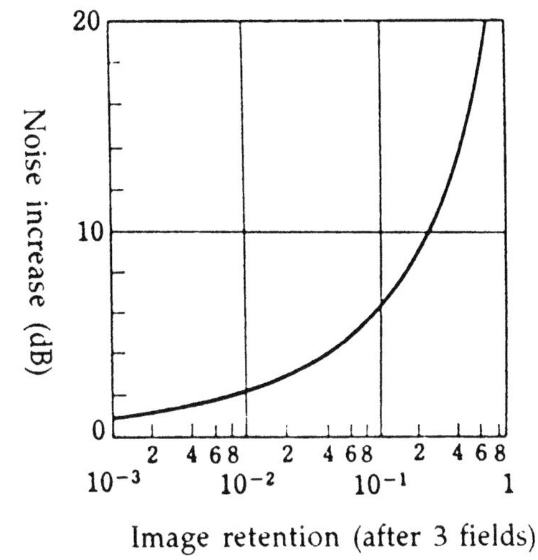

FIGURE 2.21. Noise increase caused by after-image compensation.

This can be realized easily by using a field delay memory. The increase in noise resulting from this compensation for after-image is shown in Figure 2.21. Because the noise increase caused by the compensation is not much compared with the after-image value of the actual camera tube (a few percent after 3 fields), the noise increase poses no problem with respect to SN ratio. However, after-image is equivalent to a low pass filter in the temporal frequency region. The frequency characteristic of after-images, which is about equal to the low pass filter based on the capacitance effect (analogous to 1/60-second shutter of a photographic camera), is a factor in the degradation of dynamic resolution.

(3) Sensitivity

(a) Physical Sensitivity. The sensitivity of a Hi-Vision camera can be defined in terms of physical sensitivity and effective sensitivity. Physical sensitivity is defined as the quantity of incident light necessary to obtain the reference output image signal level in the imaging of a white object (with a reflectance of 90%). It depends almost completely on the photoelectric conversion film of the camera tube. Hi-Vision camera tubes use saticon or Plumbicon film just as current standard TV camera tubes do, and so their sensitivity is essentially the same as that

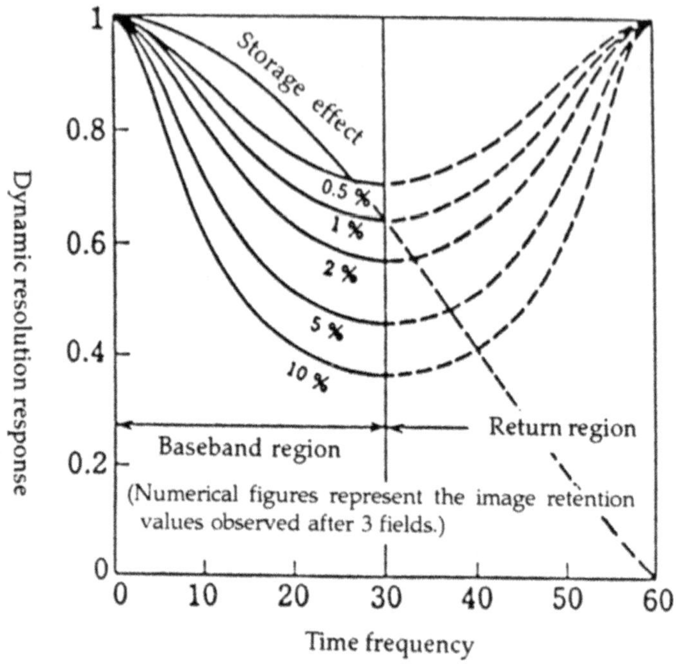

FIGURE 2.22. Frequency characteristic of after-image.

of camera tubes of standard TV cameras. In a strict sense, a Hi-Vision camera has a slightly lower sensitivity than the standard TV camera because its camera tube signal current is larger than the standard TV camera to improve the SN ratio, and because Hi-Vision, with a broader screen (aspect ratio of 16:9), has a lower utilization of light.* This difference is not very large—about half an f-stop, or 1.5 times as much in terms of incident light.

(b) Effective Sensitivity. Effective sensitivity is defined as the light quantity necessary for obtaining a good Hi-Vision image quality. Effective sensitivity depends not only on the photoelectric conversion film characteristics, but also on the characteristics of the optical lens. As described before, resolution deteriorates on currently available lenses if the f-stop is smaller than 2.8. This limit on the performance of currently available optical lenses limits the effective sensitivity. This problem, which can be solved by improving the lens characteristics, is not an essential problem.

The depth of field of the lens limits effective sensitivity. When an object is in focus, the depth of field is defined as the distance in front of and behind the object also in focus. The brighter a lens is, or the smaller an f-stop is, the shallower (smaller) the depth of field is. An extremely bright lens (that is, a lens with an extremely small f-stop) can only focus on a part of the object. If the object is a close-up shot of a man, the man may have only his nose in focus, with his eyes out of focus. In such a case, the high resolution of Hi-Vision is of no use. Thus there is a lower limit to f-stops that can actually be used.

Figure 2.23 shows the depth of fields of Hi-Vision and the current standard TV. In the figure, A, B and C represent the objects, and the A′, B′ and C′ are the images of the objects. If the lens is focused on point A, the light from each object passes through the areas shown with solid lines to the images. The light from point

*A screen with an aspect ratio of 16:9 has a smaller area than a screen with an aspect ratio of 4:3 provided that the diagonal length of these screens are equal. As a result, the total signal charge quantity stored in the former is smaller than in the latter.

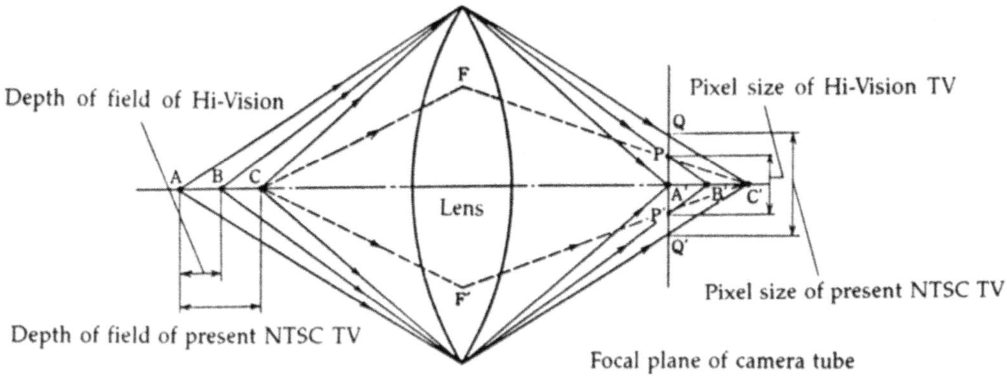

FIGURE 2.23. Depth of field.

A forms an image on the camera tube plane. Light from B and C, however, is dispersed over the PP′ or QQ′ regions. These images are not sharp. However, if the dispersion does not exceed the size of one pixel, the lack of sharpness does not pose a problem because the resolution of the camera depends on the size of the pixel of the camera tube. If the dispersion size is larger than a pixel, then the lack of sharpness will determine the resolution of the camera tube.

Let us define depth of field as the distance between two objects that would cause a dispersion equal in size to a pixel. Because Hi-Vision pixels are about half the pixel size of standard TV, the depth of field should also be about half as shown in Figure 2.23. To obtain a depth of field as deep as that of standard TV, the light coming from point C should be limited to the FF′ shown in the center area. This means that the lens aperture needs to be half the size, in other words 2 f-stops down from a fully open aperture. The quantity of light from the object that reaches the camera tube is reduced to one-fourth. Thus a Hi-vision camera needs four times as much light on an object as a current standard TV camera. This figure, added to the decrease in physical sensitivity described above, means that a Hi-Vision camera requires six times as much illumination as a standard TV camera. Thus the effective sensitivity of Hi-Vision cameras is lower than the standard TV camera, and the limit on effective sensitivity imposed by depth of field is quite basic.

(4) SN Ratio

(a) Preamplifier Noise. SN ratio is the most essential item in camera design. However, it is not necessarily easy to obtain a good SN ratio for Hi-Vision cameras because of their broad image signal band. The predominant noise in present television cameras is the thermal noise of FET used in the first stage preamp. The SN ratio decreases at a rate of 3/2 of the required signal band width B. This decrease in the SN ratio occurs because the noise band broadens (proportional to $B^{1/2}$) and because, as the FET has a capacitive input impedance, the input voltage decreases in inverse proportion to signal frequency, causing an equivalent increase in the noise component (proportional to B). In other words, the result is triangular noise. Assuming that the camera tube signal currents are equal, the SN ratio of a Hi-Vision camera with 30 MHz bandwidth should be worse than a standard television camera with a 4.2 MHz bandwidth, by 25.6 dB ($=20 \log(30/4.2)^{3/2}$).

Because the noise in a Hi-Vision image looks different from the noise in a standard TV image, a Hi-Vision camera does not require the same SN ratio as that of the standard camera. Since the noise on a Hi-Vision image is relatively minute, it is more difficult to see than on a standard TV image. Thus the SN ratio on a Hi-Vision camera can be set at a lower level than that of a standard camera. This difference is said to be about 10 dB. But even after taking this into consideration, the SN ratio of a Hi-Vision camera is at least 10 dB worse than on a standard camera (25.6 − 10 = 15.6).

(b) Shot Noise. In addition to FET noise, photon shot noise is also a basic noise. Although the photon shot noise is rarely considered to be a problem at present, it will surely be recognized as a problem in the future when improvements have been made in FET and other areas. Even if a noise-free FET were developed, the SN ratio obtained under the present signal current level (0.4 μA) would be 46 dB. Incidentally, photon shot noise is a flat spectrum (noise amplitude being constant regardless of frequency). However, the noise in an electrical signal output from a camera tube, influenced by the camera tube resolution, generates a spectrum with a lower high-frequency component. This spectrum, after undergoing the previously described edge compensation, returns to a flat spectrum. The theoretical limiting value mentioned above refers to the signal after the edge compensation.

Because the amplitude of shot noise is proportional to the square root of the incident light quantity, the limiting value of the shot noise in the SN ratio is also proportional to the square root of the incident light quantity. If follows that an incident light quantity above a certain level is required to maintain a level of SN ratio acceptable for Hi-Vision. Although the detection limit of the shot noise in Hi-Vision has not been directly measured, it can be estimated in the following manner. Shot noise is large in the bright areas of the image screen and small in the dark areas. In other words, shot noise is dependent on the signal level. However, because the gamma processing circuit in the camera has a characteristic in which the gain is high in the low signal level (dark area), and low in the high signal level (bright area), the shot noise does not have the signal level dependence with output from the gamma processing circuit.* This means that the shot noise output from the camera (with an edge compensation effect) can be treated like regular transmission line noise which does not have any level dependence. However, it should be noted that shot noise compressed by gamma correction is equivalent to transmission line noise which is about 6 dB smaller. For example, the above-mentioned shot noise with an SN ratio of 46 dB is equivalent to a flat noise with an SN ratio of 52 dB that is mixed into the transmission line.

The detection limit of flat noise is said to be about 45 dB in terms of SN ratio. This level corresponds to about 39 dB of shot noise, which, in terms of incident light quantity given signal current value, is 80 nA. If a FET without noise were developed, and if the camera SN ratio only depended on shot noise, the reference signal current of a camera tube could be lowered to this level. Because the reference signal current for the current Hi-Vision camera is 0.3–0.4 nA, this is about a fivefold improvement of sensitivity. The photoelectric conversion film of a saticon or the like has a quantum efficiency of 30 to 40% (G light). If this can be improved to 100%, a further threefold improvement could be achieved. Together with the improvement from the lowered signal current mentioned above, the total improvement in sensitivity would be about 15 times. However, improvement in sensitivity beyond this level reduces the SN ratio due to the shot noise limit. Accordingly, improvement in the sensitivity of a Hi-Vision camera has a theoretical limit of 15 times the present level, that is, about 2000 lux at an F-stop of 11.

(5) Registration

Registration error can cause not only image deterioration in the form of color misalignment, but also resolution deterioration in the luminance signal synthesized from RGB signals. For Hi-Vision, which is characterized by its high resolution, the deterioration of resolution needs to be minimized.

Figure 2.24 shows the calculated relationship between registration error and luminance signal resolution. The horizontal axis is the size of registration error expressed in terms of the distance between scanning lines, while the vertical

*If the signal level is x, the shot noise is $nx^{0.5}$ (n is a proportional constant). Assuming the gamma circuit output to be y, then $y = (x + ny^{0.5})^r$. If the noise component is small, y is approximated by $y = x^r + nrx^{r-0.5}$. If $r = 0.5$, then $y = x^r + 0.5n$. Thus the shot noise has no signal level dependence. Further, the noise level decreases by half.

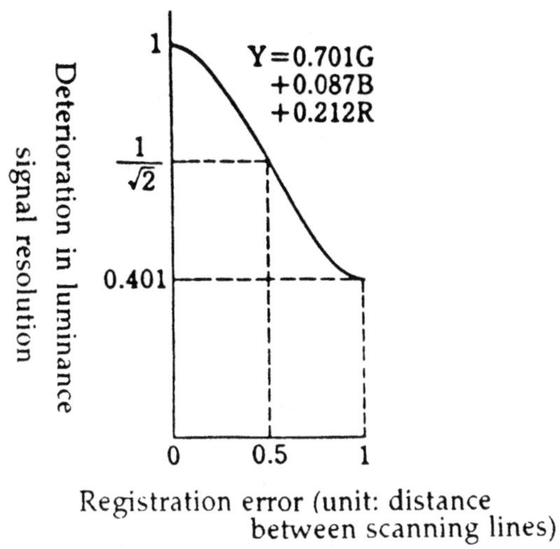

FIGURE 2.24. Deterioration of resolution resulting from registration error.

axis is the deterioration in resolution of the maximum number of TV resolution lines that can be transmitted by Hi-Vision (1035 TV lines). While the deterioration in resolution depends on the type of registration error and so does not always occur in this manner, the figure, which shows the worst case, indicates that the resolution of a camera could deteriorate to 40% due to a registration error even when the resolution of the camera tube is 100%. The complete absence of registration error, while desirable, is impossible to realize. If the deterioration in resolution due to registration error were less than −3 dB, the allowable registration error would be less than one-half the distance between scanning lines.

2.2.2 Hi-Vision Camera Technology

A Hi-Vision camera, which works according to the same principles as a standard camera, does not have to depend on a completely different technology. However, some of the techniques are unique because of its broad bandwidth and high precision.

(1) High Sensitivity Camera Tube

Because the effective sensitivity of a Hi-Vision camera is limited by the depth of field as discussed above, the Hi-Vision camera needs to have a camera with a higher sensitivity to overcome this handicap. Two types of high sensitivity camera tubes have been developed. One new type of tube makes use of an improvement of the existing saticon film by adding a large quantity of the sensitizer tellurium (Te). Tellurium, a recognized sensitizer, had not been used to improve the performance of commercial cameras because the simple addition of Te tended to cause burned-in images (a pattern of the image remains after pointing the camera at an object for too long). However, recent technological advances have made it possible to improve the sensitivity of the camera tube without burn-in. This is made possible by controlling the amount and the distribution of Te in the saticon film. Te, which is effective in increasing sensitivity to long wavelength light (red light), can double the sensitivity to red light, while increasing sensitivity to green and blue light by 1.5 times and 1.1 times, respectively.

The other type of new high sensitivity camera tube is called the HARP tube (High gain Avalanche Rushing amorphous Photoconductor). This is an entirely new technology that amplifies the signal charge within the photoconducting film. In this technology, holes excited by incident light are made to avalanche by means of a high electric field applied to a photoelectric conversion film (HARP film). The avalanche factor varies widely with the size of the electric field (target voltage). Figure 2.25 shows some measurements of the target voltage of the HARP tube and its output signal current. At a target voltage of around 150 V, avalanching does not occur and the tube shows a sensitivity level comparable to a saticon. However, at 200 V the signal current begins to increase, and at 240 to 250 V is approximately ten times this level. The avalanche factor continues to increase with the target voltage, but a sharp increase in dark current (the avalanche of thermally excited holes in the absence of incident light) makes the use of the camera tube impossible in this region.

Because the avalanche effect is accompanied by almost no noise generation, the HARP tube is a high sensitivity, low noise camera tube. The

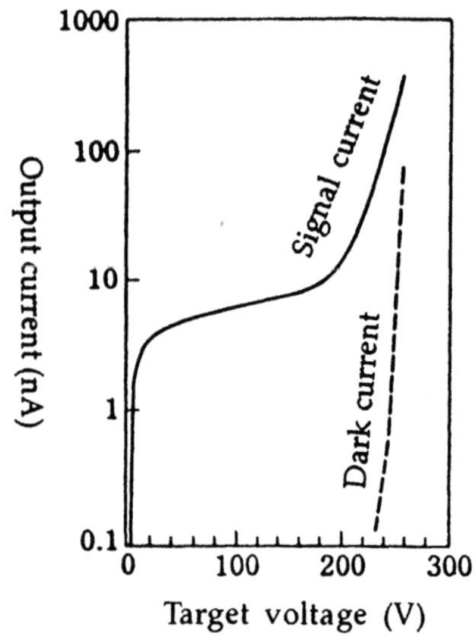

FIGURE 2.25. Voltage-current characteristics of HARP tube target.

tenfold increase in sensitivity realized by the HARP tube is close to the theoretical limit.

(2) Preamplifier

The noise characteristics of the preamplifier depend on the FET used for the first stage amplification. In recent years, gallium arsenide (GaAs) FETs have been adopted for this purpose. Although developed as ultrahigh frequency signal amplifying devices, GaAs FETs had not been used as a first stage amplifier despite their low noise characteristic because of 1/f noise (a noise whose power density increases inversely with frequency) at frequency levels of several tens of MHz and below. However, recent advances in manufacturing techniques have made it possible to limit their corner frequency (the frequency within which 1/f noise appears) to 10 MHz or less.

As with camera tubes, when amplifying an original current signal using a voltage amplifying device that has a capacitive input impedance, the input voltage decreases in proportion to the frequency. To compensate for this decrease, the preamplifier is designed to increase the amplification factor (6 dB/octave) in proportion to frequency. Thus if the noise generated by the amplifying device is flat, then the preamp output noise is triangular. If 1/f noise is included, since the noise from the amplifying device is constant if the corner frequency is exceeded, the amplitude of the noise increases in proportion to frequency as with triangular noise. But for frequencies below the corner frequency, the output amplification factor is 3 dB/oct due to the −3 dB/oct characteristic of the noise from the amplifying device and the frequency char-

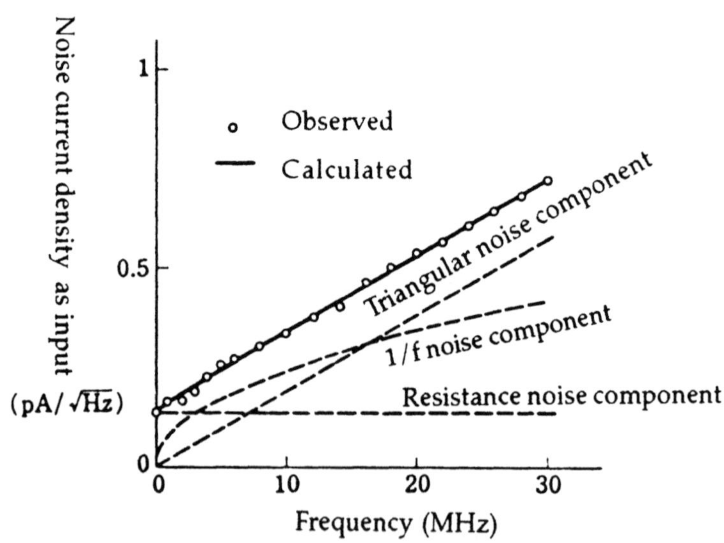

FIGURE 2.26. Noise characteristics of a GaAs FET preamplifier.

acteristic of the amplification factor (6 dB/oct). In other words, the noise amplitude is proportional to the square root of the frequency.

Figure 2.26 shows an example of noise spectra of a preamplifier with a GaAs FET. The SN ratio of a GaAs FET preamplifier is better than that of a silicon FET by about 6 dB.

(3) Camera Cables

The transmission loss of a multicore camera cable is approximately 100 dB/km at 30 MHz. This makes long distance transmission difficult; a cable length of approximately 200 meters is the practical limit. For longer camera cables, optical fiber is used to transmit the R, G, and B signals in parallel over three fibers. Technologies have been developed to automatically correct the levels on these channels and to multiplex control signals from the CCU (Camera Control Unit) over a single fiber.

(4) Registration Correction Circuit

There are two major types of registration errors. The first type, called a static error, is the error that cannot be removed by camera adjustments. The second type is called a dynamic error, and occurs when the camera registration shifts from its adjusted state while in operation.

A circuit called DRC (Digital Registration Correction) is very effective in correcting static error. It divides the screen into a grid with several hundred points so that registration can be adjusted individually at each point. The circuit has achieved a registration precision of less than four-tenths of the space between scanning lines. To avoid the tremendous workload and time required to manually adjust several hundred points, as well as human error, the DRC has been made fully functional by means of an automatic setup function that performs these adjustments automatically.

In the actual operation of the camera, the registration will vary with changes in temperature, geomagnetism, and focal length of the zoom lens. To correct these dynamic errors, many cameras are equipped with a method of detecting the values of these variable factors and controlling the corrective waveforms. However, as this is a feed-forward control technique, it is limited in the precision of its corrections and its reproducibility. In addition, with this technique the control data has to be changed whenever the lens or camera tube is exchanged because these affect the dynamic error.

To solve this problem, a method has been developed that detects registration errors using only the image signal. Because this detection technique makes possible registration monitoring at all times during camera operation, when a registration error is caused by some factor during operation, a feedback loop is formed to reduce the error to zero. Portable cameras with this function have already been developed.

2.3 TELECINE

2.3.1 Telecine for Hi-Vision

So far, three types of telecines have been developed for Hi-Vision—laser, FSS (flying spot scanner), and camera tube.

(1) Laser Telecine[9]

A laser telecine reads the film image directly with a high luminance laser beam converged on a micro spot. This system is capable of obtaining a high SN ratio and high resolution image.

A photograph and basic configuration of a laser telecine are shown in Figures 2.27 and 2.28, respectively. The scanning beam source for the R channel is an He-Ne laser, while those for the G and B channels are an Ar^+ laser and an He-Cd laser, respectively. Power fluctuations and noise from the three laser beams for R, G and B are reduced by feedback loops with an Acousto-Optical Modulator (AOM). The three beams are then synthesized into one beam with Dichroic Mirrors (DM). The synthesized laser beam is given a horizontal deflection by a rotating polygon optical deflector,[10] and goes through an auxiliary deflector before it converges on the film surface. The scanning of the film with the laser beam is performed by sequential one-line scanning. The vertical reflection is performed by the film which is running continuously. The auxiliary deflector performs fine adjustments of the scanning position of the laser beam in the vertical direction in responding

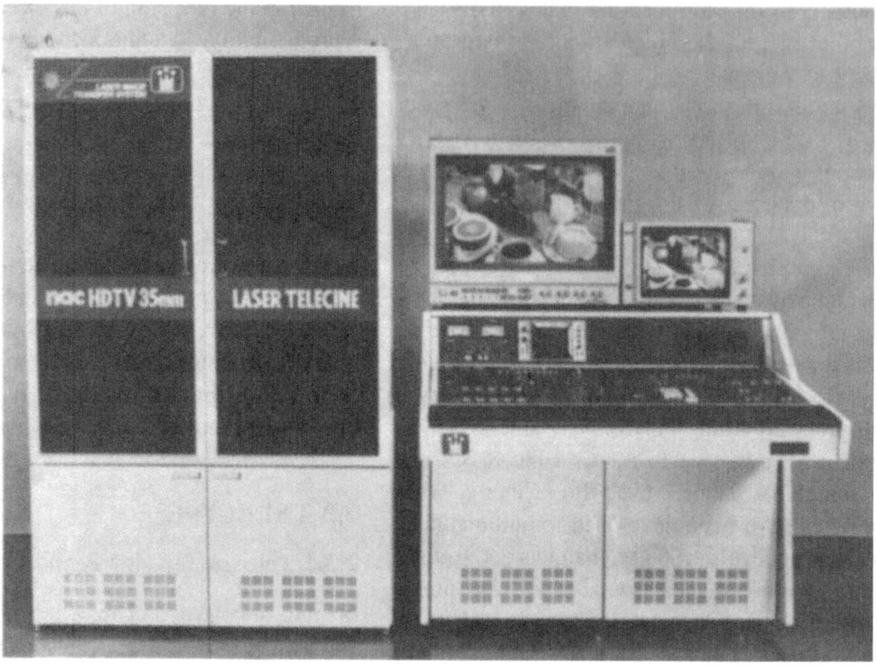

FIGURE 2.27. A laser telecine.

to the effective screen height of the image on the film and the direction and speed of the film. The laser beam that has penetrated through the film is separated back to R, G, and B beams, which then are converted by photomultipliers to electrical signals. The R, G, and B image signals then receive gamma correction and other color corrections, and are converted to digital signals before they are sent to the system conversion unit. In the system conversion unit, they go through conversions for sequential-interlace scanning, aspect ratio, and the number of images per second. The aspect ratio conversion is performed with beam deflection for the vertical direction, and D-D conversion for the horizontal direction. The conversion of the film's 24 frames

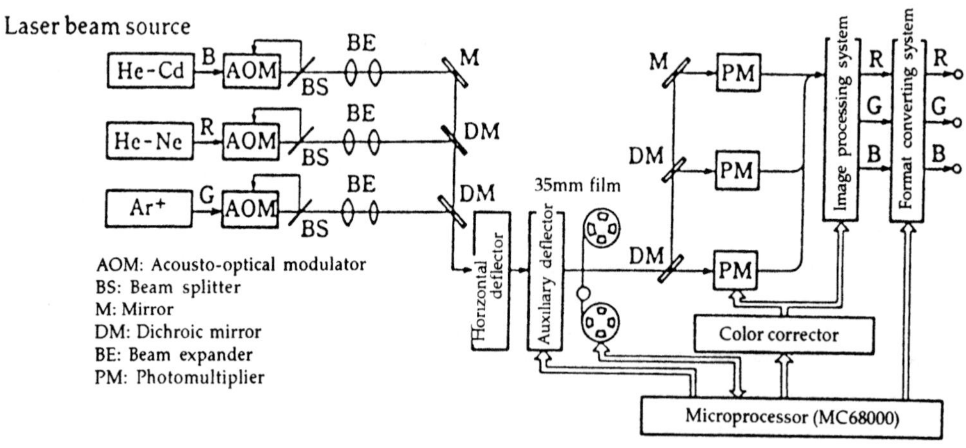

FIGURE 2.28. Basic configuration of a laser telecine.

per second into television's 60 fields per second is done using frame memory, and in addition to the 2-3 system, a system based on motion compensation has also been developed.[11]

(2) FSS Telecine[12]

In the FSS system, an electron beam from an electron gun is used to form a raster on a CRT (Cathode Ray Tube), and the light emitted from the fluorescent screen transfers the image onto the film. Panning and zooming is made possible by varying the size and position of the raster on the CRT. Because there is only one deflected beam, no registration error is caused.

Figure 2.29 is a photograph of an FSS telecine, and Figure 2.30 shows the basic configuration of an FSS telecine. In this telecine, a capstan drives the film continuously. The light that has passed through the film is separated into R, G, and B colors by dichroic mirrors. The colors are then converted into electrical signals with photomultipliers. In the electrical unit, the CRT after glow correction, shading correction, gamma correction, and other types of image processing steps are executed. This telecine, like the laser telecine, uses frame memory in converting the number of images per second.

(3) Camera Tube Telecine—The Saticon Telecine[13]

The camera tube telecine system stops the film at every frame and takes its picture. The film is advanced with a 2-3 advancing method, wherein the camera shoots a frame twice before the frame is advanced, and then shoots the next frame three times before it is advanced. The conversion of the number of the images per second is done by repeating the process. In this respect, the camera tube telecine system is markedly different from the laser telecine and FSS telecine, which depend on a continuous tape driven system and a frame memory system.

This telecine is called a saticon telecine because it uses an electromagnetic converging, static reflecting (MS) 1-inch saticon (H-4187) in its camera tube. Shown in Figures 2.31 and 2.32

FIGURE 2.29. FSS telecine.

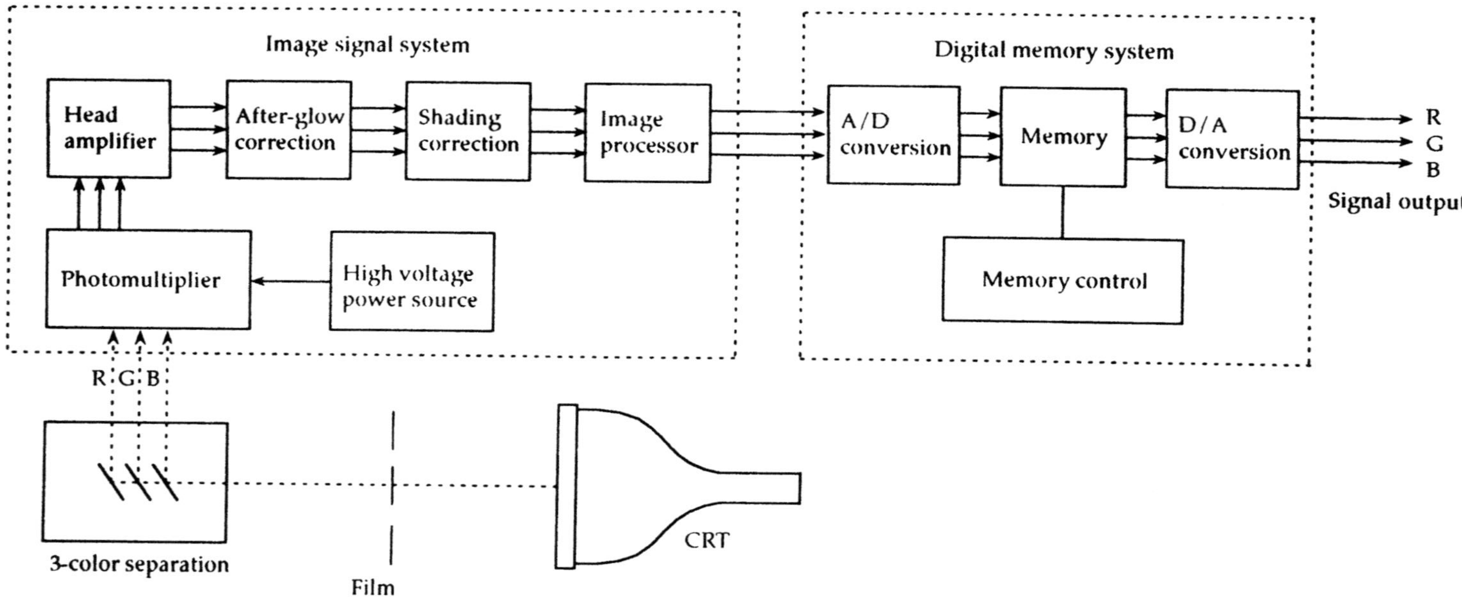

FIGURE 2.30. Basic configuration of FSS telecine.

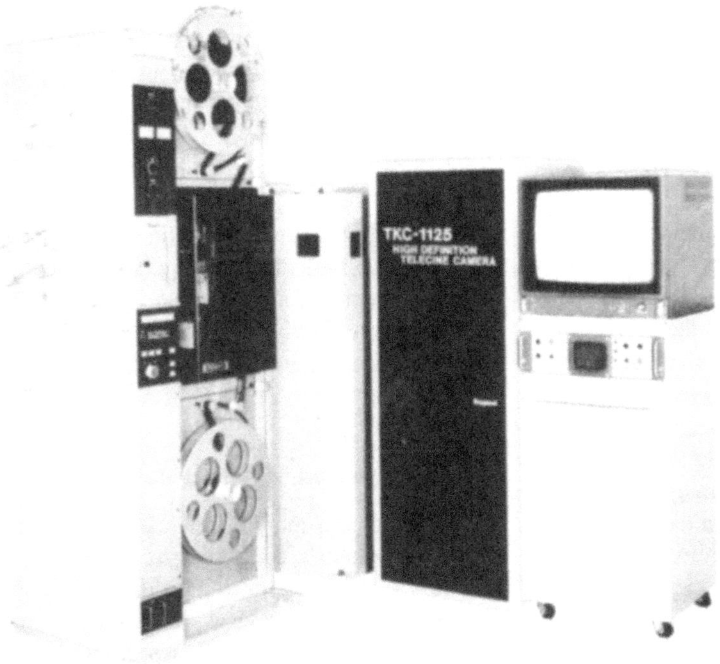

FIGURE 2.31. Saticon telecine.

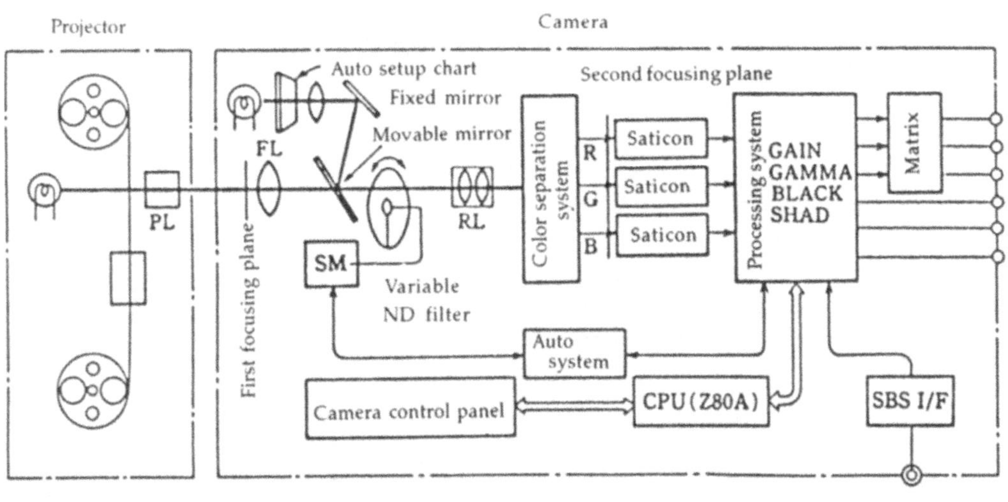

FIGURE 2.32. Basic configuration of saticon telecine.

are a photograph of and a basic configuration of a saticon telecine. The telecine consists of two sections—the camera's main body and the film projector unit. The image projected by the film projector is received by the first focusing plane of the cameras main body. The image then passes through a field lens, variable ND filter, and relay lens before it is separated into R, G, and B by a color-separating optical system with a prism, so that the images are formed on the second imaging plane. Then the R, G, and B signals are converted by the saticon telecine into electrical signals which receive various image processing treatments such as shading, gain balancing, gamma balancing, etc. before they are output. In this system, the image flutter in the film projector section is limited to less than 0.015%, or smaller, in both horizontal and vertical directions, by fixing of the film with registration pins.

2.3.2 Film Size

While there are various sizes of movie films such as 16 mm, 35 mm, 70 mm, the unit area resolution and granularity of the film itself are the same regardless of the format. It follows that the sharpness and graininess of the image on the screen increases with the film size provided that the screen size is the same. Figure 2.33 shows the resolutions of representative movie films in use today, using film size as a parameter. The actual resolution can be much lower than is shown by the curves, depending on the camera, exposure, development process, and duplicating process. Incidentally, the resolution of a 35 mm film shown in a movie theater is said to be 700 to 800 TV lines at best.[14] For Hi-Vision, 70 mm film is desirable in terms of image quality, and 16 mm film is inadequate. Although 35 mm film is not really satisfactory in image quality, it is suitable if we consider the abundant supply of program materials and the restrictions in the equipment size in program production.

2.3.3 Conversion of Aspect Ratio

Today's 35 mm movie films can be classified by the projector's aperture size. As shown in Table 2.4, these apertures have different aspect ratios. There are two ways of converting these film aspect ratios to the 16:9 aspect ratio of Hi-Vision. In one method, the film image spans across from the left to right border, and the upper and lower margins are either cut off or blackened. In the other, the film image fills the screen vertically from the upper to lower edge, and the left and right sides are either cut or blackened. To do either of these, one of the following methods is employed:

- Change in film scanning beam deflection width,
- Change in image size using a zoom lens,
- Digital D-D conversion.

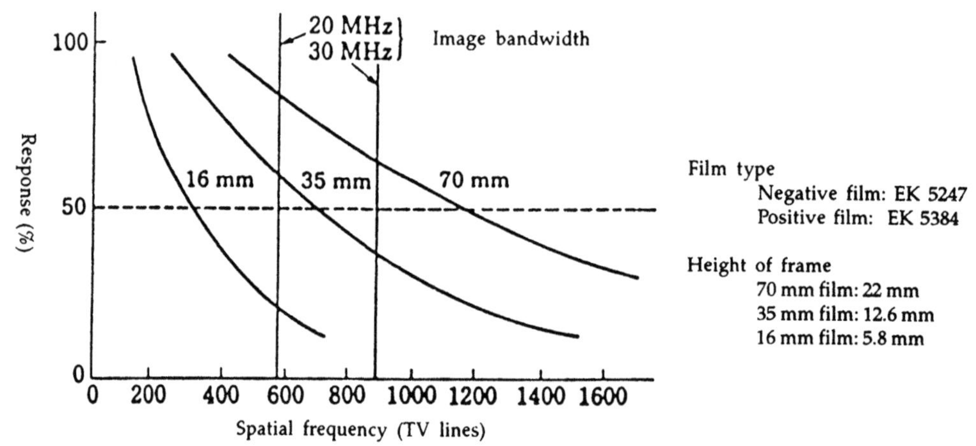

FIGURE 2.33. Resolution of various movie films (product of negative and positive film).

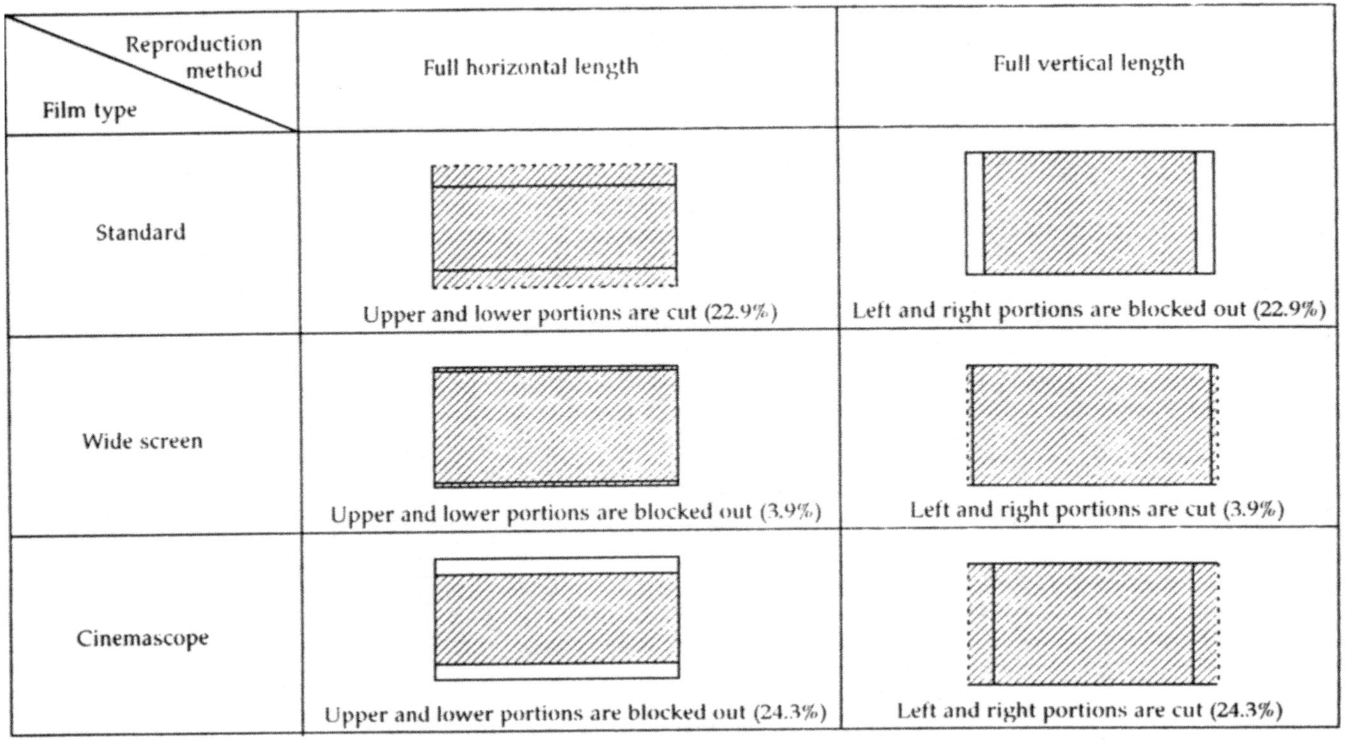

The frame shown with bold lines indicates the Hi-Vision format. The hatched area is the film frame.

FIGURE 2.34. Aspect ratios.

TABLE 2.4. Aperture sizes of 35mm projectors.

Type	Width (mm)	Height (mm)	Aspect ratio
Standard	21.00	15.30	1.37 : 1
Wide screen	21.00	11.35	1.85 : 1
Cinemascope	21.31	18.16	2.35 : 1

2.3.4 Conversion of Number of Images per Second

Movie film usually runs at a rate of 24 frames per second. To convert this rate to the Hi-Vision rate of 60 fields per second, one of the following techniques can be employed:

- 2-3 system, in which a frame is converted into 2 fields and the next consecutive frame into 3 fields,
- Linear interpolation type insertion system that uses insert filters,
- Motion compensation insertion system in which position insertion is performed with motion vectors.

(1) 2-3 System

The 2-3 conversion process is shown in Figure 2.35. Assume Q_k (where k is an integer) is the input signal before conversion sampled at 24 Hz, and P_m (where m is an integer) is the signal after conversion sampled at 60 Hz. The converted output signal P_k is

$$\begin{array}{lll} P_0 = Q_0 & P_1 = Q_0 & P_2 = Q_0 \\ P_3 = Q_1 & P_4 = Q_1 & \\ P_5 = Q_2 & P_6 = Q_2 & P_7 = Q_2 \\ P_8 = Q_3 & P_9 = Q_3 & \end{array} \tag{2.5}$$

Figure 2.36 shows an example of the realization of this process by the use of frame memory. The frame memory is written into at 24 Hz synchronously with the input signal from the film. The read-out is done at 60 Hz by synchronizing with the TV signal. In this system, judder (unnatural movements) is caused because the same frame is repeated two or three times.

(2) Linear Interpolation Type Insertion System.

Figure 2.37 shows the conversion process for a linear interpolation with two taps from the inserted filters. In this case, the output signal after conversion, P_m which is the sum of the products of impulse response h_n (where n is an integer) and input signal Q_k, is expressed by the following equations.

$$\begin{aligned} P_0 &= h_0Q_0 + h_5Q_1 \\ P_1 &= h_{-1}Q_0 + h_4Q_1 \\ P_2 &= h_{-2}Q_0 + h_3Q_1 \\ P_3 &= h_{-3}Q_1 + h_2Q_2 \\ P_4 &= h_{-4}Q_1 + h_1Q_2 \\ P_5 &= h_0Q_2 + h_5Q_3 \end{aligned} \tag{2.6}$$

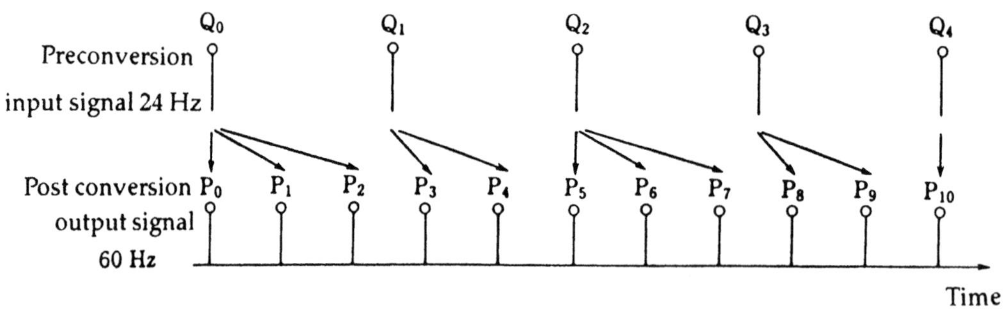

FIGURE 2.35. Conversion process in 2-3 method.

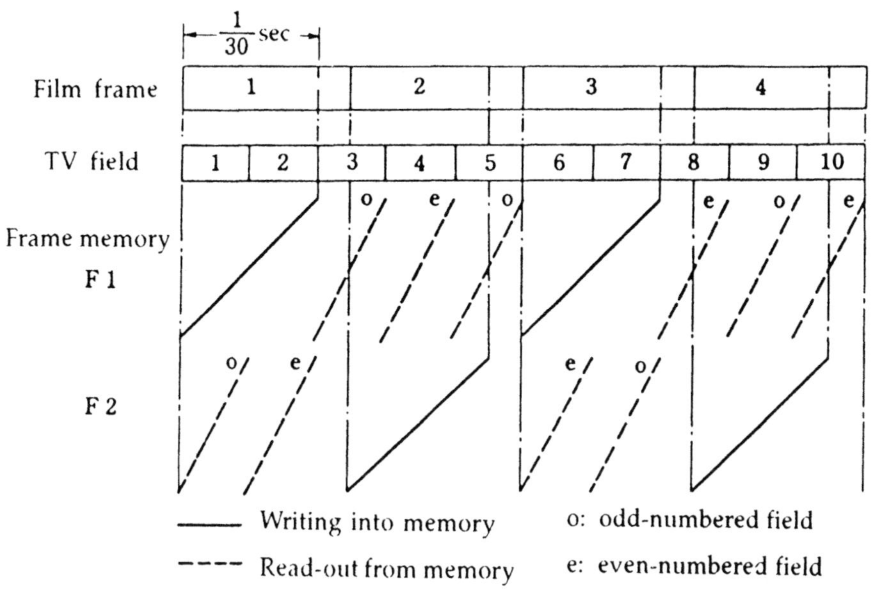

FIGURE 2.36. Example of 2-3 conversion with frame memory.

The frequency spectra shown in Figure 2.38 compares frequency spectra of the 2-3 method and linear interpolation insertion when converting an image in which the sine wave moves horizontally each frame (24 Hz) to a 60 Hz field frequency signal. In the spectra, the flicker components formed at the multiples of 12 Hz have caused interference in the form of jitters. Linear interpolation insertion decreases the energy of jitter components, but causes moving objects to blur.

(3) Motion Correcting Insertion System

Figure 2.39 shows examples of the conversions by the 2-3 method and motion correcting insertion of a scene of a moving car from a 24-frame

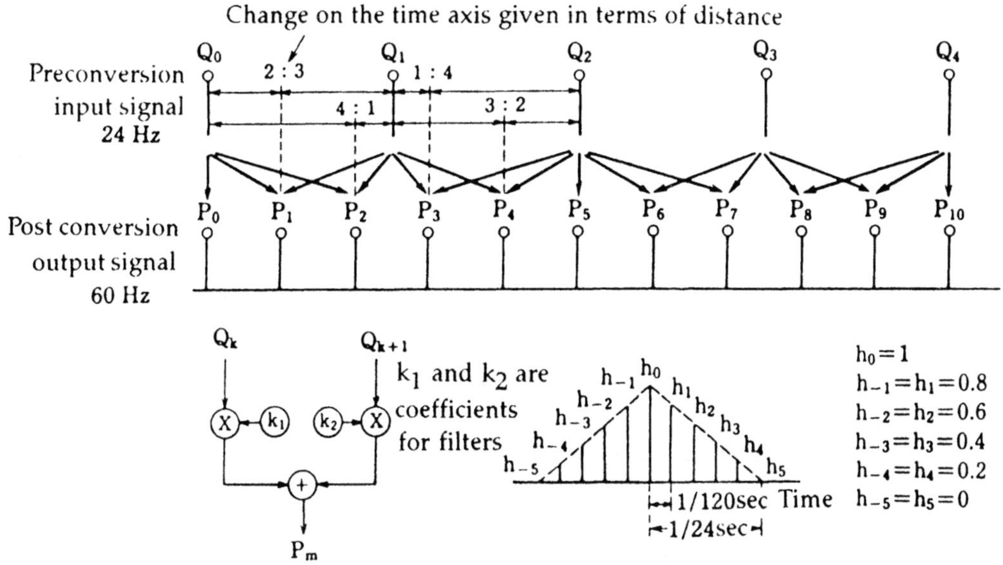

FIGURE 2.37. Conversion process according to linear interpolation method.

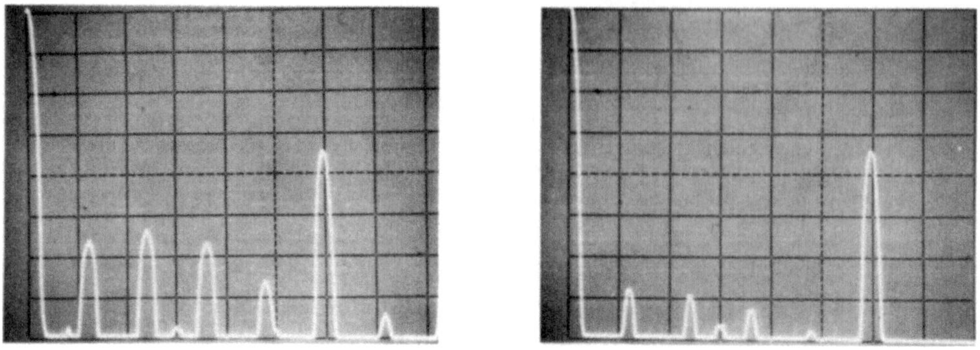

FIGURE 2.38. Frequency spectra of converted images.

signal to a 60-field signal. In the images converted by the 2-3 method, the motion is discontinuous, and the vehicle's movement is unnatural (due to judder). But with the images converted by the motion correcting insertion system, the car moves smoothly.

Figure 2.40 shows the conversion process using motion correcting interpolation. In this conversion, the motion vector detected from the input signals before conversion, Q^k and Q_{k+1}, are converted by calculations based on the distances of sample points on the time axis after conversion. The image location is then moved by that amount in the horizontal and vertical directions to obtain an insertion frame. In Figure 2.40, for example, an insertion frame P_4 is formed from input signals Q_1 and Q_2 in the following manner. Suppose that a motion vector **V** has been detected from previous frame Q_1 and present frame Q_2. Because the ratio between the distance in time between P_4 and Q_1 and distance in time between P_4 and Q_2 is 3:2, an insertion frame P can be obtained by correcting the position of the previous frame Q_1 by a magnitude of 2/5 of the motion vector **V** and the present frame Q_2 by $-3/5$, obtaining the weighted av-

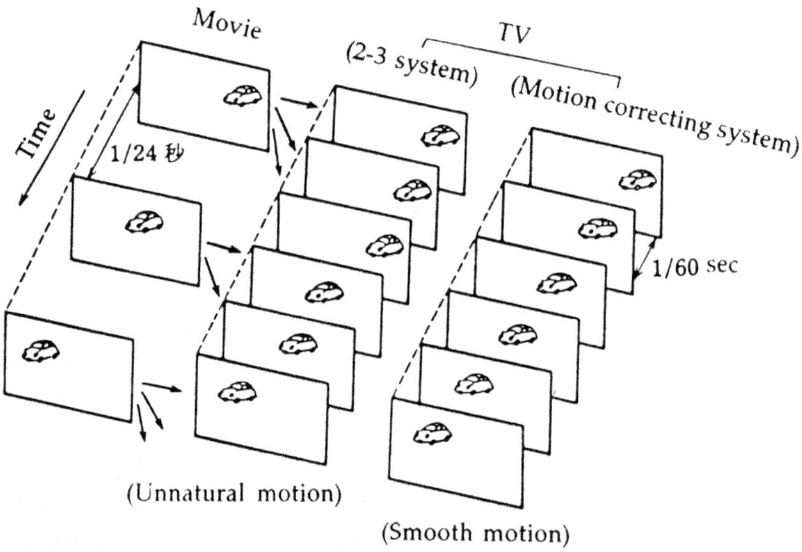

FIGURE 2.39. Frame frequency conversions by means of 2-3 method and motion compensating insertion method.

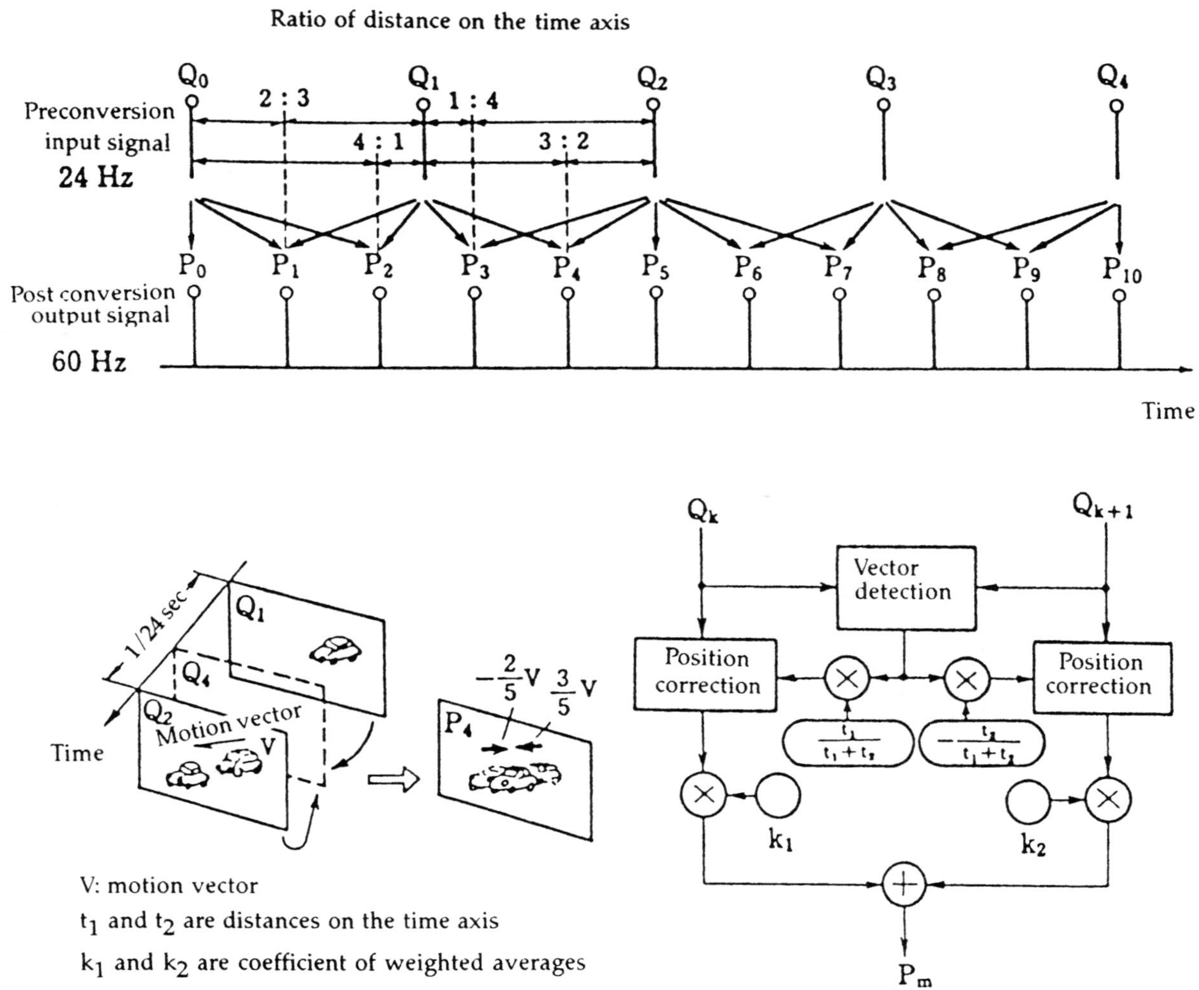

V: motion vector

t_1 and t_2 are distances on the time axis

k_1 and k_2 are coefficient of weighted averages

FIGURE 2.40. Conversion process in motion compensation insertion method.

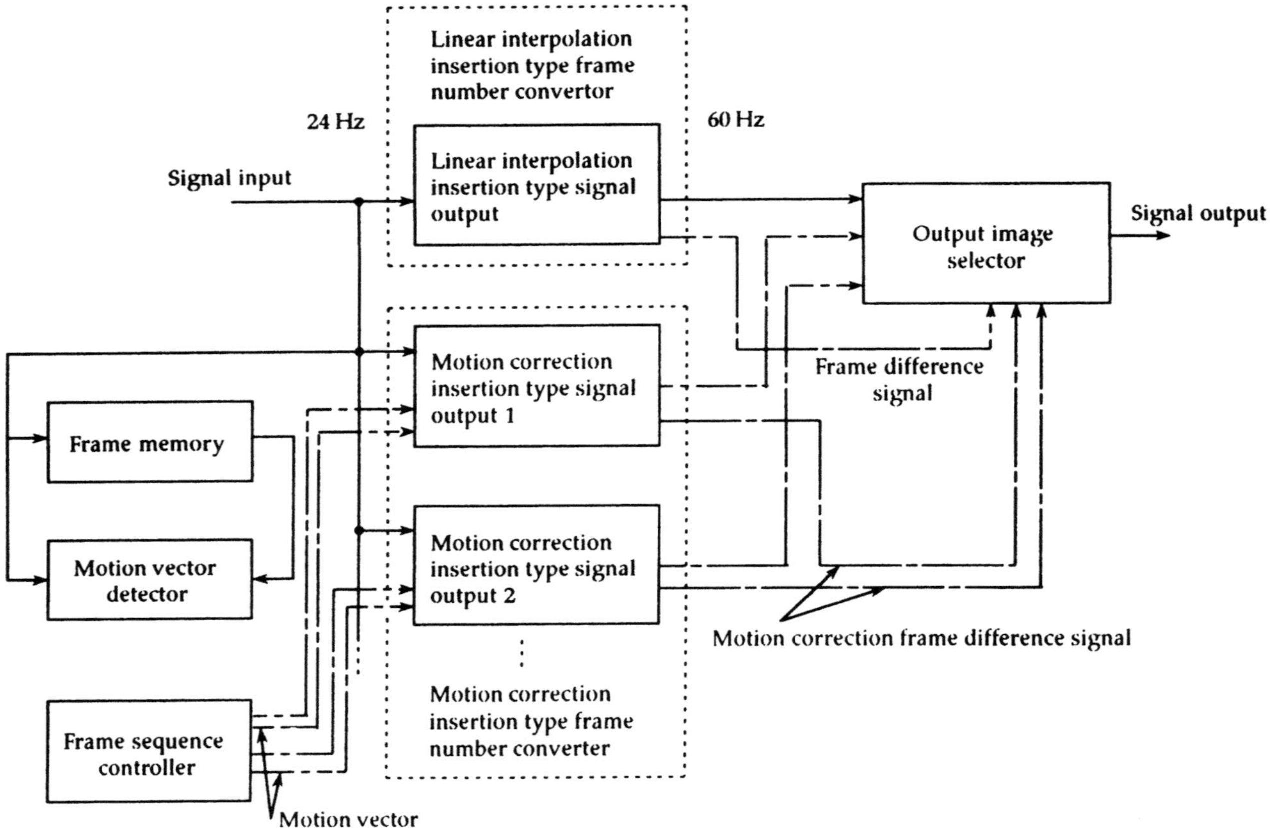

FIGURE 2.41. Example of a motion compensating frame conversion.

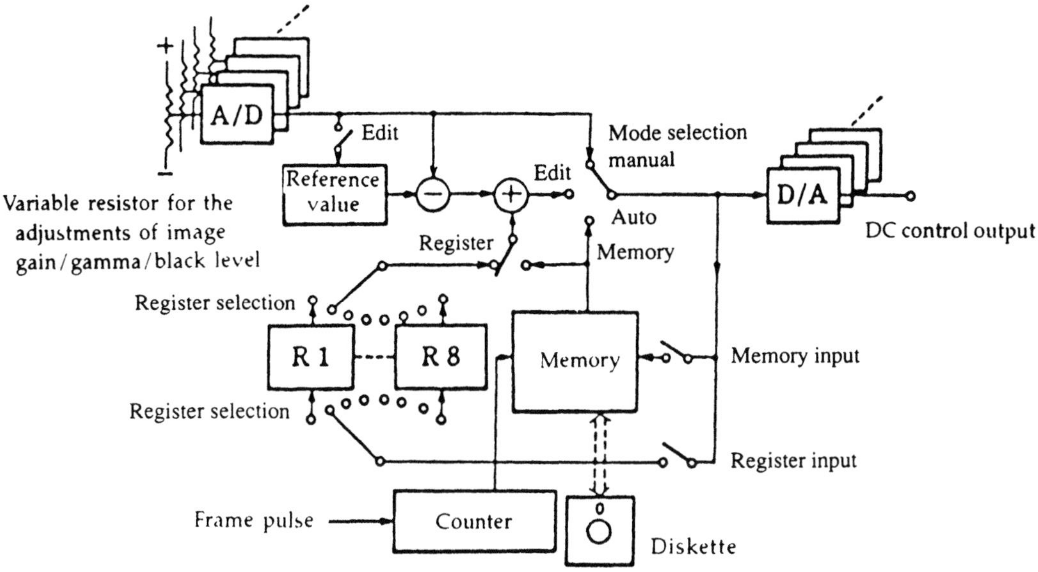

FIGURE 2.42. Compensation data flow in the color corrector.

erage of the two signals, and performing the motion correction.

In ordinary images, the actual motions on a screen will be complex and variable in directions and magnitudes. It will be practically impossible to correct exactly the motions of all the moving parts. A sensible solution would be to combine motion correction insertion with the 2-3 method or linear interpolation insertion. Such a configuration is shown in Figure 2.41. Although the use of a large number of motion vector detectors would produce more reliable results, the actual number of detectors will be one to four because of the restrictions in actual hardware configuration. Images obtained by the motion corrections based on motion vectors are combined with the images obtained by the 2-3 method or linear interpolation insertion through pixel by pixel selections to form optimized insertion frames.

2.3.5 Color Correction

The image density range, tone, and color balance, of film images can vary in different movies, or even between scenes. Thus a telecine must have a color correction function that sets the optimum reproduction conditions for each scene.

Figure 2.42 shows an example of correction data flow in a color corrector. The color corrector is able to adjust image gain, gamma, and black level for R, G, and B channels, either individually or in common for all the channels. The correction data which an operator adjusts while watching the monitor and frame count number are entered in the memory. Because several major scenes can appear repeatedly, it is convenient if the correction data once stored can be read out repeatedly whenever they are needed. Registers R1 to R8 are provided to meet this need. If necessary, the corrector is able to give fine adjustments to the data of certain cuts. In performing the fine adjustments, the data in the registers are used as references. By turning the adjustment knob, the adjusting data are added to the data in the register in differential operation. The differential operation, which is applicable to adjust the memory output, is convenient when it becomes necessary to revise the correction data which has already been incorporated in a color correction data table.

2.3.6 Movie Production

One industrial application of Hi-Vision is in movie production. The technology associated with this application is called electrocinematography. Electrocinematography is a technology that edits material videotaped with Hi-Vision cameras with such techniques as chromakey syntheses and special effects, and then converts the results to a movie film format. Compared with optical processing, this technology decreases movie production time. Further, it makes possible the use of complex processes.

Telecine equipment is also being used in movie production. Such use occurs after high speed or low speed shooting with a film camera. The telecine is used to convert the film to video signals and then to subject them to chromakey synthesis and special effect processing. However, this application requires the solution of a number of problems, including correcting picture blur caused during the shooting, development, or telecine reproduction; matching the tone and color between the film and the video camera material; and optimizing the conversion from a 24-frame system to a 60-field system and back to the 24-frame system.

REFERENCES

1. Ogusu, "Development of Camera Tubes for High Definition Television," *NHK Technical Report* Vol. 25, No. 8, pp 37–50 (1982).
2. Kokusu, "Diode Gun Camera Tubes," *Journal of the Institute of Television Engineers of Japan (ITEJ)*, Vol. 34, No. 8, pp. 34–38 (1980).
3. Tanioka et al. "Increasing the Sensitivity of a Photoconductive Target," *ITEJ Technical Report,* Vol. 10, No. 45, ED-87-1 (1987).
4. Yamazaki et al., "UV Characteristics of an Amorphous Se Avalanche Target," *ITEJ Proceedings,* 2–8 (1988).
5. Tanioka et al., "Red Light Sensitizing Amorphous Se Avalanche Photoconductive Target," *ITEJ Proceedings,* pp. 2–7 (1988).
6. Nishida et al., "Wide Dynamic Range HDTV Image Sensor with Aliasing Suppression," *IEEE, ICCE Digest of Technical Papers,* pp. 60–61 (1988).
7. Nagano et al., "2-Million Pixel High Resolution CCD Image Sensor," *ITEJ Technical Report,* ED 87-174, pp. 45–50 (1988).
8. Manabe et al., "A 2-Million Pixel CCD Imager Overlaid with an Amorphous Silicon Photoconversion Layer," *IEEE, ISSCC Digest of Technical Papers,* pp. 50–51 (1988).
9. Hirabayashi, Motoki, "A Laser Telecine for Hi-Vision," *ITEJ Technical Report,* OPT 229 (Sept. 1986).
10. Masuko, Kameyama, Kutsuzawa, Naraoka, "Development of a Rotating Polygon Optical Deflector," *NHK Technical Journal* Vol. 32, No. 1 (1980).
11. Hirabayashi, Nojiri, Tanaka, "Motion Compensating Frame Number Conversion for Laser Telecine," *ITEJ Proceedings,* pp. 9–8 (1988).
12. Osato, Brittain, "Flying Spot Telecine for Rankcintel HDTV," *ITEJ Technical Report,* OPT 88-1 (Jan. 1988).
13. Iijima, Aizawa, "Development of a Saticon Telecine for Hi-Vision," *ITEJ Technical Report,* OPT 221 (March 1986).
14. Kaiser et al., " Resolution Requirements for HDTV Based upon the Performance of 35mm Motion-Picture Films for Theatrical Viewing," *SMPTE,* Vol. 94, No. 6 (1985).

3

Transmission

Yuichi Iwadate, Yozo Utsumi, Yoshimichi Ohtsuka, Kenji Oumaru, Keiichi Kubota, Toshiyuki Takegahara, Taiji Nishigawa, Akio Yanagimachi, Takehiko Yoshino

3.1 MUSE TRANSMISSION SYSTEM

3.1.1 The MUSE System

The most accessible means of realizing Hi-Vision broadcasting in the future is with satellite broadcasting. Satellite broadcasting is capable of providing nationwide TV service both effectively and economically. Of the eight channels in the 12 GHz band assigned to Japan for direct satellite broadcasting, three channels on broadcast satellite BS-3 are being allocated for NTSC as well as for Hi-Vision television. As for terrestrial VHF and UHF frequencies, they are simply not available, and even if they were, there would be problems with the development timeline and economic feasibility involved in developing a nationwide broadcasting network. On the other hand, a nationwide cable or fiber optic network would be difficult to realize because it would necessarily be limited to cities or localities.

The Hi-Vision signal bandwidth is about five times wider than that of the current NTSC system, which is 4.2 MHz. Thus about 20 to 25 MHz is needed to transmit Hi-Vision signals. However, as the available frequency spectrum is limited, broadcasting these signals will require band compression.

Japan's direct broadcasting satellites have a channel bandwidth of 27 MHz and use Frequency Modulation (FM). With FM, the signal bandwidth must be about 1/3 of the carrier bandwidth to accommodate a sufficient frequency deviation (see Section 3.3). Thus to broadcast a Hi-Vision program on one satellite channel, the signal bandwidth must not exceed 9 MHz. To compress the signals to this bandwidth requires a highly advanced band compression method. The compression method must be able to allow the reception of weak signals and at the same time be as simple as possible.

The MUSE system[1] was developed for Hi-Vision broadcasting over a single satellite channel. It compresses the baseband signal to 8.1 MHz by a relatively simple technique without deterioration of image quality. The audio signal is broadcast either in mode A for 4-channel stereo or mode B for high quality 2-channel stereo. Both modes A and B are digitally transmitted by multiplexing during the vertical blanking period of the video signal.

As the name suggests, MUltiple sub-Nyquist-Sampling Encoding (MUSE) is a multiple subsampling encoding system. It performs subsampling twice to compress the video signal to a bandwidth of 8.1 MHz.

Figures 3.1 and 3.2 are block diagrams of a MUSE encoder and decoder. While the encoder and decoder consist of digital signal processing circuits, the transmission lines carry analog

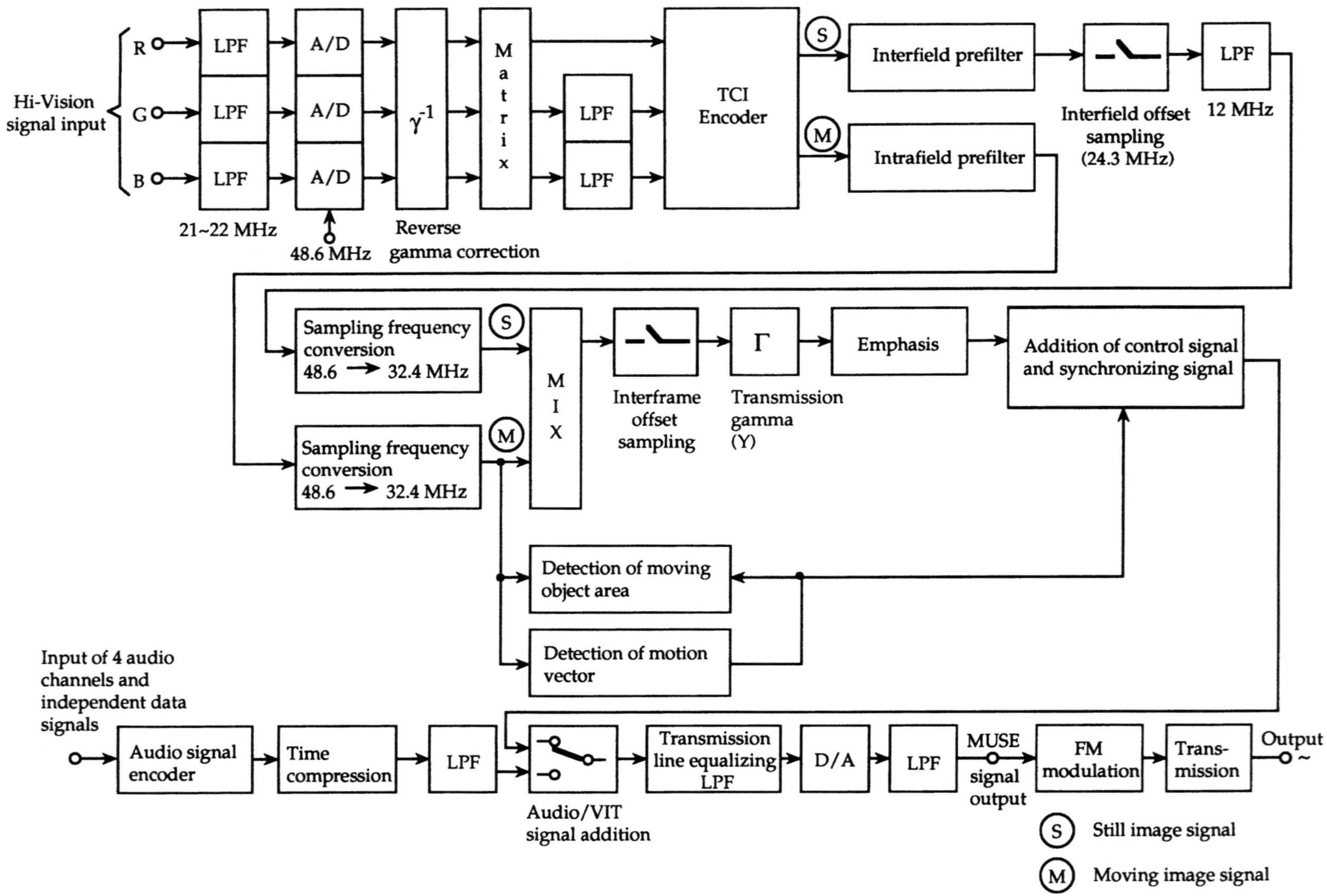

FIGURE 3.1. MUSE encoder configuration.

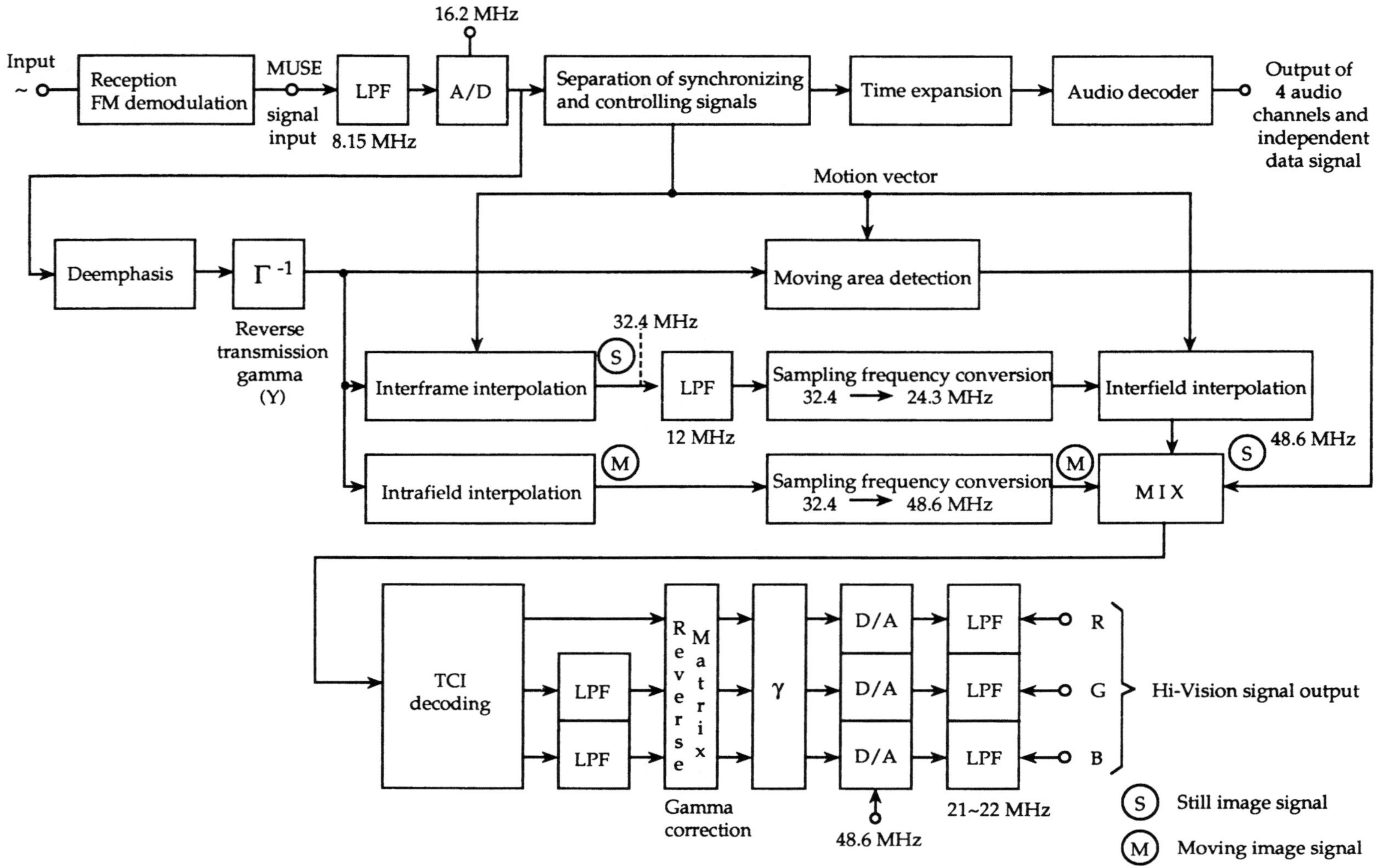

FIGURE 3.2. MUSE decoder configuration.

sampling. The digital signal processing performed by MUSE is a three-dimensional subsampling consisting of TCI conversion, motion detection and processing for still and moving images, and motion vector detection and correction. In addition, pseudo-constant-luminance transmission has been adopted to deal with problems related to band and noise balance between the luminance (Y) and chrominance (C) signals. The technical details of this system are explained in the following sections.

3.1.2 MUSE Transmission Signal

(1) Video Signal

The MUSE transmission signal is based on a TCI format, which performs time division multiplexing of the Y-C signal. The signal is then subjected to field offset, frame offset, and line offset-sampling and compressed to 8.1 MHz, and then augmented with synchronization, audio, independent data, and control signals. The MUSE signal as displayed on a monochrome monitor is shown in Figure 3.3. Figure 3.4 shows a detailed pixel-by-pixel and line-by-line description of the MUSE signal.

The sampling frequency for the MUSE transmission signal is 16.2 MHz. One line (1H) is sampled at 480 points, of which 11 points are assigned to HD (horizontal synchronizing signal), 94 points to C signals, and 374 points to the Y signal. Between Y and C, a gap of 1 CK (clock period) is inserted to prevent signal interference between them. Two C signals are time-compressed to 1/4, and the R-Y and B-Y signals are line sequentially multiplexed to odd-numbered lines and even-numbered lines, respectively.

The bandwidth of the C signal (as a time ratio) is 1/8 that of the Y signal due to the 1/4 time compression and line sequential multiplexing. Thus the C signal is easily influenced by transmission line noise when time-expanded in the decoder. To solve this problem while taking into consideration the balance in SN ratios for Y and C, the signal processing is based on the constant luminance principle (described later), and the transmission level for C is set 3 dB higher than for Y. Since C is multiplexed line sequentially, a 1H delay for Y corresponds to a 2H delay for C, and after two-dimensional filtering, line sequential insertion and other signal processing including vertical signal processing, the C signal lags behind Y For this reason, the C signal is set to precede the Y signal in the encoder by 4H.

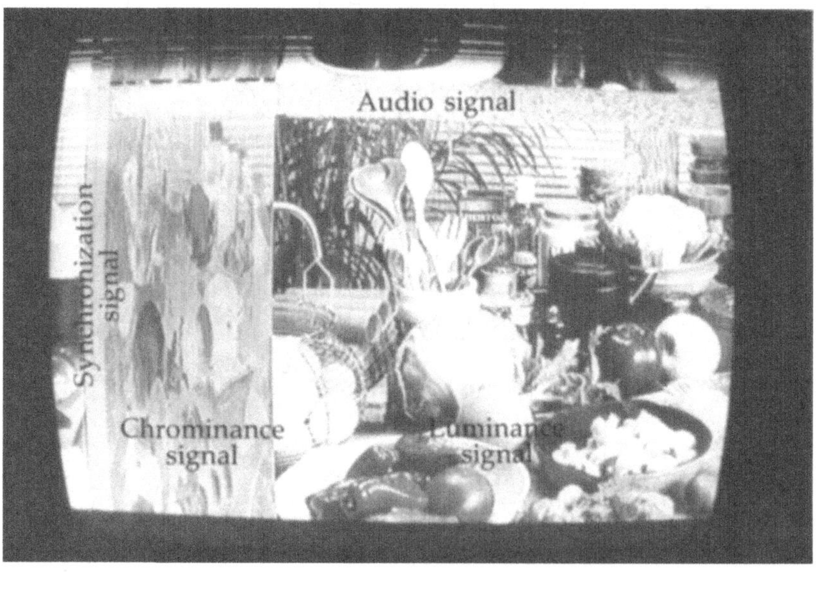

FIGURE 3.3. Allocation of MUSE transmitting signals displayed on monochrome monitor.

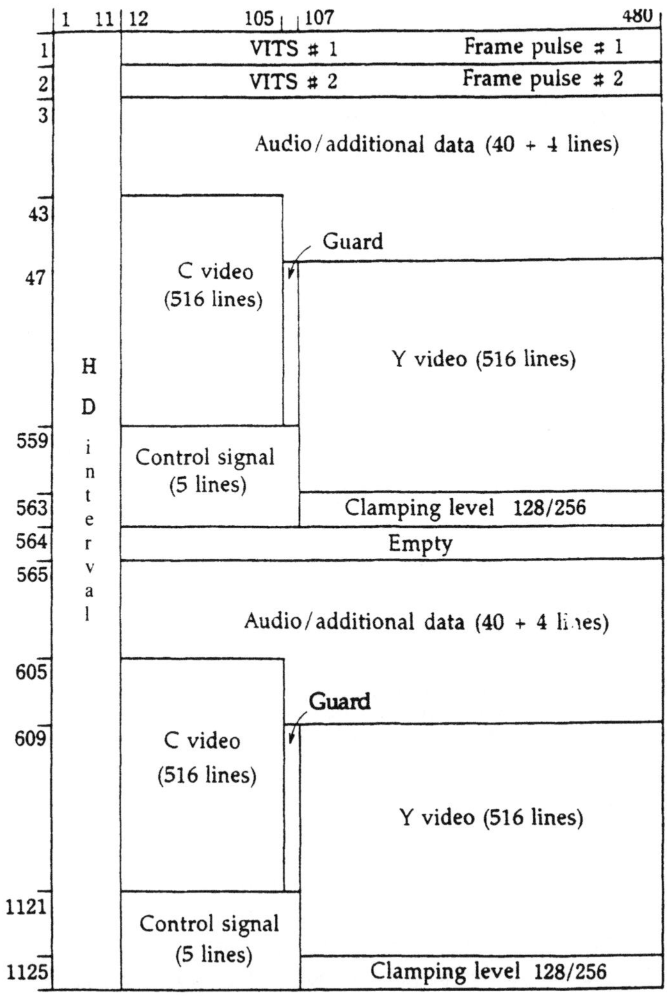

FIGURE 3.4. Form of the MUSE transmission signal.

During the vertical blanking period, audio and independent data (1350 kb/s) are multiplexed in the form of digital signals. Also multiplexed are frame pulse signals for vertical synchronization, clamp level signals for defining the neutral level of the C signal and for the AFC (Automatic Frequency Control), VIT (Vertical Impulse Test) impulse signals to equalize the transmission line, and signals for controlling the motion vector subsampling phase.

(2) Synchronization Signal

Because the MUSE system is based on the analog transmission of sampled signals, it is necessary to accurately recreate the sample clock in the decoder. As will be described later, the slightest shift in the resample clock phase can cause distortion in the waveform, resulting in ringing disturbances on the screen. To meet this strict requirement and maintain the correct phase, a vertical synchronization signal or frame pulse and a horizontal synchronization signal waveform (both shown in Figure 3.5) are used.

The frame pulses are square waves that have 100% of the video signal, which inverts at every fourth 16.2 MHz clock interval. The vertical synchronization signal is inserted every frame by using two lines whose waveform polarities

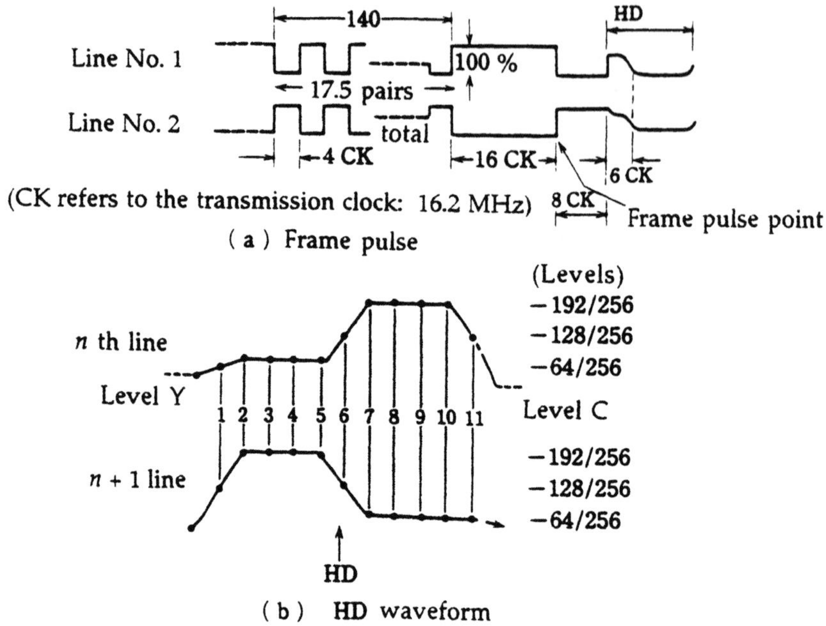

HD waveform is reset after the completion of line inversion and frame pulse transmission. (The HD for line no. 3 is rising.) Pixels #1 and #11 may be in the HD waveform or video signal. Preferably the average value of the HD waveform and video signal is taken.

FIGURE 3.5. Synchronizing signal waveforms (waveforms of frame pulses and HD).

are inverted with respect to each other. In general, the upper and the lower lines of a video signal are highly correlated. The probability that a video signal would include an inverted waveform the same shape as the frame pulse is extremely low, and even if this were to happen, it is unlikely that this condition would continue over several dozen frames. Therefore we can assume that this frame pulse waveform is able to synchronize frames accurately. Further, frame pulses can be used for automatic level control. This signal processing is indispensable for nonlinear emphasis and nonlinear processing based on pseudo constant luminance transmission, which is described later.

The horizontal synchronization signal has 50% of the video signal, and its waveform polarity is inverted every other line. The leading edge as well as the trailing edge of the waveform comprise a half cycle of a 4.05 MHz sine wave (4.05 MHz is 1/4 of the resampling clock frequency). The reason that a conventional negative polarity synchronization signal was not adopted was because our aim was to improve the SN ratio by the amount of the synchronization signal. We adopted the polarity inversion for every other line to distinguish the signal from the video signal just as with the frame pulses, and to cancel out the direct current component and avoid the distortion effect of even numbered higher harmonic waves. Although odd-numbered and even-numbered harmonics are both distorted, the odd-numbered harmonic distortion does not affect the phase changes in clock regeneration.

In the decoder, PLL (Phase-Locked-Loop) is used to reproduce the 16.2 MHz resampling clock from the synchronization signal (HD). The PLL residual phase error needs to be reduced enough to meet the resampling conditions. In the MUSE standard, the HD resampling phase is set so that the mid-range of the HD waveform can be resampled. Assuming that HD is sampled by a 16.2 MHz clock, as in Figure 3.6, and that

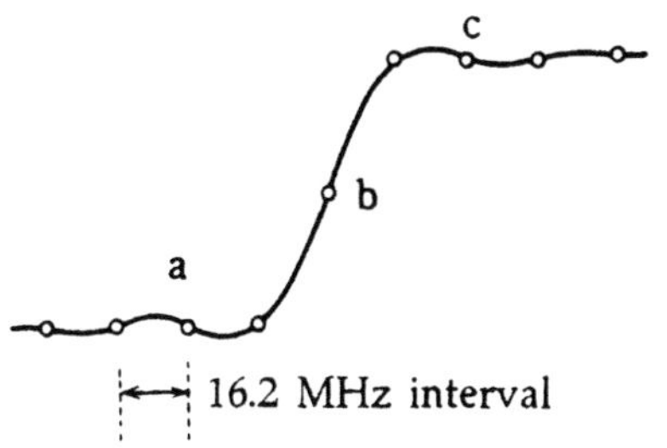

FIGURE 3.6. HD waveform and PLL control.

points a, b, and c are obtained by selecting every other sampled point, the phase error is calculated by the following expression. The sampling phase is controlled by

$$[\text{Phase error}] = \pm\left(b - \frac{c + a}{2}\right) \quad (3.1)$$

so that the PLL phase error can be reduced to zero—that is, level b can be made equal to the average value of levels a and c. However, the plus and minus signs are reversed every 1H depending on the polarity of the HD waveform.

3.1.3 Analog Transmission of Sampled Values

MUSE transmission is based on the analog transmission of sampled values using PAM (Pulse Amplitude Modulation), and is the first television transmission method in the world to adopt this method. In fact, the adoption of this technique was instrumental to MUSE. The basic requirement for the analog transmission of sampled values is to sample accurately without mutual interference between the encoder's D-A conversion and the decoder's A-D conversion processes. The conditions that preclude interference between sampled values (hereafter the sampling conditions) are known as Nyquist's first theorem.[2] As shown in Figure 3.7 (a), these conditions are satisfied when the point symmetry of the frequency characteristic of the transmission path is half of the sampling frequency (i.e., 8.1 MHz), and the group delay characteristic is uniform within the band. The frequency characteristic described above is called the "−6 dB roll-off characteristic."

The resampling conditions can be expressed in a time series as follows. In Figure 3.7(b), sampled impulses are represented by a circle and analog transmission by a solid line. The signal is resampled and returned to an impulse (circle) when the ringing frequency is equal to one-half of the sampling frequency, that is, when the zero-crossing point of the ringing is resampled. All signals are the overlapping of such impulses separated from each other by time. This means that all sampled values are freed from waveform interference once an impulse meets the resampling conditions. This type of response is given by the −6 dB roll-off characteristic.

If the −6 dB roll-of characteristic shown in Figure 3.7 is not realized, the resultant waveform interference shows up as ringing disturbances on the screen. Should the transmission line characteristics be particularly bad, an automatic equalizer is employed. This is a variable digital filter (transmission line equalizing LPF) that operates at the doubled oversampling rate of 32.4 MHz. The transmission line equalization uses VIT signals with impulse waveforms multiplexed into the vertical blanking period. The frequency characteristic and group delay characteristic of transmission lines can be found by measuring the impulse responses during decoding. The results are fed back to the variable digital filter so that the resampling conditions can be satisfied. Automatic equalization with VIT signals is applicable not only when the transmission path involves satellite broadcasting, but also when other types of media such as CATV, VTR, and video disk are used.

3.1.4 Band Compression for MUSE

(1) Three-Dimensional Subsampling

Along with the analog transmission of sampled values, another basic technique of the MUSE system is three-dimensional subsampling.

Before discussing three-dimensional subsampling, let us consider a two-dimensional rhombic lattice sampling pattern and its two-dimensional frequency spectrum as shown in

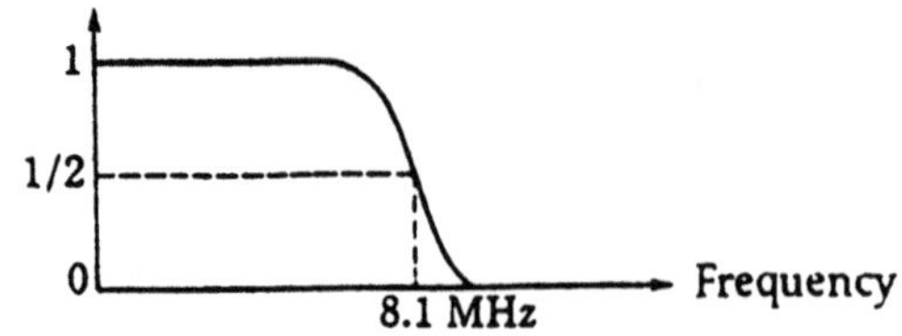

(a) Expressed as frequency

Resampling conditions are satisfied if the frequency characteristic is point symmetric at 8.1 MHz, and if the group delay characteristic is constant within the band.

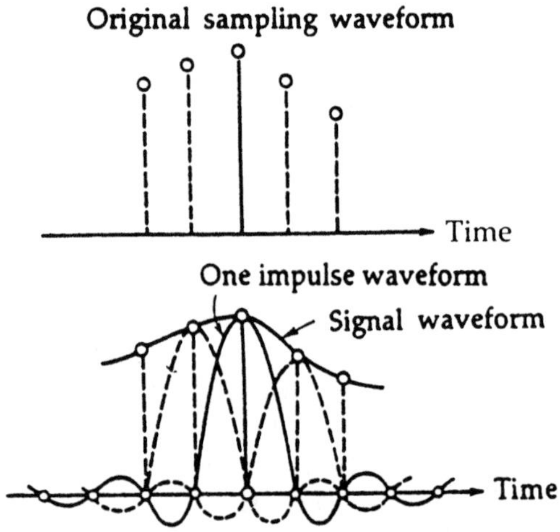

(b) Expressed as time series

The impulse waveform completely reverts to its original form when the zero crossing point of the ringing is resampled.

FIGURE 3.7. Resampling conditions.

Figure 3.8(a). (The horizontal and vertical sampling frequencies have both been normalized to 1.) The two-dimensional frequency spectrum of this pattern is shown in Figure 3.8(b). As the double circles indicate, the sampling frequency is $\sqrt{2}$ from the origin. Since the sampling theorem states that the pass band is one-half of the sampling frequency, the two-dimensional frequency does not produce aliasing distortion in this sampling pattern in the diamond-shaped area. Within this area, the real frequency region is the hatched area. In other words, while horizontal and vertical bandwidths up to the sampling frequency can be transmitted, the diagonal transmission bandwidth is $1\sqrt{2}$.

MUSE performs subsampling twice. Following the two-dimensional subsampling, subsampling is done in the time (frame) dimension. Shown in Figure 3.9 is the MUSE signal processing for the luminance signal (Y) and its relationship to subsampling patterns. While the initial sampling patterns is a square lattice with a frequency of 48.6 MHz, the signal processing is different for the stationary and moving portions of the image, and so there are two separate processing paths.

In processing the stationary portion of an image, the first process is field offset subsampling at a clock rate of 24.3 MHz, in which phase inversion occurs at every field. The subsampling

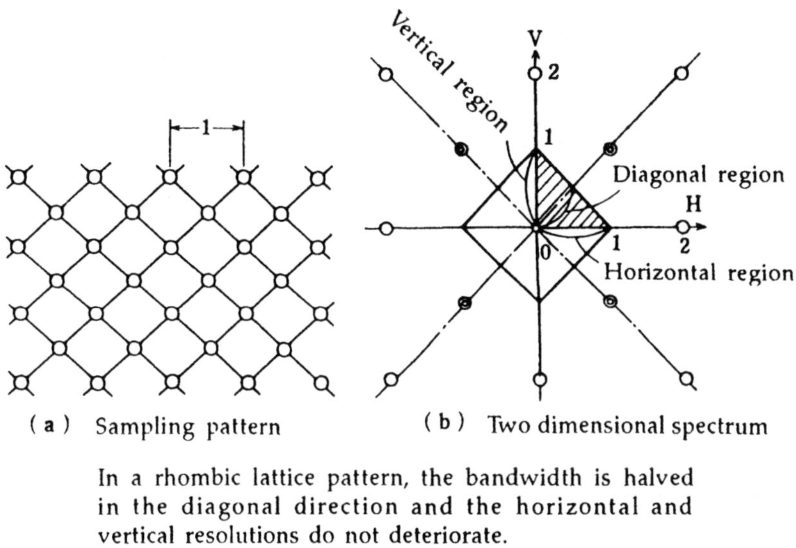

FIGURE 3.8. Sampling pattern and two-dimensional passing region.

pattern is the same as the two-dimensional subsampling described above. In this case, however, due to interlacing, one TV screen image (1 frame) is composed of 2 fields. Next, the signal is passed through a 12 MHz LPF and a subsampling signal is inserted, and after the sampling pattern is returned to the initial form (the sample value inserted here is different from the initial sample value), the sampling frequency is converted to 32.4 MHz. Finally, frame offset subsampling is carried out at an interframe inversion clock rate of 16.2 MHz. This sampling is performed in a rhombic lattice pattern in the direction of time. With this pattern, a pixel which is a sampling point in one frame will not be one in the next frame.

Shown in Figure 3.10 are the transmittable temporal and spatial regions and the aliasing spectrum in subsampling for the Y signal of the stationary portion of the image after processing. In the figure, the vertical frequency is the spatial frequency expressed in TV lines based on 1125 TV lines per frame. The hatched areas in the figure represent the transmittable region. However, the hatched region defined with a broken line in the 20 to 24 MHz range shows no visible improvement despite the fact that it is a transmittable region. Considering the aliasing for the noise reduction, this region should be eliminated. Before subsampling is begun, a prefilter is needed to prevent the aliasing spectrum and the spectrum of the original signal from overlapping. The prefilter's passing band can also be considered the transmittable region. As the horizontal-temporal spectrum in Figure 3.10 indicates, except for the low region of 0 to 4 MHz, the transmittable region for a stationary area is halved to 7.5 Hz.

For signal processing of moving images, field offset subsampling is not performed. Instead, the first process is band limiting and is performed at 16 MHz. Next, the sampling frequency is changed to 32.4 MHz. Finally, line offset subsampling is performed at 16.2 MH clock. This line offset subsampling actually is the application of the previously described two-dimensional rhombic lattice subsampling limited to the area within the field. The transmittable region căn be found by converting the 1 for the horizontal and vertical frequencies in Figure 3.8 to 16.2 MHz and 1125/2 TV lines, respectively. In other words, the transmittable region of the moving image is one-half of the transmittable region of the stationary image. Because subsampling is not performed in the temporal dimension, the transmission characteristic is flat in the direction of time.

As described above, MUSE compresses the

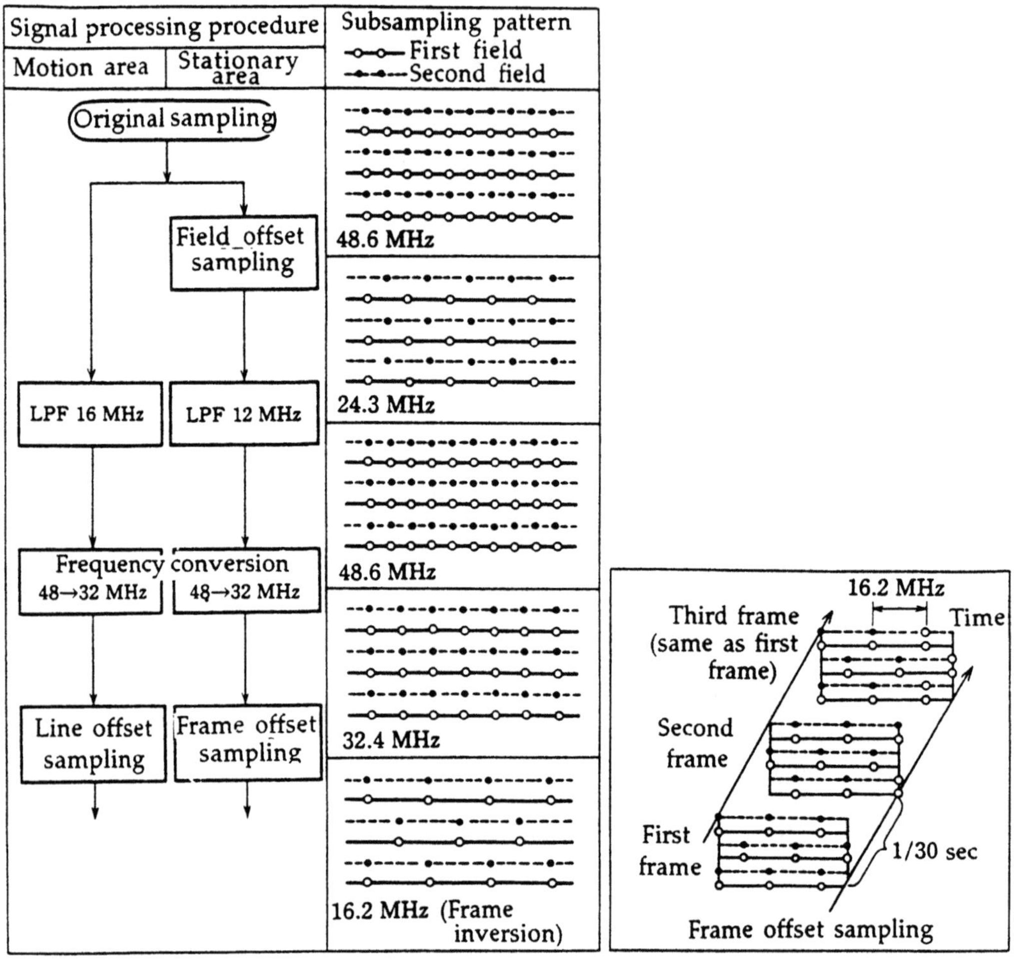

FIGURE 3.9. Luminance signal processing procedure and subsampling pattern in the MUSE encoder.

video signal to 8.1 MHz by subsampling twice for stationary parts of an image and once for moving parts. The unique feature of this system is the fact that subsampling compresses the signal not by one-half and then one-half again, but by one-half and then two-thirds by inserting a frequency conversion to 32.4 MHz. The system has the following advantages. First, there is no aliasing component in the low frequency region. This is an important element of TV signal processing because of the concentration of signal energy in the low frequency region, and the more visible deterioration and disturbance in image quality in the low frequency region. Second, as noise is heavily distributed toward the high frequency region, the absence of the aliasing of high frequency noise into the low frequency region is advantageous. Another feature of this system is the absence of the aliasing of interframe offset sampling in the region below 4 MHz. This means that motion detection in the decoder is possible using one frame difference with a component less than 4 MHz, as described below. This feature makes the system capable of accurate motion detection.

(2) Signal Processing of Stationary and Moving Images and Motion Detection

Figure 3.11 shows the Y signal processing scheme in the MUSE encoder and decoder. In the en-

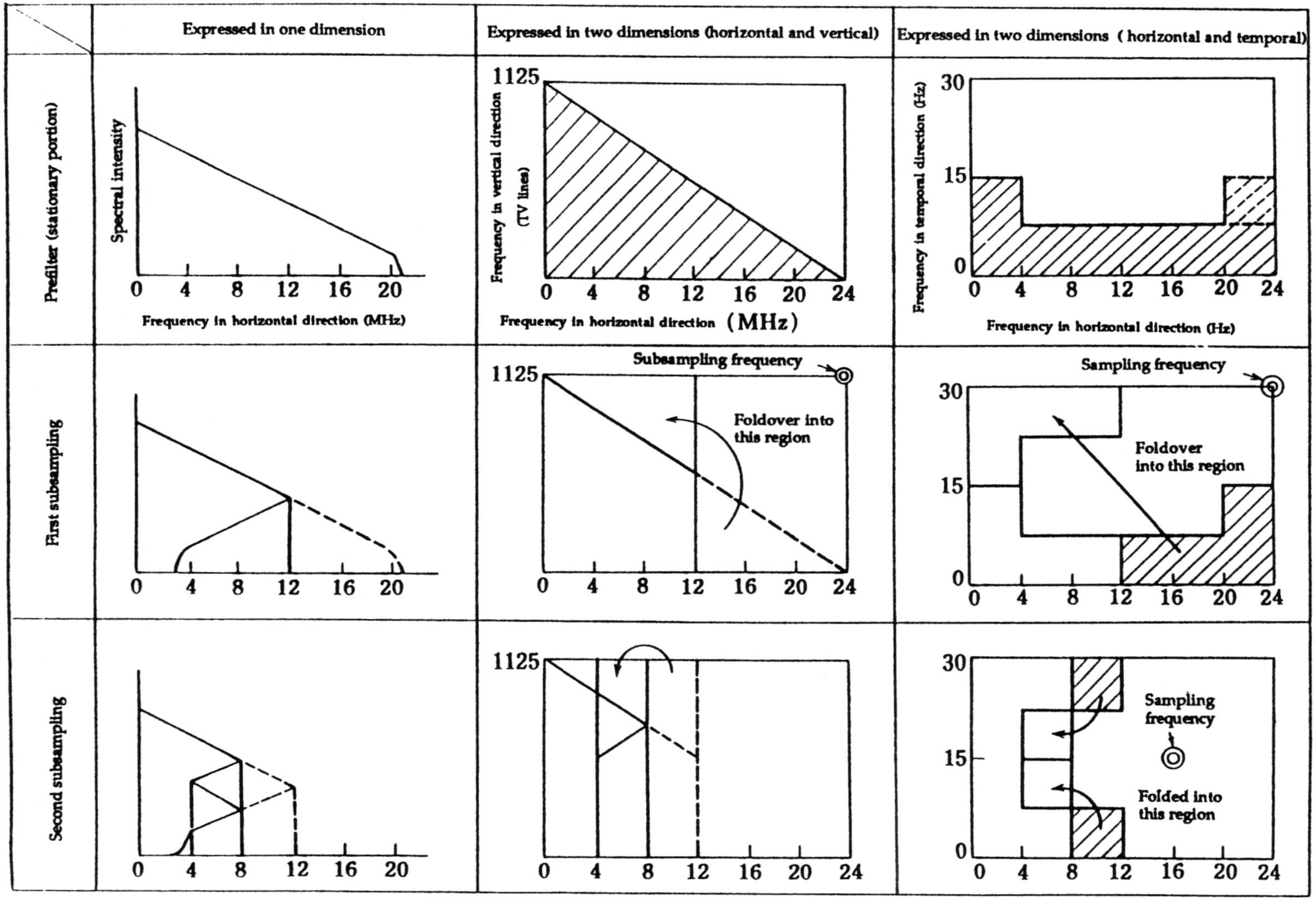

FIGURE 3.10. The transmittable spatial and temporal regions and spectrum of the aliasing luminance signal.

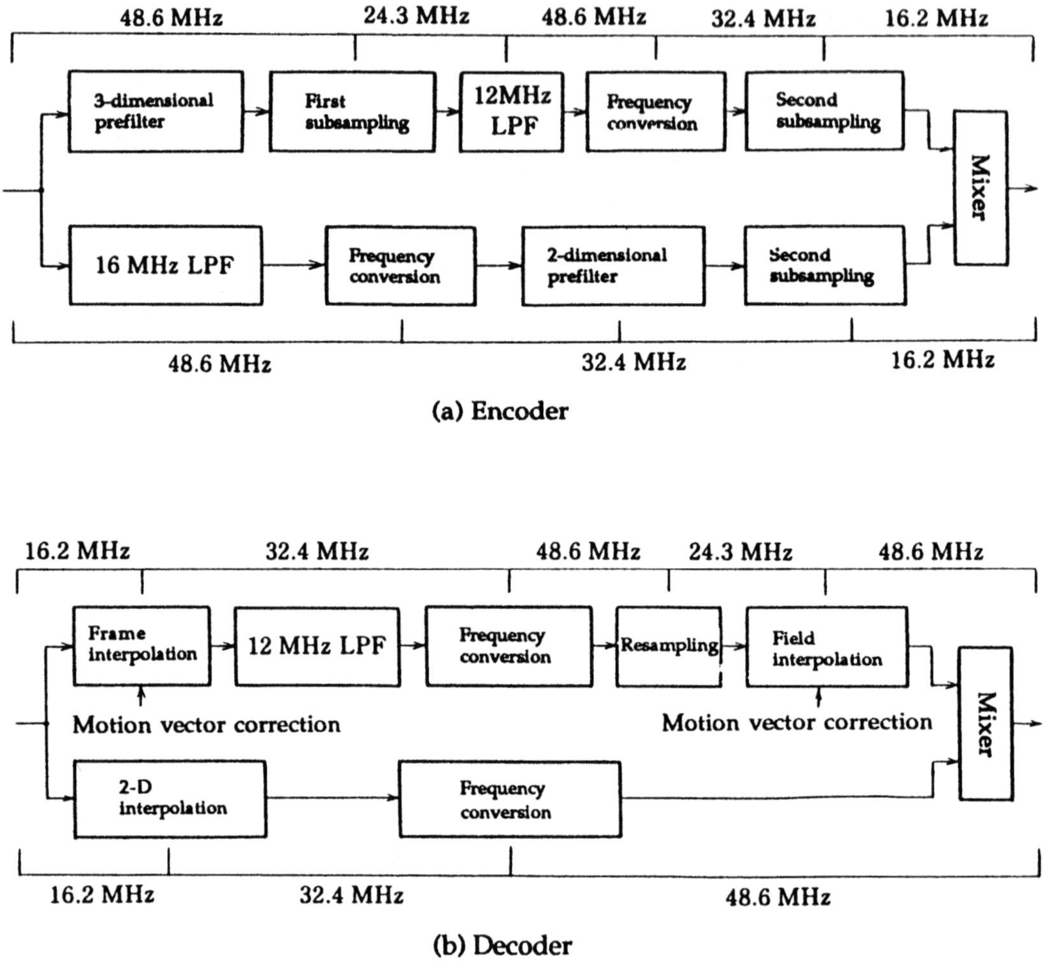

FIGURE 3.11. Signal processing for Y signal.

coder, input signals are separated into a stationary component (top) and a moving component (bottom) to be processed according to the procedures described above. The signals separately processed are them mixed together in a mixer. The mixing ratio is determined by the amount of motion detected in individual pixels. In other words, if a pixel is stationary, the mixer lets a stationary signal pass through, and if the pixel is moving, a motion signal passes through. However, the mixer does not switch between stationary and moving paths, but considers the intermediate amounts of motion and performs the mixing in 16 levels to avoid the generation of switching noise.

The Y signal processing in the decoder generally reverses the encoding process. In this case, again, the signals are separated into stationary (top row) and moving (bottom row) components, and are appropriately recombined in the mixer. In the initial stage of stationary component processing, frame insertion (interpolation) of the signal is performed. This is the insertion of frame offset sampling (secondary subsampling) that fills the pixels that are dropped out by the subsampling with the signal from the preceding frame. Next, the unwanted signals are filtered out with a 12 MHz LPF, and the sampling frequency is up-converted to 48.6 MHz. Then resampling is performed with the same clock used for the field offset subsampling, and the aliasing signal is returned to its original state.

Finally, stationary processing is concluded with the field insertion of the signal. The field insertion inserts (interpolation) signals to fill the sampling points omitted by the first subsampling using the signals from the current and preceding fields. This process corresponds to the LPF that removes the diagonal high frequency components, and its pass band is shown by the hatched region in Figure 3.8(b).

The signal processing for the moving component is completed with the field interpolation by means of a two-dimensional LPF and upconversion of the sampling frequency to 48.6 MHz. The interpolation filter has the characteristics shown in Figure 3.8(b), with normalized values for the horizontal and vertical directions being 16.2 MHz and 1125/2 TV lines.

While the encoder detects motion by using the difference between the preceding frame and the present frame, the decoder does not have access to the preceding frame due to the frame offset subsampling, and so cannot use simple frame differences. However, as mentioned before, the MUSE signal does not have interframe offset subsampling aliasing in the region below 4 MHz. This means that one frame differences can be used for motion detection using the sub-4 MHz component. In practice, however, motion detection in the region below 4 MHz fails to detect the motion of small objects. To compensate for this inadequacy, motion detection by 2-frame difference (which use the sample point two frames previous) is also employed.

(3) Motion Vector Detection and Correction

The MUSE system also features motion vector detection and correction capabilities.[3] As described previously, the transmittable bandwidths of stationary and moving components are different. The horizontal limiting resolutions of the stationary and moving components are 24 MHz and 16 MHz, respectively. In actual practice, with a relatively simple filter and no special contrivance, the limits are 20 MHz for a stationary system and 13 MHz for a moving component. Although the resolution of moving objects is lower than that of stationary objects, this is not a problem because human visual acuity is also lower with moving objects, and because TV cameras also have a lower resolution with respect to moving objects due to a capacitance effect. However, when the camera pans and tilts to follow a moving object, the low resolution becomes more conspicuous because the eyes follow the object with the camera movements. In this case, it would be preferable to use the stationary signal processing as much as possible. To accommodate this need, motion vector detection and motion vector correction techniques have been incorporated in MUSE.

In performing the motion vector detection and correction, the direction and magnitude of movement in the image are detected in the encoder (motion vector detection). These values then are transmitted to the decoder as control signals. In the decoder, the location of the signal from the preceding frame is shifted by the value of the motion vector (motion vector correction) before the signal is inserted for the interframe interpolation. By following this procedure, it becomes possible to apply stationary signal processing to moving images.

Because the detection and correction of motion vectors is performed to save the loss of resolution resulting from panning and tilting the camera, one motion vector is used for each field. The detection of motion vectors is done by the pattern matching method described below.

The pattern matching technique of motion vector detection calculates the correlation between the preceding frame image and the present frame image while the preceding frame image is being shifted in small increments. The motion vector is the amount of the shift when the correlation is maximized. Assume the video image level at the two-dimensional location of a particular pixel x to be $A^N(x)$, and the video image level in the preceding frame in which a shift of v has occurred with respect to x to be $A^{N-1}(x - v)$. The correlation is given by the following equation.

$$D(v) = \sum_{x \in \mathbf{F}} f\left\{ |A^N(x) - A^{N-1}(x - v)| \right\} \quad (3.2)$$

where F is the set of all the pixels in one field. This quadratic function is the usual definition of a correlation function. However, when a small object with large level changes moves across a flat image, it is necessary that the frame difference not determine the motion vector. To satisfy this requirement, we employed the following form of the function.

$$f\{a\} = \begin{cases} 0, & a \le d \\ [K\sqrt{a - d}], & a > d \end{cases} \quad (3.3)$$

where d is a micro threshold level for avoiding the effect of noise, K is a constant that brings about the agreement in bit number between a and $f\{a\}$. [] indicates the omission of fractions. In this case, the motion vector is obtained as a shift quantity that minimizes the value of $D(v)$ with respect to all the values of shift quantity v.

Although F has a range of one field, the detection of the motion vectors of all the pixels requires an enormous amount of hardware. In terms of accuracy, an F with a few thousand pixels is sufficient. This setup reduces the frame memory requirement for motion vector detection to a few Kbytes, and makes low speed signal processing possible. It also reduces the complexity of the hardware.

3.1.5 Pseudo Constant Luminance Transmission

The cathode ray tube of a receiver shows a nonlinear relationship between electrical signal input and light output. This relationship is called the gamma characteristic. To make receivers more economical, gamma correction is done by the camera, and the gamma-corrected RGB signals undergo a matrix conversion into Y and C signals. In the course of this process, crosstalk occurs between Y and C signals. For example, a part of the luminance component may become part of the chrominance signal, and since the chrominance signal band is narrower than the luminance band, image details would be lost in areas with highly saturated colors. Or transmission line noise received by the chrominance signal may be converted to the luminance signal. Because the chrominance signal is time-expanded four times in the decoder, the transmission noise frequency is reduced to 1/4 its original value and becomes more conspicuous. This type of crosstalk between Y and C signals does not occur when the image is achromatic, but becomes more prominent as the color levels of the image increases.

The crosstalk between Y and C signals can be avoided if the signal transmission is carried out in a linear system instead of a gamma system. In such a setup, gamma correction must be carried out on the receiver side. In this method, the constant luminance principle holds.[4]

MUSE has incorporated the constant luminance principle because of the reasons described above. However, the introduction of the constant luminance principle without modification generates the following problems:

1. Insufficient precision of quantization of resolution at the 8-bit level,
2. Poor SN ratio with respect to Y in the low level (dark) region,
3. With respect to C, insufficient bit precision and poor SN ratio in the low saturation region.

Figure 3.12 shows the configuration of nonlinear signal processing in MUSE. Because gamma correction of the camera output is standard, an inverse gamma correction is carried out in the MUSE encoder. Gamma correction for the display is done on the decoder output. These gamma and inverse gamma corrections must have inverse characteristics at the encoder and decoder. In order to solve problem (1), the gamma and inverse gamma characteristic curves in the encoder and decoder have gentle slopes; they are not corrected to complete linearity. Thus, the matrix and reverse matrix are not perfectly linear signal systems and show gentle gamma characteristics. Because they do not exactly meet the constant luminance principle, the system is called the pseudo constant luminance principle. To deal with problem (2), the Y signal is transmitted after applying a transmission gamma and black level expansion. Problem (3) is solved by applying an LPF on C signals, followed by nonlinear correction of a low level signal expansion.

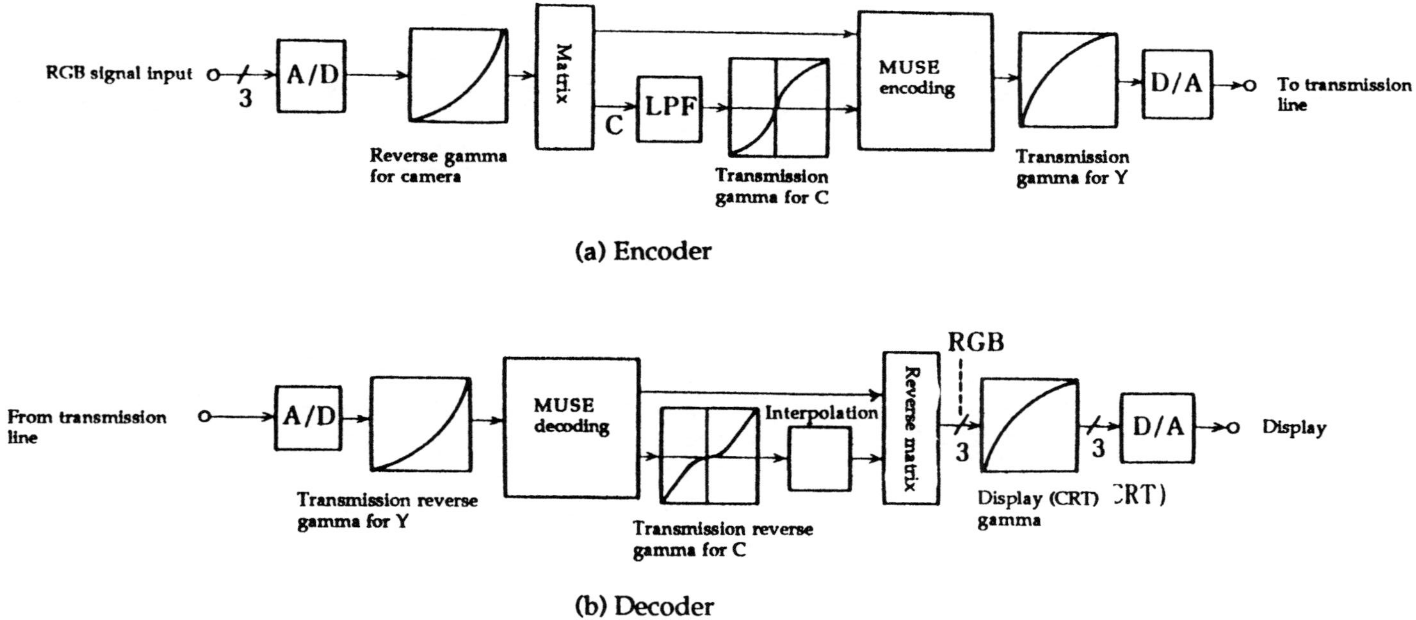

FIGURE 3.12. Nonlinear signal processing in the MUSE system.

In addition, C signals are increased by 3 dB because 100% saturation as shown in the color bars are seldom generated for ordinary images. These nonlinear signal processing methods were set after comprehensively examining the balance between the SN ratios of Y and C signals and the resolution of highly saturated areas.

3.1.6 Application of the MUSE System

In transmission experiments with MUSE using broadcast satellite BS-2, a satisfactory image was received with a 75 centimeter reception antenna (December 1986). A VSB-AM ground broadcasting experiment on UHF conducted in the U.S., and a U.S.–Canada transmission experiment with a communications satellite that covered the whole North American continent (October 1987) have also proven the effectiveness of the MUSE system.

A MUSE receiver build from currently available ICs involves a considerable amount of hardware and consumes a large amount of power. However, if the receivers are produced in large quantity using ICs developed for MUSE, the problems related to hardware size and economy will be solved. Development is under way for application specific ICs for the MUSE system.*

Peripheral equipment for the MUSE system for home use, such as VCRs, video disks, and a MUSE/NTSC converters[5] have been built on an experimental basis. A simplified MUSE encoder for a video camera has also been built and its image quality has been confirmed. MUSE-NTSC converters are being developed as low cost adaptors so that conventional TV sets can receive Hi-Vision broadcast programs.

3.2 AUDIO SIGNAL TRANSMISSION SYSTEM FOR MUSE

3.2.1 Audio Signal Transmission for Hi-Vision

The MUSE system is able to compress the Hi-Vision video signal bandwidth, which exceeds 20 MHz, to about 8 MHz. But as the video signal bandwidth still takes up most of the satellite transmission bandwidth, it is not possible to transmit audio signals on the subcarrier dedicated for the purpose as is done with conventional TV. An audio signal transmission method that does not increase the transmission bandwidth is time-division multiplexing within a blanking period of the video signal. Because the horizontal blanking period is too short for this, the vertical blanking period is used.

Multiplexing the digitized audio signal can be done in either of the two methods shown in Figure 3.13. The first method, called RF time division multiplexing, compresses the signal in the time axis to fit within the vertical blanking period, then directly modulates the carrier and transmits it during the vertical blanking period by switching to and from the video carrier. In the other method, the digitized audio signal is time-division multiplexed in the baseband during the vertical blanking period of the video signals.

RF time-division multiplexing has ample transmission capacity because it uses a modulating method that is appropriate to the amount of information being transmitted. However, this method requires a highly sophisticated technique that is capable of the stable reception of RF signals in burst mode. Baseband multiplexing is inferior in terms of its smaller transmission capacity for VCRs, disks, and CATV, but is better suited for Hi-Vision audio signal transmission in terms of receiver stability and cost.

3.2.2 Compression Encoding of Audio Signal

(1) Bit Rate and Compression for Audio Signal

The bit rate for audio signal transmission was set at 1.35 Mbit/s for several reasons. At this rate, the error rate is sufficiently low even if the reception CN ratio deteriorates. In addition, the rate is a simple integer ratio of the MUSE video signal clock frequency, and is compatible with the standard international audio sampling frequencies of 48 kHz and 32 kHz. Since the rate is about 65% of the transmission bit rate for conventional TV satellite broadcasting of 2.048

*See Appendix for recent developments.

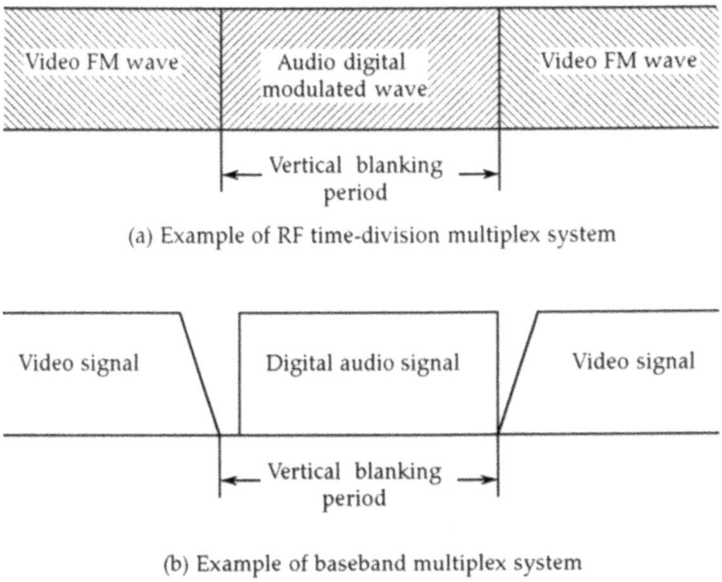

FIGURE 3.13. Multiplexing methods of audio signal.

Mbit/s, the bandwidth needs to be compressed further by about 40%.

The large screen size and high image resolution of Hi-Vision provide far more realistic images than does conventional television. To further enhance this image effect with sound, MUSE has been designed with 4-channel stereo broadcasting.

To transmit the audio signal for 4 channels within the vertical blanking period of the Hi-Vision video signal, the data from one sampling made at a sampling frequency of 32 kHz must be compressed to about 8 bits. In addition, because the SHF band used in satellite broadcasting is extremely sensitive to attenuation caused by rainfall, the compression must be resistant to bit errors and have the shortest possible time delay.

In encoding the audio signal for Hi-Vision transmission with the requirements described above, we adopted a two-stage compression technique—a bandwidth compression that uses the correlations in the sound signals and prevents signal deterioration, followed by a near-instantaneous compression that takes into consideration the human auditory sense.

The bandwidth compression uses differential PCM, which transmits only the difference in successive audio signals. Differential PCM compression works well for signals that only have a low frequency component because the differential value of successive signals are small and highly correlated, thus producing a small quantization number and high rate of compression. However, as Figure 3.14 shows, when the signal has a high frequency component, the differential values also become large and exceed the original quantization number. To solve this problem, we developed DANCE (Differential PCM Audio Near-instantaneous Compressing and Expanding), which uses the near-instantaneous compressing and expanding technique

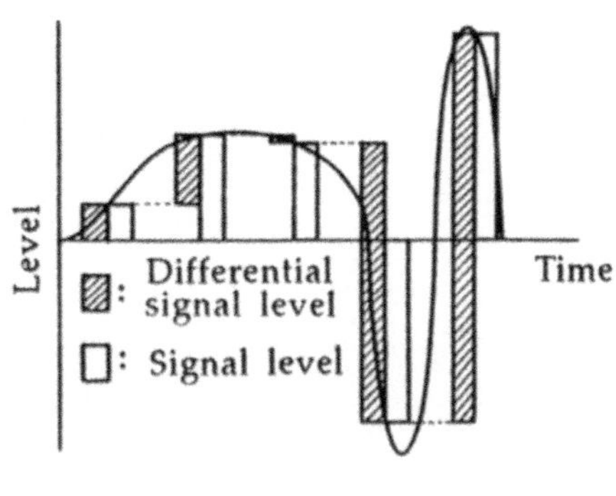

FIGURE 3.14. Original and differential signal levels.

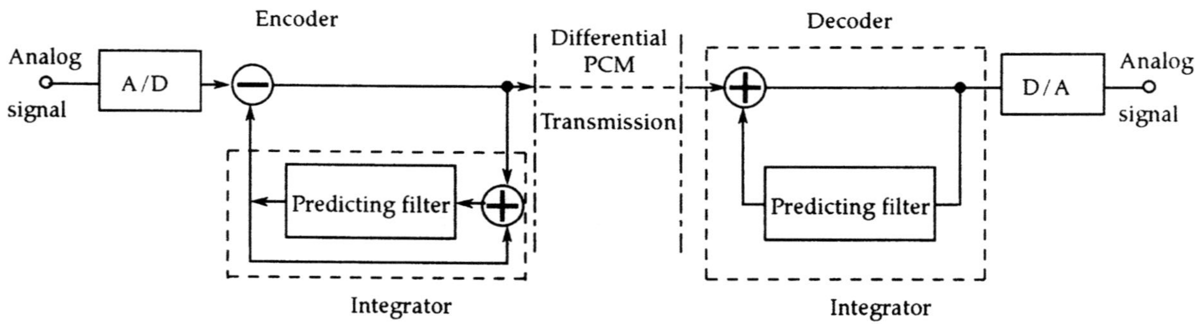

FIGURE 3.15. Principle of the predictive encoding system.

currently used in conventional satellite television.[6] This technique allows the setting of two transmission modes, Hi-Vision Mode A, which compresses a uniformly quantized, 15-bit signal sampled at 32 kHz to 8 bits, allowing for the transmission of 4 channels, and Hi-Vision Mode B, which compresses a 48 kHz 16-bit PCM signal to 11 bits so that 2 audio channels can be transmitted.

(2) Differential PCM Audio Near-Instantaneous Compressing and Expanding (DANCE)

The differential coding is a kind of predictive coding system that utilizes the correlation of sound signals along the time axis. Shown in Figure 3.15 is a configuration of the predictive coding system.

In this system, the value of the present signal is used to predict the size of the next signal. The transmission of the prediction error, which is the difference between this predicted value and the actual value, is generally referred to as differential PCM.[7]

A differential PCM signal can be encoded with a smaller quantization range that can a PCM signal. A large sound signal quantized into a PCM code with 16 bits will fit into a range averaging 12 bits. With differential PCM, most sound signals will fit into a 9-bit range, and on average the quantization range can be reduced by 2 to 3 bits compared to PCM. However, when the frequency components of the signals are high, the quantization number becomes larger than the original PCM signal.

Figure 3.16 shows the dynamic range of differential PCM at a sampling frequency of 48 kHz. The noise level assumes a certain constant value that depends on the quantization level of the PCM signal before the differential is taken. The maximum reproducible level depends on the bit count of the differential PCM. The maximum transmissible level of 11-bit differential PCM is shown in the figure as line 1. This level corresponds to a 10-bit level at 24 kHz (one-half the sampling frequency) and 11 bits at 8 kHz (1/6 the sampling frequency). As the frequency becomes lower, the dynamic ranges increase. Although the correlation is low in the high frequency samples, signals in the medium and low regions, which undulate very slowly, have high correlations and small differential val-

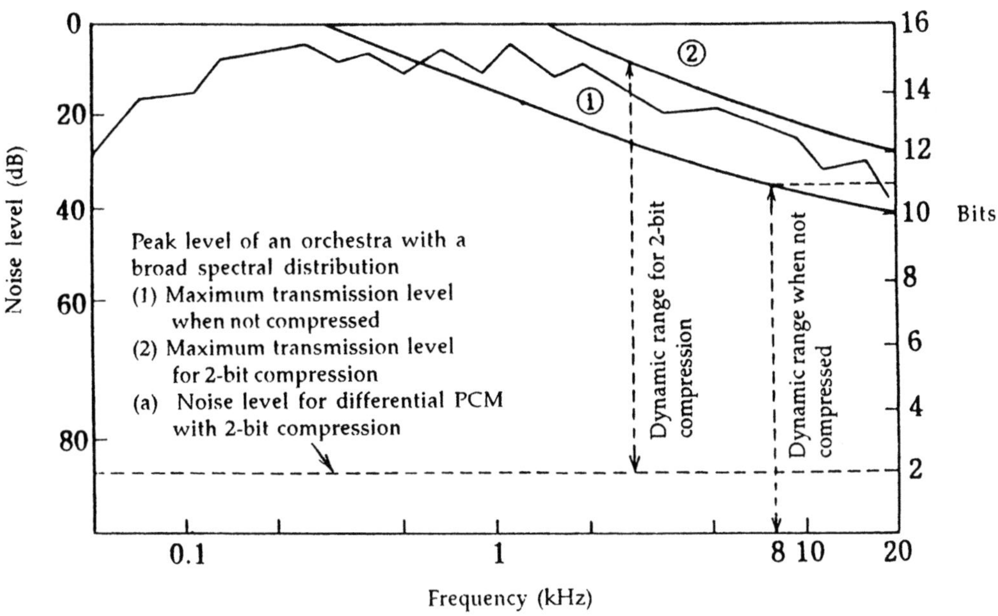

FIGURE 3.16. Dynamic range of near-instantaneous differential PCM compression and expansion at a sampling frequency of 48 Khz.

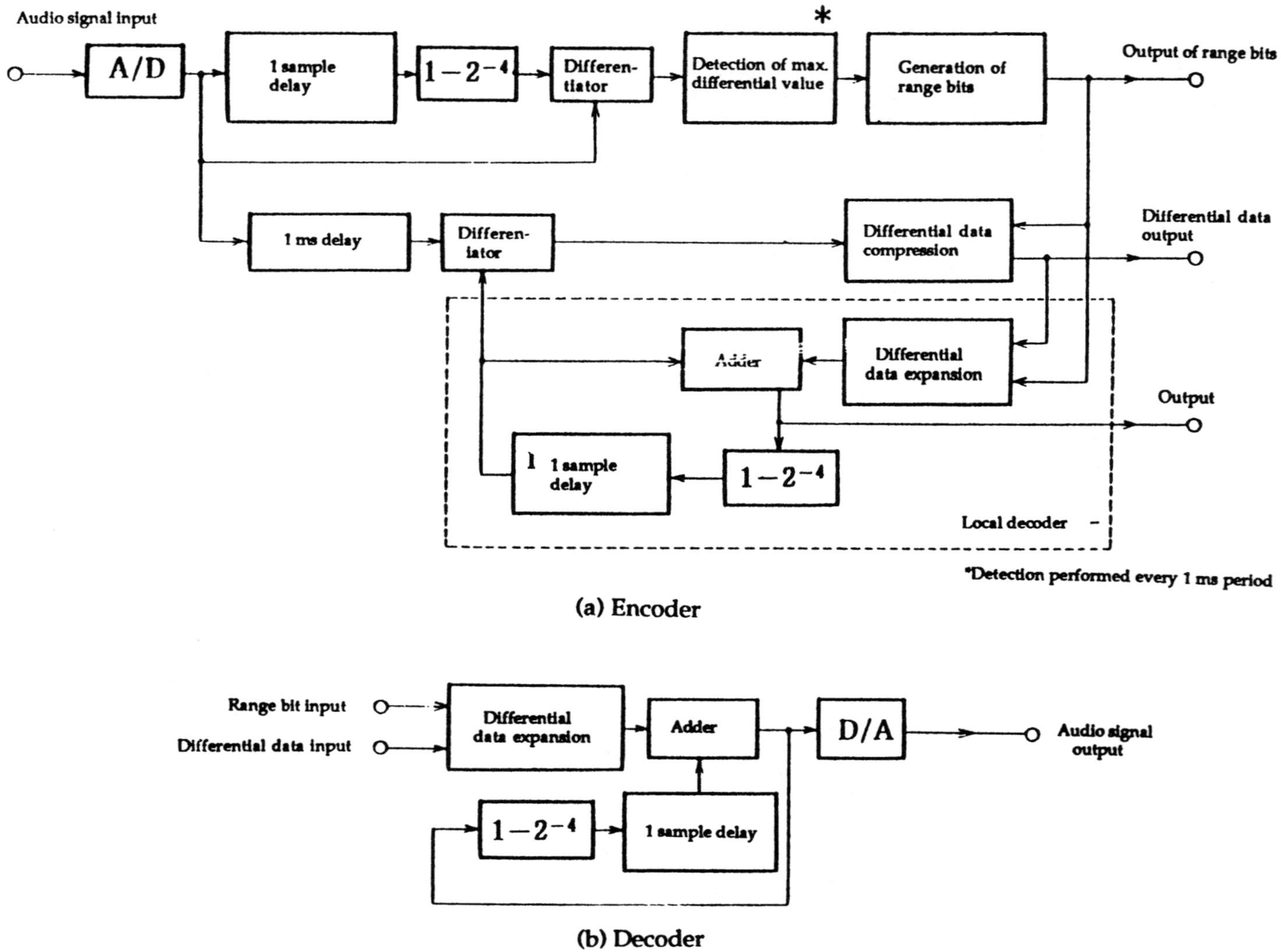

FIGURE 3.17. Near-instantaneous compression and expansion of differential PCM.

ues both before and after sampling, and are able to realize a large dynamic range with a small number of bits.

In general, sound signals in nature become smaller as their frequency increases. However, since the human auditory sense declines sharply in sensitivity as the frequency increases, differential PCM, whose dynamic range decreases as the frequency increases, conforms to this law of nature.

However, because high frequency signal levels change sharply, the differential values can be large and exceed the bit rate set for the medium and lower range. In the spectral distribution for an orchestra shown in Figure 3.16, whenever the high frequency component exceeds line number 1, and additional near-instantaneous compression is done, and depending on the size of that signal, a 2-bit compression takes place to quadruple the quantization level. As a result, the maximum transmissible level increases to line number 2, and the noise level also increases by two bits and reaches dotted line (a). The maximum noise caused by the 2-bit compression also increases to the dotted line. However, because the signal level also increases and masks the noise, the noise is less perceptible to the ear.[8]

Near-instantaneous compression needs further explanation at this point. Its operation is similar to that of a digital voltmeter that has an auto-range function. A voltmeter changes its operating range in response to the magnitude of the voltage, be it 1V, 10V, or 100V. With near-instantaneous compression, the range bit correspond to the range selection of the voltmeter, while the transmitted data corresponds to the voltage reading.

Figure 3.17 shows the configuration of the encoder and decoder for near-instantaneous compressing and expanding differential PCM for Hi-Vision. In the compressing operation, the input signal is separated into 1 ms intervals, and the range is determined by detecting the maximum differential between intervals. The configuration inside the box with the dotted lines is exactly the same as that of the decoder for the receiver. Thus the value decoded in the receiver is used as the predicted value which is compared to the next signal to calculate the differential value, thereby continually correcting the compression error.

(3) Leakage Coefficient

In demodulating differential PCM, ample measures are taken to prevent errors when the receiver performs an additive operation on the differential values to obtain the original PCM. Nevertheless, when the CN ratio deteriorates, error correction is disabled, resulting in the addition of wrong differential values which remain permanently in the adder. As Figure 3.18 shows,

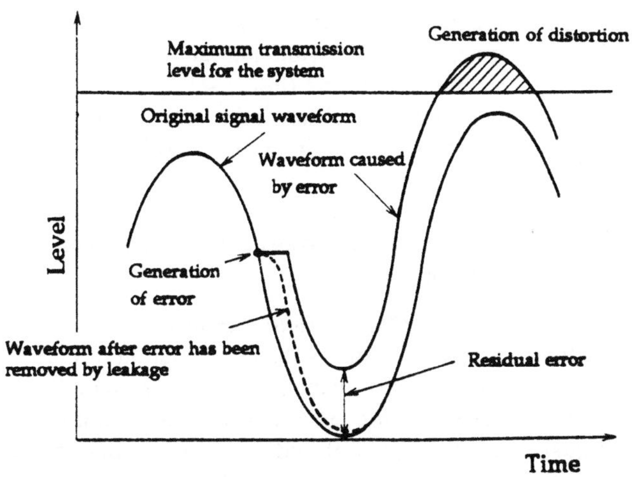

FIGURE 3.18. Distortion due to bit errors in differential PCM.

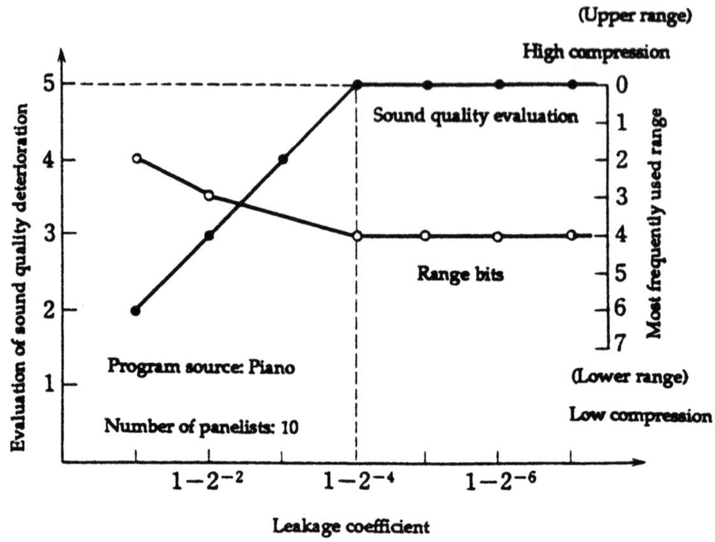

FIGURE 3.19. Relationship between leakage coefficient and sound quality.

this error not only causes a distortion of the signal waveform, but narrows the dynamic range by the amount of the error. For this reason, it is necessary to remove the error in the adder. As the configuration of DANCE in Figure 3.17 indicates, the error can be eliminated by the attenuation of the output value caused by multiplying the output from the addition in the decoder by a factor of $1-2^{-n}$. This operation is called leaking, and $1-2^{-n}$ is the leakage coefficient. A smaller leakage factor can accelerate the error correction, but it also attenuates the program signal itself, causing the sound quality to deteriorate. To prevent this from occurring, it is necessary to perform a correction in the encoder that reverses the characteristic of the decoder. However, an intense correction by leakage performed in near-instantaneous compression increases the differential value, and thus causes a deterioration in sound quality due to unnecessary compression. To optimize the process, the leakage coefficient should be as small as possible so it will remain within a range that does not cause deterioration in sound quality.[9]

Figure 3.19 shows the relationship between the most frequently used range and sound quality. The results in this figure indicate the optimum leakage factor to be $1-2^{-4}$.

3.2.3 Evaluation of DANCE Sound Quality

Compared to a uniform quantized PCM signal, the sound quality of Hi-Vision Mode A (8 bits) is equivalent to a 14.5-bit signal, while Hi-Vision Mode B (11 bits) can transmit programs with a sound quality equal to Mode B of conventional satellite broadcast television.[10] We conducted audio tests comparing the sound quality of conventional satellite broadcast television with both Hi-Vision Mode A and Mode B, and arranged the results on a scale. Six types of programs—orchestra, piano, pop music, male speaking voice, traditional Japanese music, and sounds from nature (insect sounds)—were used to evaluate the different encoding systems. The results are shown in Figure 3.20.[11]

3.2.4 Baseband Multiplex Transmission System

(1) Encoding of Audio Signals

Figure 3.21 shows the configuration of the Hi-Vision audio signal transmission system. Table 3.1 shows Modes A and B, which can transmit the two types of audio signals having official bandwidths of 15 kHz and 20 kHz.

The compression encoding of audio signals

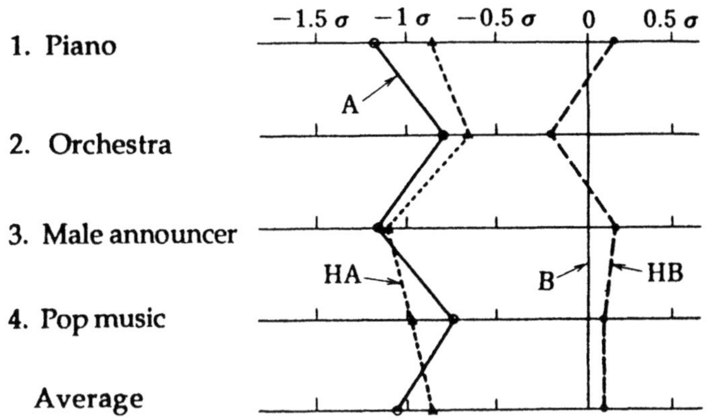

A 14/10 near-instantaneous PCM compression and expansion (conventional satellite television, Mode A)
HA 15/8 near-instantaneous differential PCM compression and expansion (Hi-Vision Mode A)
B 16-bit uniform PCM quantization (conventional satellite television, Mode B)
HB 15/11 near-instantaneous differential PCM compression and expansion (Hi-Vision Mode B)

FIGURE 3.20. Results of subjective hearing test.

in Mode A involves the generation of a differential signal from a 15-bit uniformly quantized encoded signal, which then undergoes a near-instantaneous compression. In the compression, the signal is first divided into 1ms segments. The maximum differential value in each segment is compared against the ranges shown on the horizontal axes of the graphs in Figure 3.22. Then the conversion line (range bits) is determined, and the signal is converted to a new 8-bit code.

In Mode B, a similar procedure is used to convert a 16-bit uniformly quantized signal into an 11-bit signal.

(2) Transmission Signal Format

One characteristic of digital transmission is the ability to freely change the quality of transmission, such as the number of audio channels and transmission mode, to conform with the content of the program being transmitted. It is also possible to transmit fascimile and other nonaudio signals. The transmission format has been designed to meet the diverse requirements described above.

Figure 3.23 shows the frame configurations for Mode A and Mode B. Each consists of a total of 1350 bits, including a frame synchronization signal at the beginning of each frame, a control signal that switches the mode and number of channels, audio data, and correction code. The bit rate is 1.35 Mbit/s.

(3) Interleaving

The first step in obtaining audio signal sample data is a simple 16-sample word interleaving. The purpose here is to prevent consecutive audio signals from entering the same correction block, thereby preventing consecutive word errors. By doing so, word errors that cannot be corrected can effectively be compensated.

The next step is bit interleaving. Bit errors that occur in the transmission line can develop into a burst error of several successive bits. The error correction code BCH (82, 74) used in this system can make 1-bit corrections and detect errors of up to 2 bits for any 84-bit block. But it cannot respond to a burst error any larger than this.

To deal with this problem, an interleaving transmission is performed as described below. A BCH (82, 74) block consisting of 16 rows is written into memory row by row as shown in

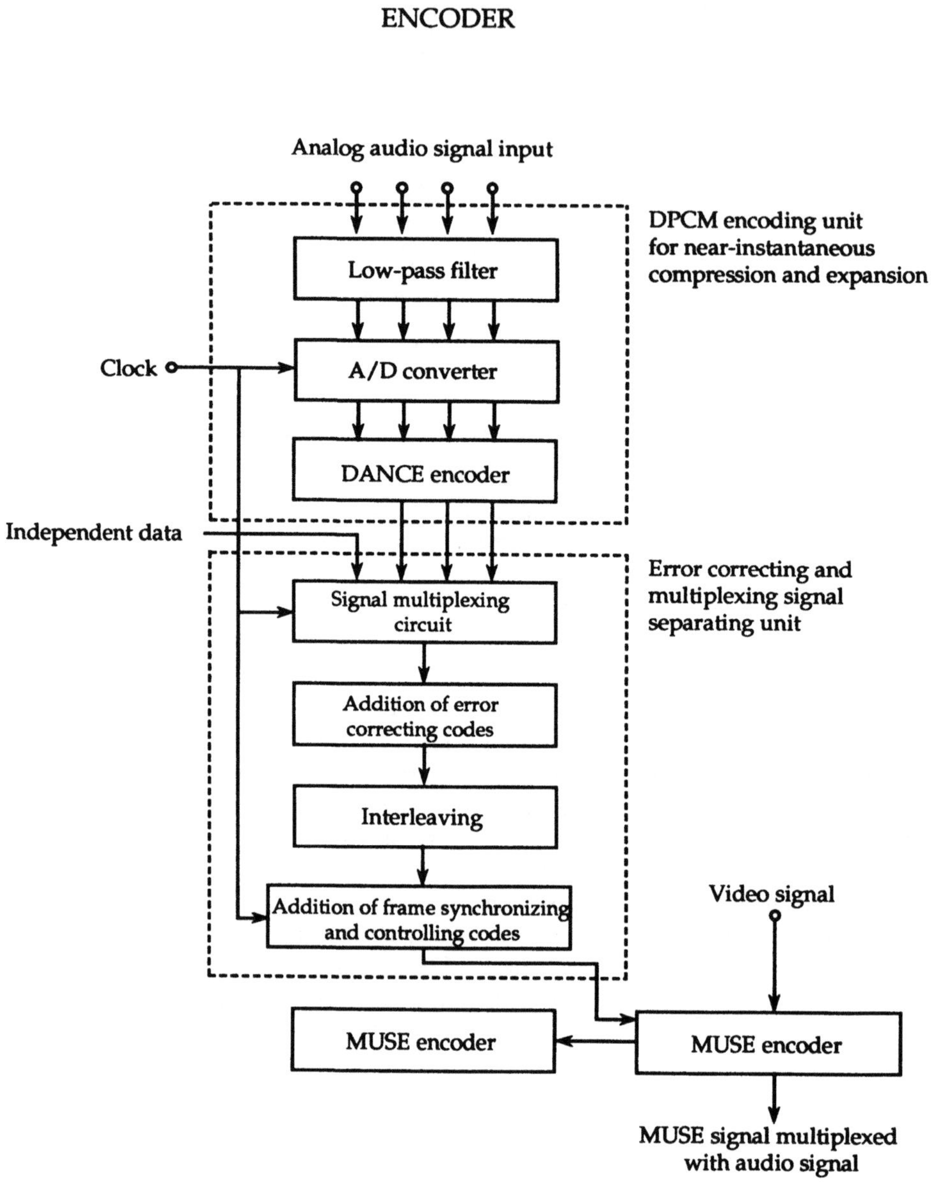

FIGURE 3.21. Audio signal transmission system for Hi-Vision.

Figure 3.24 to form a matrix of 16 rows × 82 columns. The matrix is then read and transmitted in columns starting with the leftmost column. At the receiver, the data is written into memory starting from the left column and read out in rows while making error corrections. In this procedure, a continuous burst error can be treated as random error. This technique is effective in correcting a burst error with a maximum size of 16 bits. The frame synchronizing code, composed of 16 bits, has the following pattern.

0001001101011110

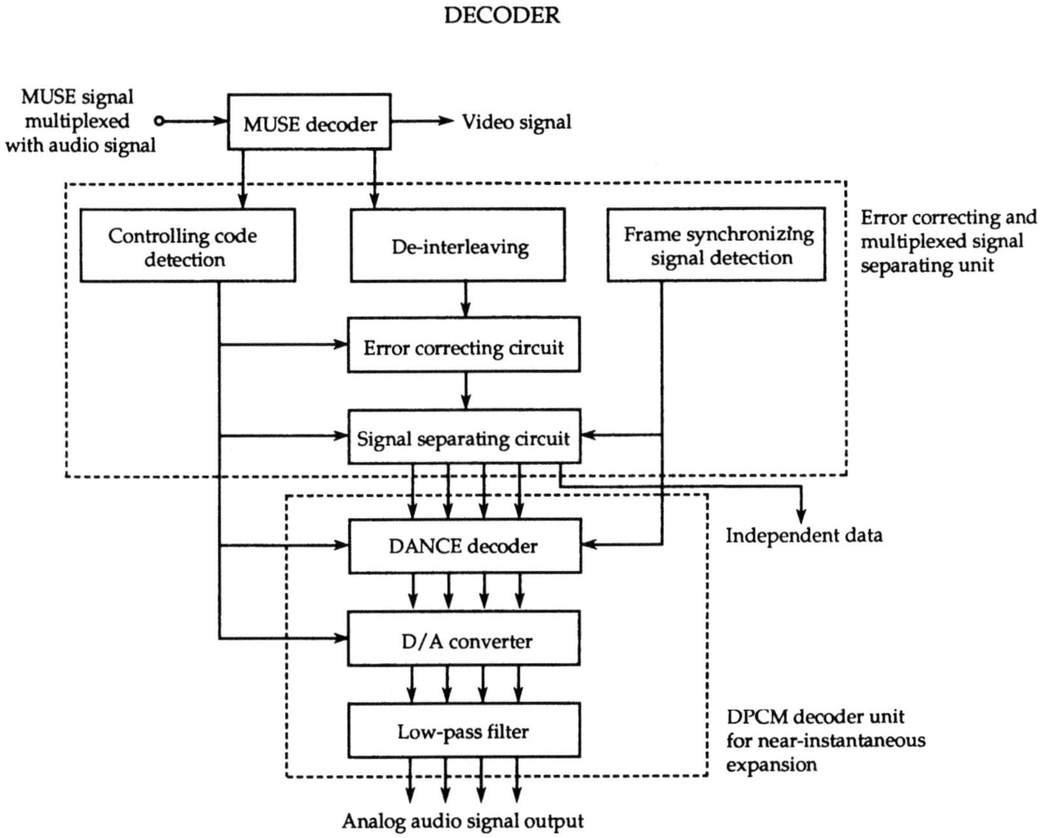

FIGURE 3.21 (continued)

(4) Error Control System

Rainfall can affect satellite broadcasting by causing the reception CN ratio to deteriorate. To secure a sufficiently high sound quality even if the reception CN ratio deteriorates, it is necessary to correct transmission code errors. For this purpose, the error correction system uses range bits with a 1-bit error correction, 2-bit error detection code based on BCH·SEC·DED (7, 3) coding. If the range bits are incorrect, the levels become irregular, resulting in a marked deterioration in sound quality. Thus the range bits with correction code are supplemented, along with the audio data and independent data, with an error correction code of BCH·SEC·DED (82, 74). This code is obtained by adding an error detection capability to the original BCH (127, 120) to obtain (127, 119) code, which is then shortened by 45 bits. The resulting code, with a block length $n = 82$, data bit length $k = 74$, test bit length $m = n - k = 8$ bits, can correct 1-bit errors out of 82 bits, and detect 2-bit errors out of 82 bits. The generator of this code is $x^8 + x^7 + x^4 + x^3 + x + 1$; and the generator for range bits is $x^4 + x^3 + x^2 + 1$. The error correction for the control code uses a majority decision method, which compares code patterns and chooses the one with more matches as being correct.

3.2.5 Multiplex Modulation into Video Signals

(1) Multiplexing Audio and Independent Data

In this process, successive 1350 kb/s of audio data and independent data undergo interframe interleaving. The data is then time-compressed

TABLE 3.1. Basic transmission parameters for audio signal of Hi-Vision television.

Transmission mode		A	B
CODING			
Audio signal bandwidth		15 kHz	20 kHz
Sampling frequency		32 kHz	48 kHz
Time of sampling		Same as in case of stereo	
Audio quantization		15 bits linear	16 bits linear
Compression / expansion		1. Near instantaneous compression and expansion of differential values of above signal into 8 bits (8 range).	2. Near instantaneous compression and expansion of differential values of above signal into 8 bits (6 range).
Leakage value		$1 - 2^{-4}$	
SIGNAL MULTIPLEXING			
Code transmission rate		1350 Kb/s	
No. of audio channels		4 channels	2 channels
Independent data transmission capacity		128 Kb/s	112 Kb/s
No. of frame bits		1350 bits	
Frame sync pattern		16 bits / frame (0001001101011110)	
Control code		22 bits / frame	
Word interleaving		16 words	
Bit interleaving		16 bits	
Error control:	Audio • data	BCH SEC •DED (82, 74)	
	Range bit	In addition to above, BCH SEC • DED (7, 3)	
	Control code	Multiple decision making through repeated sending	
TIME COMPRESSION			
Transmission interval		Vertical blanking period	
Modulation method		Ternary signal	
Code transmission rate		12.15 Mbaud	

1. Differential values of signals quantized at 15 linear bits per sample are compressed to 8 bits per sample, and 8-range control data is sent every 32 samples.
2. Differential values of signals quantized at 16 linear bits per sample are compressed to 11 bits per sample, and 6-range control data is sent every 48 samples.

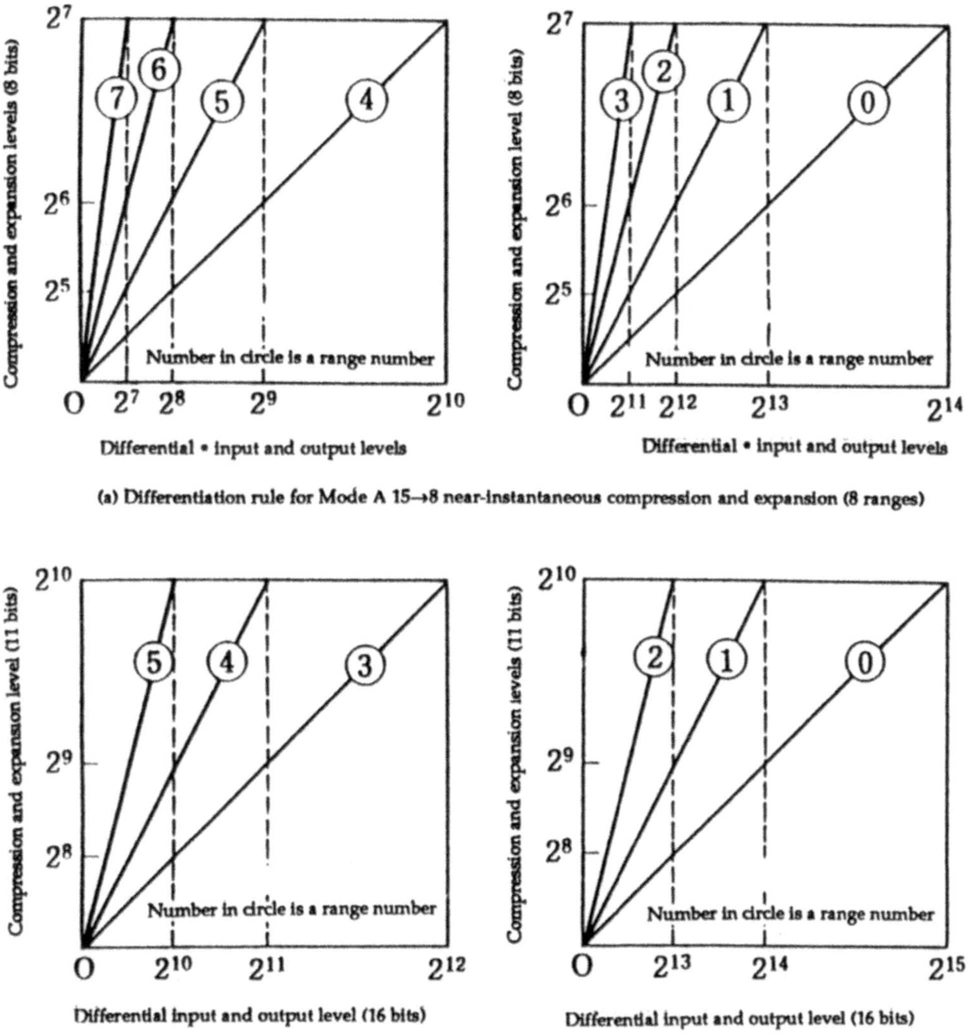

FIGURE 3.22. Range classification (8 bits).

to fit into the vertical blanking period and converted into burst signals with a transmission rate of 18.255 Mbit/s. Using binary/ternary conversion, the transmission rate is reduced to 12.15 M baud, and the signals are time-division multiplexed into the vertical blanking period.

Figure 3.25 shows the configurations of time-compressed multiplexing and time expansion separation circuits for audio and independent data.

(2) Frame Interleaving and De-Interleaving

Since the 1350 kb/s audio signal being multiplexed into the video signal is subjected to 16-bit intraframe bit interleaving, it is possible to correct 16-bit burst errors. However, because the signal is time-compressed before being multiplexed into the vertical video signal blanking period, the signal is significantly affected by noise during transmission. To prevent longer burst errors, interframe interleaving that spans

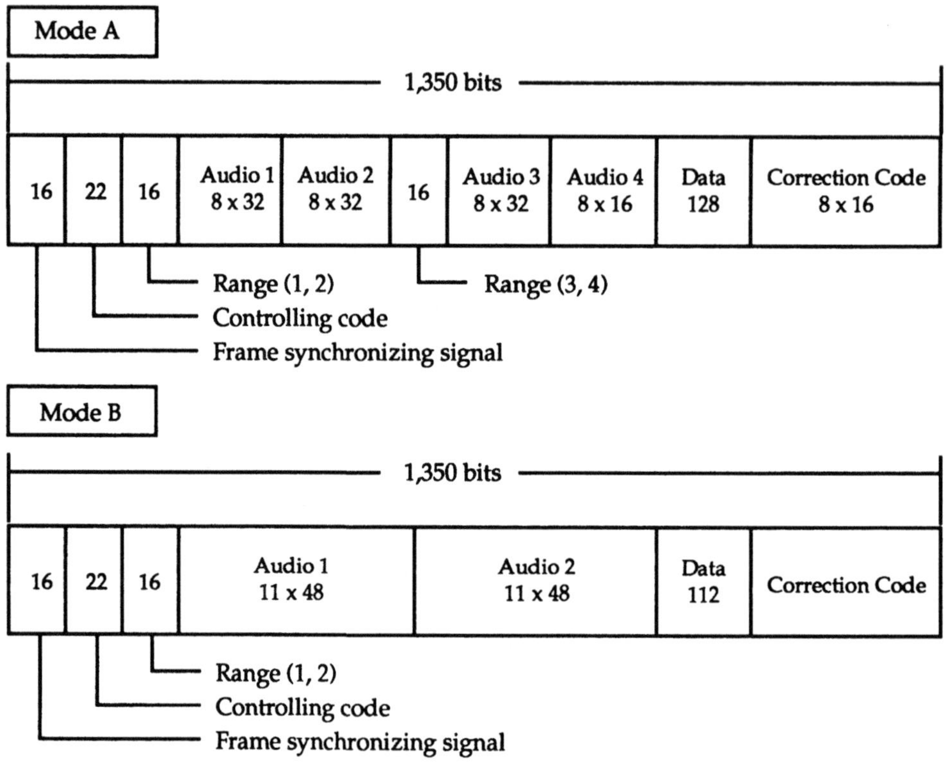

FIGURE 3.23. Configuration of audio frame.

15 audio frames has been added to the system. This interframe interleaving is able to correct burst errors of up to $16 \times 15 = 240$ bits, which is equivalent to 0.44 TV signal line.

Figure 3.26 shows the configurations for frame interleaving and de-interleaving, which are exactly alike. The configurations are realized with either fourteen 1350-stage shift registers, or 1350 clock pulse delay lines and a rotary switch. The rotary switches for interleaving and de-interleaving rotate in opposite directions, and their rotation must be synchronized with each other so that the first data of the multiplexed audio and independent data can instruct the switch positions to be P_1 and Q_1, respectively.

(3) Time-Compression Multiplexing System

The audio and independent data for one television field has a total bit count of 1,350,000/60 = 22,500 bits. This data is time-compressed and multiplexed into one vertical blanking period. The vertical blanking period has four lines with color difference signals if effectively used, and if this portion is also used effectively, 44 lines worth of audio and independent data can be multiplexed. The format of this time-compression multiplexing is shown in Figure 3.27.

Multiplexing occurs at two locations, at lines 3-46 and 565-608, totaling about 85 lines. The empty areas before and after the data bits are guard areas to prevent interference between the HD signal and the data signals, and are called gray areas.

The empty area between the data and color difference signals, also called a gray area, exists to match up the bit count being multiplexed into one field. The audio and independent data are clocked at 12.15 MHz, while the video data is

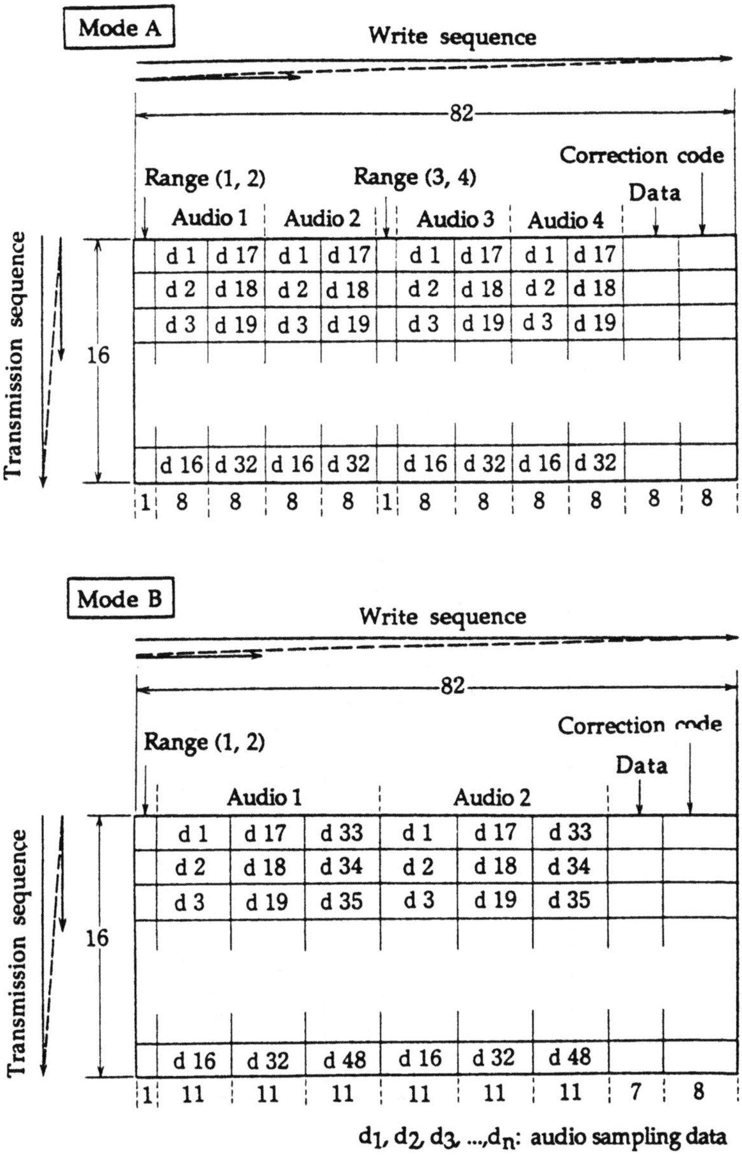

FIGURE 3.24. Bit interleaving matrix.

clocked at 16 MHz. These two clock phases are matched at sampling No. 15.

(4) Binary/Ternary Conversion

In the binary/ternary conversion, three successive bits are converted into ternary/2 baud. The conversion format and the binary/ternary relationship are specified in Table 3.2.

In this table, it is assumed that the data on the left side is transmitted earlier in time. The ternary level expresses the dynamic range of the video signal in 8 bits.

Because 3 bits can express 8 levels of data, there is one unused level (11) in the conversion from the ternary to the binary system. This unused level is called the dissipation level. An

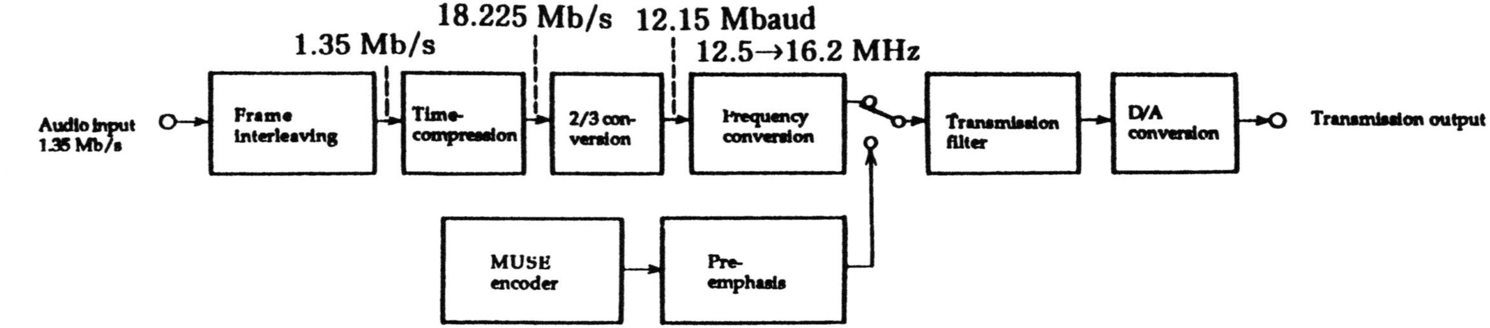

Audio/independent data time-compression multiplex unit

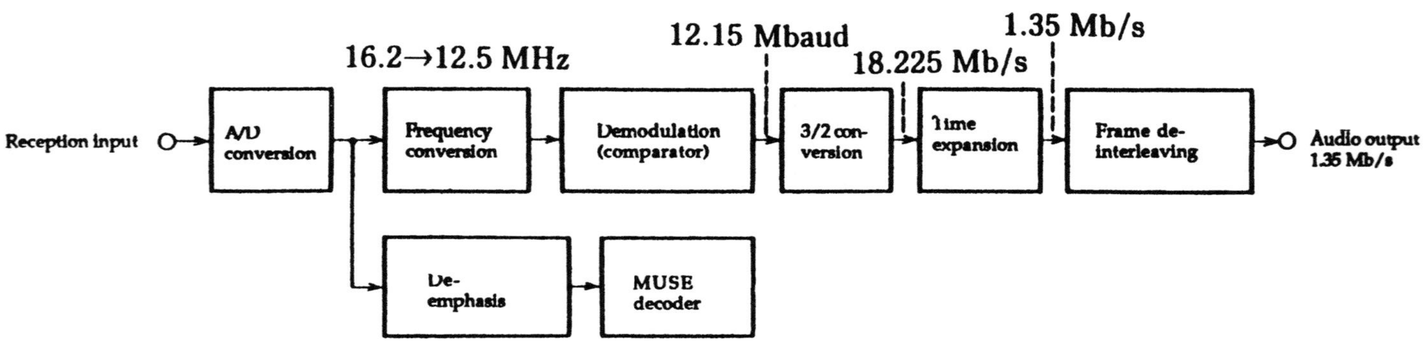

Audio/independent data time expansion separator unit

FIGURE 3.25. Time-compression expansion and multiplexing/separation of audio or independent data.

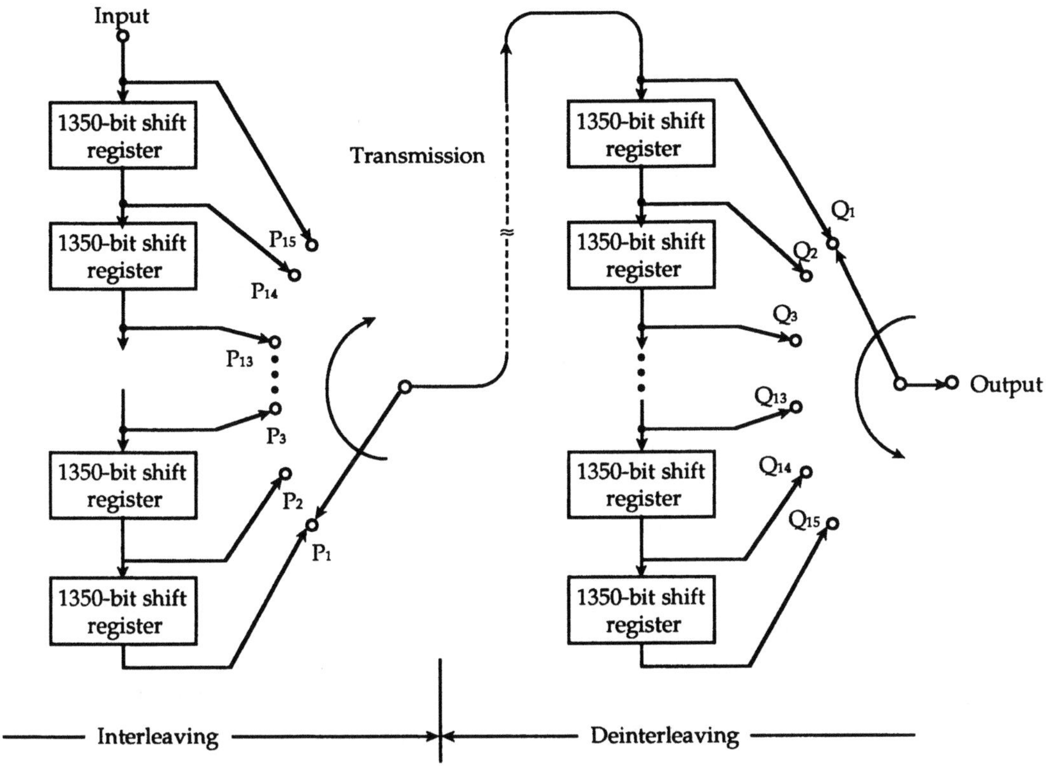

- The switch rotates in the direction of the arrow every 1.35 MHz clock.
- The switch is reset to P1 and Q1 by the first data of the vertical blanking period.

FIGURE 3.26. Configuration of frame interleaving and de-interleaving.

- Lines 3–42, 565–604

HD	Empty	Data	Empty
11	3	464 (348 baud 552 bits)	2

- Lines 43–46, 605–608

HD	Color difference signal	Empty	Data	Empty
11	94	13	360 (270 baud 405 bits)	2

The clock phases of the audio and independent data (12.15 Mbaud) and the video signal (16.2 MHz) coincide at sample No. 15.

FIGURE 3.27. Time-compression multiplexing format.

TABLE 3.2. **Binary/ternary conversion.**

Binary / 3 bits			Ternary / 2 baud	
0	0	0	0	0
0	0	1	0	1
0	1	0	1	2
0	1	1	0	2
1	0	0	1	0
1	0	1	2	0
1	1	0	2	2
1	1	1	2	1

Transmission is from left.

error caused by the detection of the dissipation level is called the dissipation error. The binary/ternary conversion format has been formulated so that a 1-baud error is not translated into a 3-bit error, and also so that the dissipation error is limited to 1 bit as much as possible.

Because the peak value of the eye pattern cannot exceed that of the video signal in Hi-Vision AM signal transmission, the ternary levels have been set so that these peak values match. When FM transmission is used as in satellite broadcasting, the audio and independent data bandwidth is narrower (6 MHz) than the video signal, and so the peak value of the eye pattern can exceed the peak value of the video signal. Reducing the degree of modulation increases errors due to the noise, noise can increase errors, but increasing the degree of modulation causes excess modulation, which also increases errors. The values shown in Table 3.3 were derived from experimental results to minimize errors.

TABLE 3.3. **Ternary levels (Unit: 1/256).**

	FM	AM
0	21 1/3	48
1	128	128
2	234 2/3	208

3.3 SATELLITE TRANSMISSION OF MUSE

3.3.1 Satellite Transmission of MUSE

The 12 GHz SHF band is used for satellite broadcasting. Of the fifteen channels shown in Figure 3.28, the eight odd-numbered channels have been assigned to Japan. If one 27 MHz channel is used for FM MUSE transmission, the frequency deviation allowed by the Carson rule* is about 10.8 MHz, and the resulting improvement with FM is 12.5 dB.

The reception CN ratio with a 75cm satellite reception antenna is about 18 dB, which means the MUSE signal reception SN ratio is an unsatisfactory 30.5 dB. To improve the SN ratio as much as possible, an additional nonlinear emphasis circuit shown in Figure 3.29 is used for FM MUSE transmission.

Through the repeated MUSE signal transmission experiments using the BS-2b, we have confirmed the possibility of Hi-Vision satellite broadcasting with MUSE signals. In addition, transmission with the Canadian Anik satellite conducted in October 1987, and a three-stage relay transmission from Nara, Japan to Brisbane, Australia using CS, INTELSAT, and

*The Carson rule states that for FM transmission of TV signals, required bandwidth = 2 (Baseband bandwidth) + (Max. frequency shift).

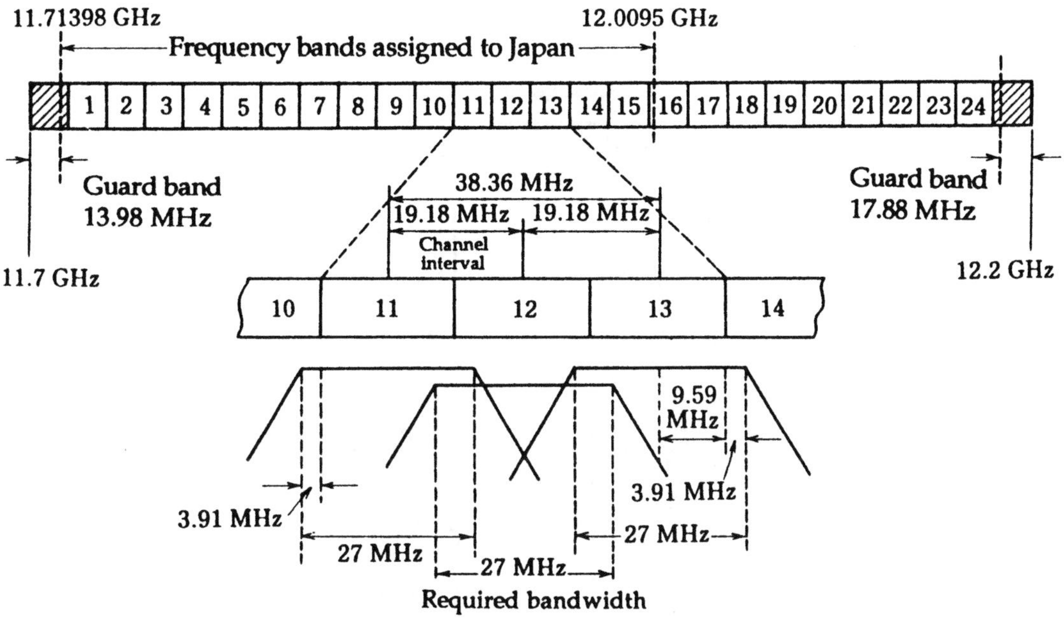

FIGURE 3.28. Channel allocation for satellite broadcasting.

AUSSAT satellites, have confirmed the possibility of domestic and international relays with communications satellites.

3.2.2 MUSE Modulation

(1) SN ratio of FM signal demodulation

Figure 3.29 shows the relationship between the CN ratio (C/N) of the FM demodulation input and the SN ratio (S/N) of the MUSE output signal, as given by the following equation.

$$S/N = \frac{B \cdot (f_d)^2}{\int_0^{f_m} f^2 \cdot D^2(f) \cdot L_R^2(f)\, df} \cdot C/N \quad (3.4)$$

where f_d: maximum frequency deviation (in direct current)

$D(f)$: transfer function of deemphasis circuit

$L_R(f)$: transfer function of reception low-pass filter (see Figure 3.30)

B: reception filter bandwidth for FM signal

f_m: baseband signal bandwidth

From Equation 3.4,

$$D(f) = 1 \text{ (without emphasis)}$$

$$L_R(f) = \begin{cases} 1 & (0 \leq f \leq f_m) \\ 0 & f < f_m \text{ (an ideal filter of bandwidth } f_m) \end{cases}$$

Then the S/N is

$$S/N = \frac{3 \cdot B \cdot (f_d)^2}{f_m^3} \cdot C/N \int I f_m \cdot C/N \quad (3.5)$$

$$I_{fm} = \frac{3 \cdot B \cdot (f_d)^2}{f_m^3} \quad (3.6)$$

The degree of FM improvement I_{fm} equals 12.5 dB if $\Delta F = 10.8$ MHz. The ratio of improvement in the SN ratio when emphasis is and is not performed is called the emphasis gain ρ_e, and is expressed by Equation 3.4.

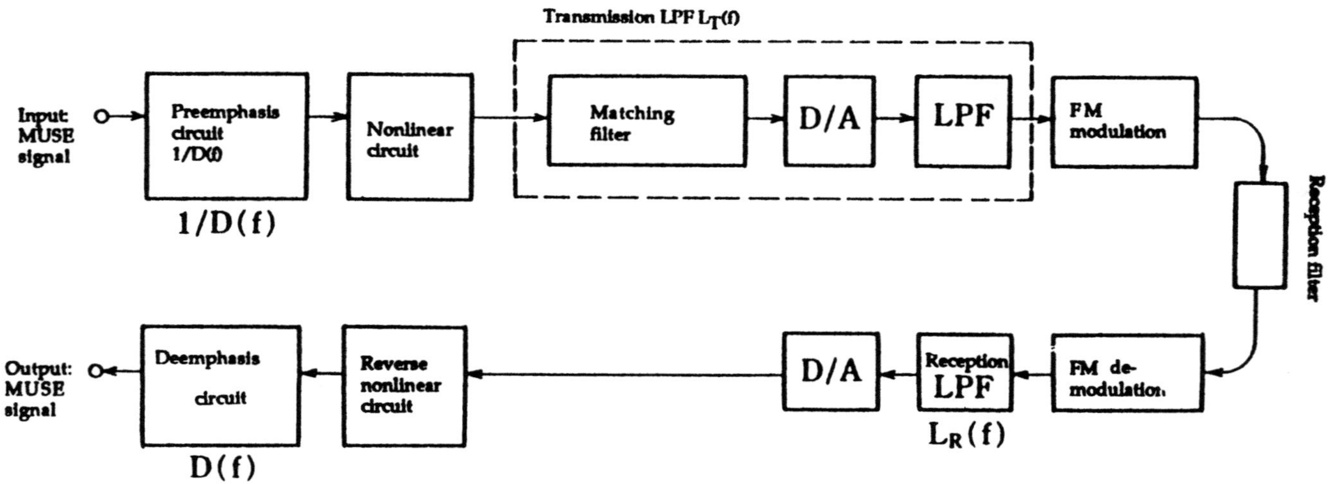

FIGURE 3.29. Block diagram of MUSE modulation/demodulation.

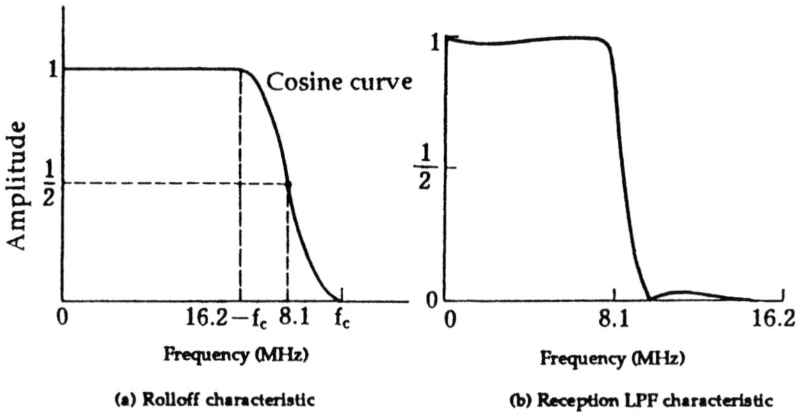

FIGURE 3.30. Overall characteristic of LPF for transmission and reception, and characteristic of LPF for reception.

$$\rho_e = \frac{\int_0^{f_m} f^2 \cdot L_R^2(f)\, df}{\int_0^{f_m} f^2 \cdot D^2(f) \cdot L_R^2(f) df} \qquad (3.7)$$

In a case when the low-pass filter for reception is an ideal filter with cut-off frequency f_m, emphasis gain ρ_e is expressed by the following equation.

$$\rho_e = \frac{f_m^3}{3\int_0^{f_m} f^2 \cdot D^2(f)\, df} \qquad (3.8)$$

With a deemphasis circuit (described in Section 3.3.3), the emphasis gain for MUSE is 9.5 dB.

(2) Characteristics of Low-Pass Filters for Transmission and Reception

When transmitting sampled values such as with MUSE, the decoder must be able to resample accurately without interference between adjacent sampled points. Interference shows up as ringing on the screen and causes image degradation.

The transmission conditions that do not cause interference between sample values are known as Nyquist's first theorem. This states that if $L_T(f)$ is the transfer function of the transmission LPF in Figure 3.29, and $L_R(f)$ is the transfer function of the reception LPF, then the overall transfer function $L_T(f) \cdot L_R(f)$ for transmission and reception must have an amplitude frequency characteristic with a point symmetric rolloff characteristic at 8.1 MHz as shown in Figure 3.30, and the phase characteristic should show a linearity within the transmission bandwidth. In many cases the rolloff characteristic is represented by cosine curves in what is called a cosine rolloff characteristic. In Figure 3.30, $100 \cdot (f_c - 8.1)/8.1$ (%) is called the rolloff percentage α.

The MUSE system uses an analog filter for the reception LPF and a digital filter called a matching filter (shown in Figure 3.29) for the transmission LPF. The cosine rolloff characteristic is distributed evenly between the transmission and reception LPFs, with each to the 1/2 power. The matching filter is adjusted so that the total system including the reception LPF characteristic has a cosine rolloff characteristic.

(3) Power Diffusion

To reduce interference between broadcast satellite services and terrestrial services that share the same frequency bands, WARC-BS and WARC-79 stipulate for the multiplexing of low frequency signals into television signals to diffuse the signal strength. The power diffusion value is determined such that the power flux density measured at a frequency bandwidth of

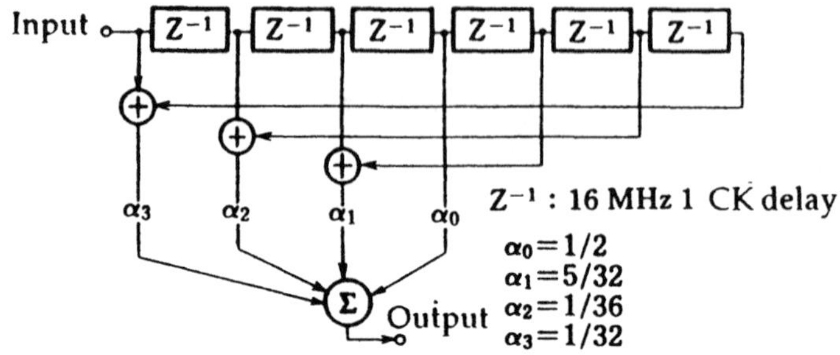

FIGURE 3.31. Deemphasis circuitry.

4 kHz should be 22 dB below the power flux density for the total frequency bandwidth of 27 MHz.

The power diffusion effect R is expressed by the following equation.

$$R = 10 \log \{(\Delta F_{p-p} + \delta f_{rms})/4)\} \quad (3.9)$$

where $\Delta F_{p\text{-}p}$ is the frequency deviation of the power diffusion signal in kHz. The frequency deviation δf_{ms} is due to the effective diffusion effect of the noise and sagging already in the television signal, and is generally about 40.

From Equation 3.9, a power diffusion of 600 $\text{kHz},_{p\text{-}p}$ is necessary for a reduction of 22 dB. The waveform of the power diffusion signal is a 30 Hz symmetrical triangular wave (which is less likely to produce flicker disturbances).

3.3.3 Nonlinear Emphasis[12]

(1) Deemphasis (Preemphasis) Circuit

To simplify the circuitry in the receiver, the deemphasis circuit is composed of digital filters as shown in Figure 3.31. In this configuration, the transfer function $D(f)$ is expressed by the following equation.

$$D(f) = \alpha_0 + 2\alpha_1 \cos(2\pi f/f_s) + 2\alpha_2 \cos(4\pi f/f_s) + 2\alpha_3 \cos(6\pi f/f_s) \quad (3.10)$$

where f_s is the MUSE transmission sampling frequency 16.2 MHz. When the values given in Figure 3.31 are used for α, the frequency characteristic shown in Figure 3.32 is obtained. The preemphasis characteristic is the inverse of the deemphasis characteristic.

(2) Nonlinear Circuit

As Figure 3.32 indicates, the preemphasis shows a tendency to increase in the high frequency region. The degree of emphasis improvement increases if a strong emphasis is used. However, as indicated by Label A in Figure 3.33, the preemphasis circuit causes considerable overshooting and undershooting. If modulated as is, the instantaneous frequency of the FM modulated wave corresponding to Label A will fall outside of the RF band region, generating an impulse noise (truncation noise) for such edge sections. The nonlinear circuit shown in Figure 3.29 is meant to prevent this problem. With the nonlinear characteristic shown in Figure 3.34 (a), the circuit compresses the signal amplitude. As a result, the instantaneous frequency of the FM modulated wave fits within the RF band. On the receiver side, the signal is returned to

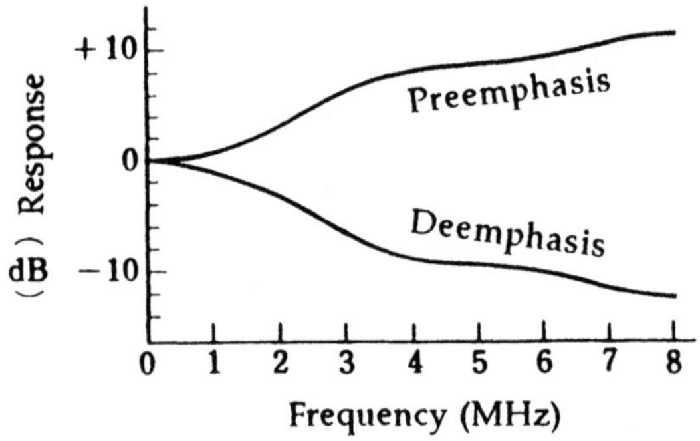

FIGURE 3.32. Emphasis characteristics.

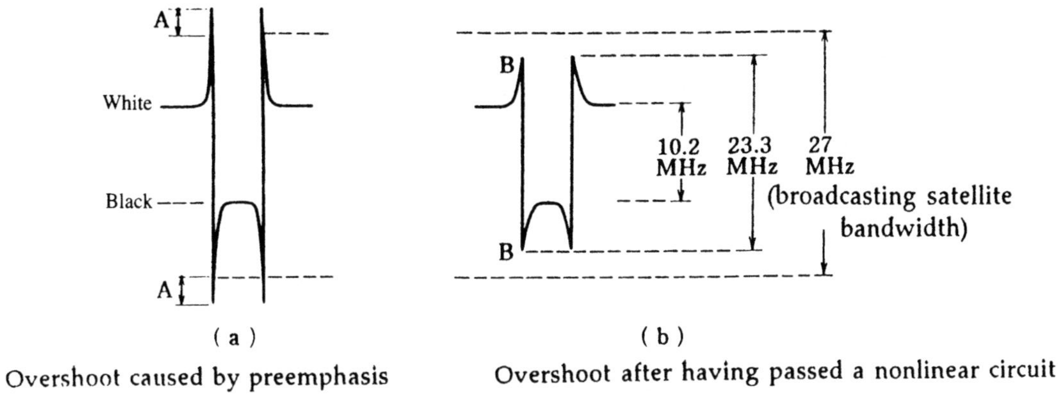

FIGURE 3.33. Suppression of overshoot by the use of a nonlinear circuit.

the original waveform by the inverse nonlinear characteristic shown in Figure 3.34 (b).

In Figure 3.34 (a), the +224 on the vertical axis is the MUSE signal's white level, and −224 (not shown) is the black level. The frequency deviation corresponding to the level difference is set at 10.2 MHz. At this setting, the frequency deviation corresponding to an output between ±512 is 23.3 MHz, which is within the bandwidth of a satellite broadcasting channel (27 MHz). When the reception CN ratio deteriorates and instantaneous noise increases, the extra bandwidth serves to abate truncation noise by accommodating peak frequency shifts from instantaneous noise such as Label B in Figure 3.33.

The inverse nonlinear circuit increases the overshooting and undershooting generated at the edges. As a result, the circuit simultaneously increases the transmission line noise of the edges.

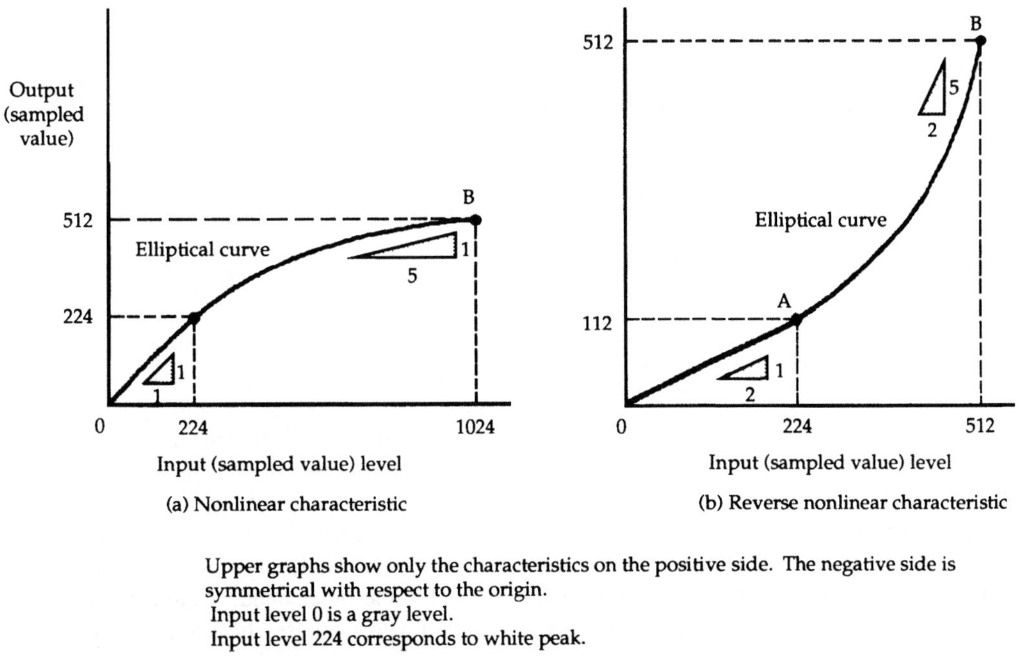

FIGURE 3.34. Nonlinear and inverse nonlinear characteristics.

However, because the noise at the edges is virtually less conspicuous, it does not cause serious image quality deterioration. On the other hand, the section of the screen with a flat luminance level receives ample emphasis, allowing the SN ratio to be improved.

3.3.4 Optimization of Modulation Parameters[13]

(1) Video Signal Modulation

Varying the modulation parameters will affect the SN ratio $(S/N)_0$ and distortion of the demodulation signal in the following manner:

1. An increase in frequency deviation f_d improves $(S/N)_0$, but increases distortion.
2. An increase in emphasis gain ρ_e improves $(S/N)_0$, but increases distortion. The reverse nonlinear circuit increases the noise in the edge area.
3. Smaller rolloff percentages are more advantageous with respect to $(S/N)_0$.

Figure 3.35 shows simulation results for waveform distortion and SN ratio resulting from simultaneous changes in the three modulation parameters mentioned above. The waveform distortion K_p is expressed in the following equation for a normal 2T pulse, based on Figure 3.36.[14]

$$K_p = \max\,(at\,/\,8T),\ 2T \le t \le 8T \quad (3.11)$$

where a: Deviation from the signal

t: Time from the center of the pulse

T: 1/2 of a half-pulse width. For a Hi-Vision bandwidth of 20 MHz, $T = 25$ ns.

Strictly speaking, the K_p for MUSE must be calculated after encoding a $2T$ pulse into MUSE, calculating the distortion, and decoding. However, the distortion from an isolated pulse with the maximum amplitude is known to have a controlling effect on K_p after decoding. Thus as shown in Figure 3.37, if $S(nT)$ is the resampling value from transmitting an isolated pulse from one MUSE signal sample, we can signify Equation 3.11 to:

$$K_p = \max\,\{(S(nT) \cdot nT_0\,/\,8T\} \quad (3.12)$$

where $n = \pm 1$ to ± 3, $T = 1\,/\,(16.2\text{ MHz})$

In Figure 3.35, the horizontal axis shows the SN improvement ratio when $(S/N)_0$ is 0 dB at $f_d = 10.8$ MHz, $\rho_e = 9.5$ dB, and $\alpha = 10\%$. Both the transmission and reception low-pass filters are assumed to have a characteristic to the 1/2 power.

In Figure 3.35, if a waveform distortion up to about $K_p = 4$ is allowed, the f_d and ρ_e combinations that produce a favorable SN ratio are $f_d = 8.8$ MHz and $\rho_e = 12$ dB, or $f_d = 12.8$ MHz and $\rho_e = 9.5$ dB. However, an attempt to obtain a large emphasis gain may end up with conspicuous noise at the edges of the image due to the reverse nonlinear circuit. Taking this problem into consideration, a combination of $f_d = 9$ to 11 MHz and $\rho_e = 10$ to 8 dB may be suitable for the purpose. In practice, because 600 kHz of frequency deviation is necessary for the power diffusion signal, f_d is determined by subtracting 600 kHz from the value obtained from the Carson rule so that $f_d = 10.2$ MHz. As for the deemphasis characteristic, the tap coefficients of $\alpha_0 = 1/2$, $\alpha_1 = 5/32$, $\alpha_2 = 1/26$, and $\alpha_3 = 1/32$ were adopted to simplify the arithmetic circuit. At these values, the emphasis gain is $\rho_e = 9.5$ dB. The rolloff percentage α is set at 10%.

(2) Audio Signal Modulation[15]

As described in Section 3.2, the digitized audio signal is multiplexed into the vertical blanking period of the video signal as a 12.15 Mbaud ternary signal. While the ternary audio signal, like the video signal, is transmitted by FM modulation, during this period, no preemphasis and deemphasis are carried out.

If the frequency deviation of the ternary audio signal is small, noise increases the bit error rate. On the other hand, a large frequency deviation also increases bit error rate due to distortion. In Figure 3.38, the bit error rate is measured for changes in frequency deviation. As

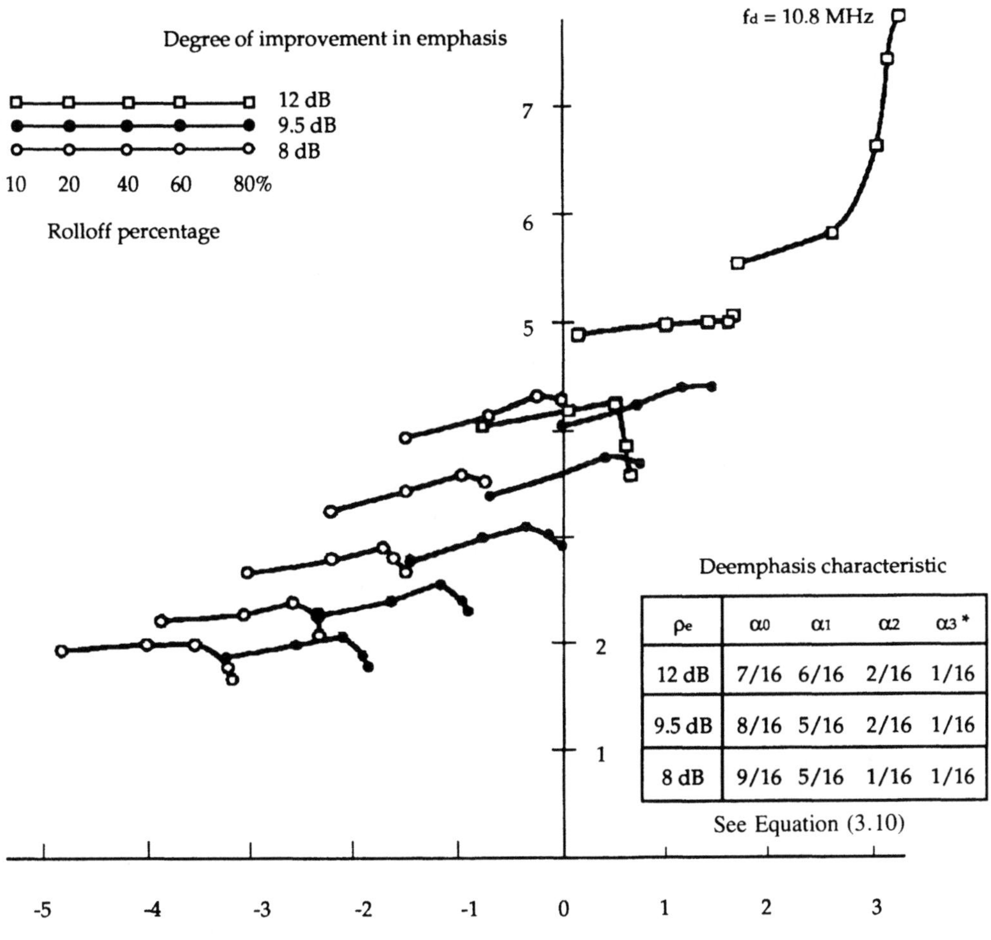

ρe	α0	α1	α2	α3 *
12 dB	7/16	6/16	2/16	1/16
9.5 dB	8/16	5/16	2/16	1/16
8 dB	9/16	5/16	1/16	1/16

FIGURE 3.35. Changes of waveform distortion K_p and SN ration with varying modulation parameters.

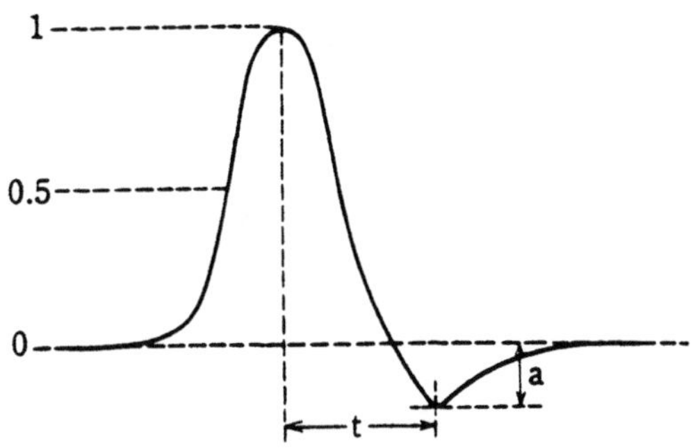

FIGURE 3.36. Graph for calculating K_p of a 2T pulse.

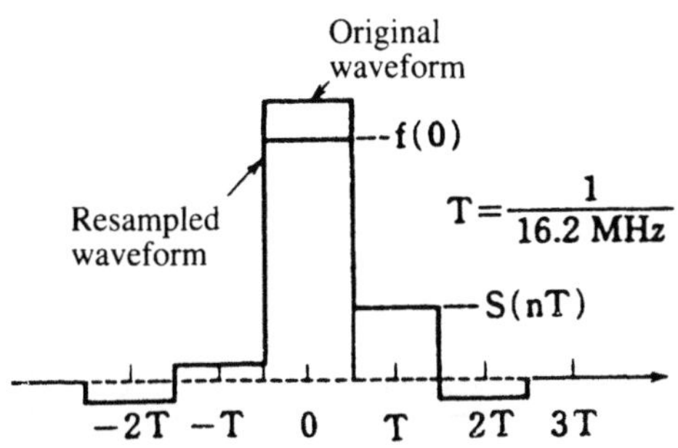

FIGURE 3.37. Sample wave corresponding to one isolated pulse of one MUSE signal sample.

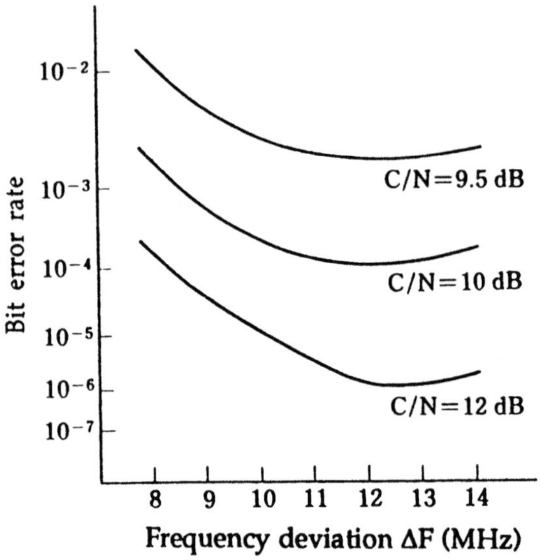

FIGURE 3.38. Bit error rate versus audio frequency deviation.

the figure indicates, the bit error rate is minimized over the broad range of frequency deviations from about 9.5 MHz to 12 MHz. Considering the increase in threshold noise when the CN ratio is low, we set the frequency deviation of the ternary audio signal at 9.76 MHz.

3.3.5 Transmission Experiment with Broadcasting Satellite BS-2

The modulation parameters for MUSE signals discussed thus far for satellite broadcasting are shown in Table 3.4. In addition, the circuit design for transmitting with broadcast satellite BS-2 is shown in Table 3.5. This circuit design is for reception in Tokyo, and assumes a rainfall attenuation level that exceeds 99% of the hourly levels in the worst month of the year.

Figure 3.39 plots the relationship between

TABLE 3.4. MUSE signal modulation parameters.

Item	Parameter
Occupied frequency bandwidth	27 MHz
Baseband bandwidth	8.1 MHz (-6dB, 10% cosine rolloff)
Modulation	FM
Modulation polarity	Positive
Power diffusion signal	600 kHz_{p-p} 30 Hz
VIDEO	
Frequency deviation	10.2 MHz_{p-p}
Deemphasis characteristic D(f)	$D(f) = \frac{1}{2} + \frac{5}{16}\cos(2\pi f/f_S) + \frac{1}{8}\cos(4\pi f/f_S) + \frac{1}{16}\cos(6\pi f/f_S)$ where f_S: 8.1 MHz
Emphasis gain	9.5 dB
Nonlinear characteristic	See Figure 3.34
Reverse nonlinear characteristic	
AUDIO	
Ternary signal frequency deviation	9.72 MHz_{p-p}

TABLE 3.5. Design of satellite transmission circuit for MUSE signal.

TWT power	100W
Frequency band	12 GHz
Channel width	27 MHz
Baseband bandwidth (−6 dB)	8.1 MHz
Modulation	FM
Frequency deviation	10.2 MHz
Effective radiating power	56.5 dBW
Satellite antenna gain	39 dB
Miscellaneous loss	2.5 dB
Free space loss	205.6 dB
Attenuation due to rainfall	2 dB *
Diameter of receiving antenna	0.75 meter **
Gain of receiving antenna	38 dB
Noise power of receiver	-129.7 dBW
Receiver noise index	2.0 dB (170K)
Antenna input noise	120K
Reception CN ratio	16.6 dB
FM improvement	11.9 dB
SN ratio (p–p/rms)	28.5 dB
Emphasis gain	9.5 dB

* 99% of the time.
**Efficiency is 70%. CN ratio is about 2 dB higher in fair weather. With the BS-3, the CN ratio improves an additional 2 dB.

the reception CN ratio and the demodulated video signal's SN ratio (prior to deemphasis) for an actual transmission from BS-2. Figure 3.40 plots the relationship between the reception CN ratio and audio signal bit error rate (before error correction). In Figure 3.39, the diameter of the reception antenna and the reception CN ratio are shown. Because the measurement was conducted in fair weather, the CN ratio exceeds the design value in Table 3.5 by about 1.5 dB.

Figure 3.41 shows video and audio quality evaluations as a function of the reception CN ratio.[16] If the standard of service is level 4, the CN ratio is 14 dB for the video signal and 9 for the audio signal.

The experimental results discussed above confirm that Hi-Vision satellite broadcasts can be received at a high level of quality with a small household antenna.

3.3.6 Transmission Experiments Performed Overseas

(1) Experiment With an Anik Satellite[17]

An experimental transmission was demonstrated over the Ku band of the Canadian domestic communications satellite Anik-C2 in October 1987 and the HDTV Colloquium in Ottawa, Canada.

A diagram of the experiment is shown in Figure 3.42. In this experiment, a mobile trans-

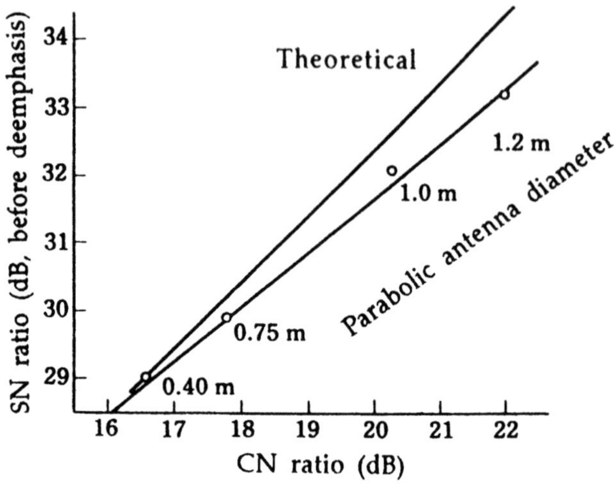

FIGURE 3.39. SN ratio of demodulated video signal versus reception CN ratio.

mitter and a 4.5 m antenna were installed in the colloquium hall in Ottawa, and MUSE signals were transmitted via the Anik-C2. At the same time, MUSE-T signals (MUSE-Transmission, a 16.2 MHz bandwidth signal for archive material that has undergone only interfield offset sampling; see Section 3.4.4) were transmitted to Ottawa via the Anik-C2 from a mobile station in Toronto. Further, the signals from the Anik-C2 were received on Long Island for a second stage relay into the United States over the RCA-K1 satellite.

As the transponder had a bandwidth of 54 MHz, MUSE signals were band-limited to 27

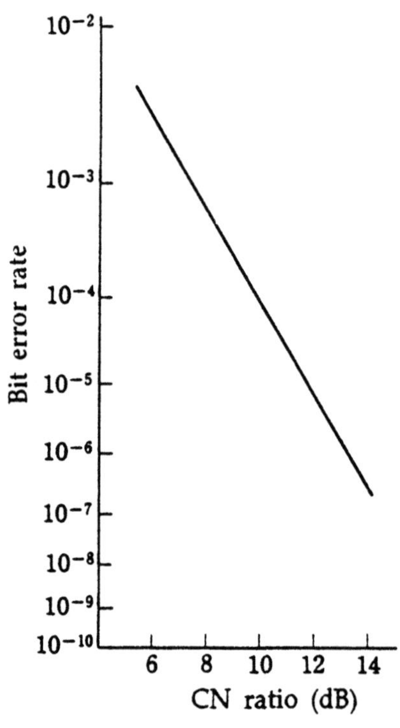

FIGURE 3.40. Audio bit error rate (prior to error correction) versus reception CN ratio.

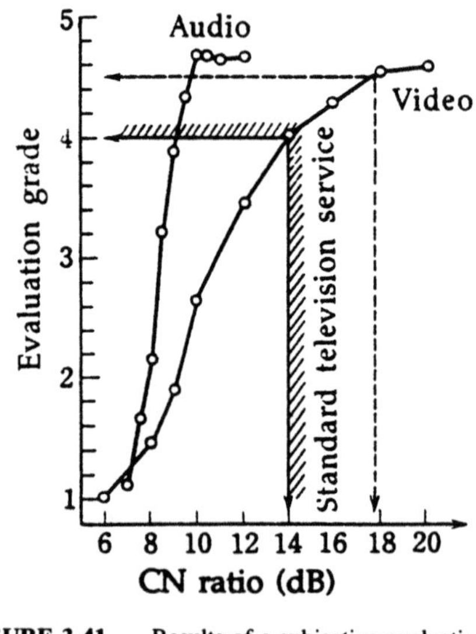

FIGURE 3.41. Results of a subjective evaluation experiment.

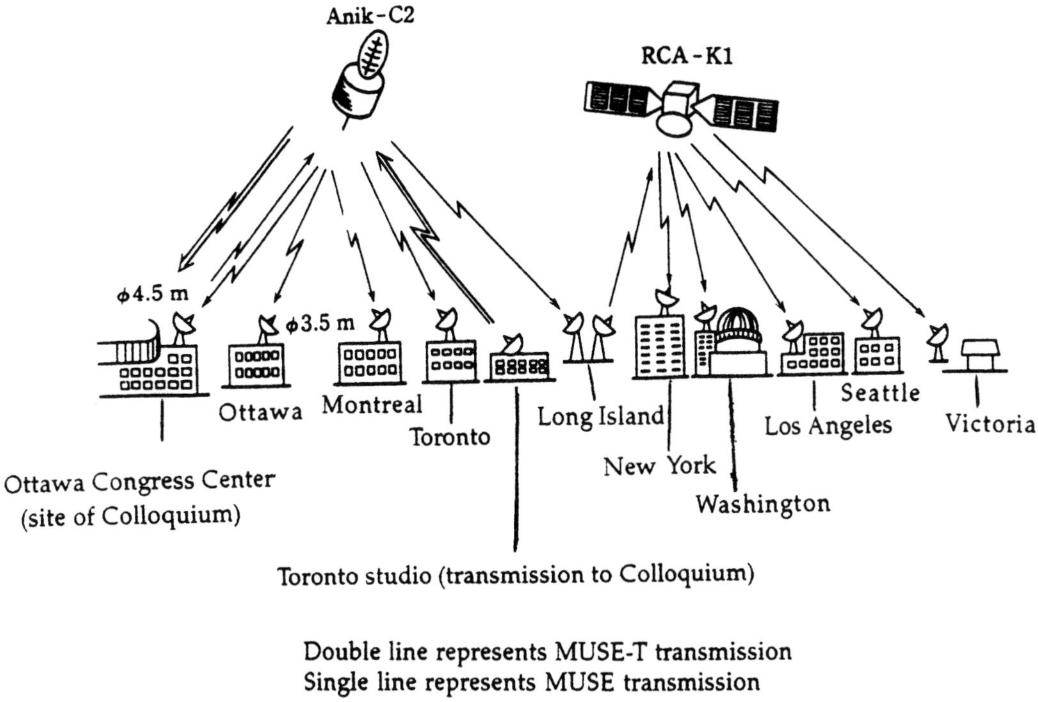

FIGURE 3.42. MUSE transmission experiment using an Anik satellite.

MHz and MUSE-T signals to 54 MHz. Table 3.6 shows the parameters and circuit design values for the Anik-C2 satellite. Characteristics such as reception CN ratio, video SN ratio, and audio bit error rate mostly agreed with their design values, and the image quality was satisfactory as well.

This experiment was highly successful in demonstrating the possibility of using MUSE signals and communications satellites for Hi-Vision relay broadcasting, and for using MUSE-T for transmitting program material.

(2) Transmission from Japan to Australia

In July 1988, a Hi-Vision transmission involving a three-stage satellite relay was performed from the Silk Road Fair in Nara, Japan to the International Leisure Fair in Brisbane, Australia.

The configuration of the transmission system is shown in Figure 3.43. The first stage was a domestic relay from an NTT mobile station located at the Silk Road Fair site to the KDD Ibaraki earth station using the C band of the Japanese domestic communications satellite CS-2. The second stage was an international relay on the C band of the INTELSAT Pacific satellite from Ibaraki to the OTCA earth station in Sydney. The final stage was a Ku band transmission with the Australian domestic communications satellite AUSSAT from Sydney to Brisbane. In the transmission between Sydney and Brisbane, both the transmission and reception were done from mobile stations. The modulation parameters for satellite broadcasting in Table 3.4 were used except for a frequency deviation of 9.8 MHz, which was changed because the power diffusion had been set to 1 MHz_{p-p}. The calculated CN ratios for each relay and for the composite of the total system are shown in Table 3.7.

The transmission result of 17.6 dB for the CN ratio is close to the calculated value. The image reception was about equal to reception with the BS satellite, with little waveform distortion and a high image quality. The success in the three-stage satellite relay proved that international and domestic relays of Hi-Vision sig-

TABLE 3.6. Circuit design of the transmission experiment with the Anik satellite.

Satellite	Anik-C2 (110.0 degrees west)	
Transponder	Channel 4H (MUSE) Channel 8H (MUSE-T)	
Uplink		
Earth station EIRP *	76.4 dBW	
Uplink propagation loss	207.2 dB	
Satellite G/T	7.0 dB/K	
Uplink circuit $(C/N)_0$	103.8 dB-Hz	
Downlink		
Satellite EIRP	49.4 dBW	
Downlink circuit propagation loss	205.7 dB	
Earth station G/T (4.5m diameter)	28.3 dB/K	
Downlink circuit $(C/N)_0$	100.6 dB-Hz	
Overall $(C/N)_0$	98.9 dB-Hz	
	MUSE	MUSE-T
CN ratio	24.6 dB	21.6 dB
FM improvement	12.4 dB	11.8 dB
Emphasis improvement	9.5 dB	9.5 dB
Unevaluated SN ratio	46.5 dB	42.9 dB

* Effective Isotropically Radiated Power.

TABLE 3.7. CN ratios for each relay segment.

Route	CN ratio
Nara to Ibaraki (CS)	20.9 dB
Ibaragi to Sydney (Intelsat)	24.1 dB
Sidney to Brisbane (AUSSAT)	21.9 dB
Overall	17.3 dB

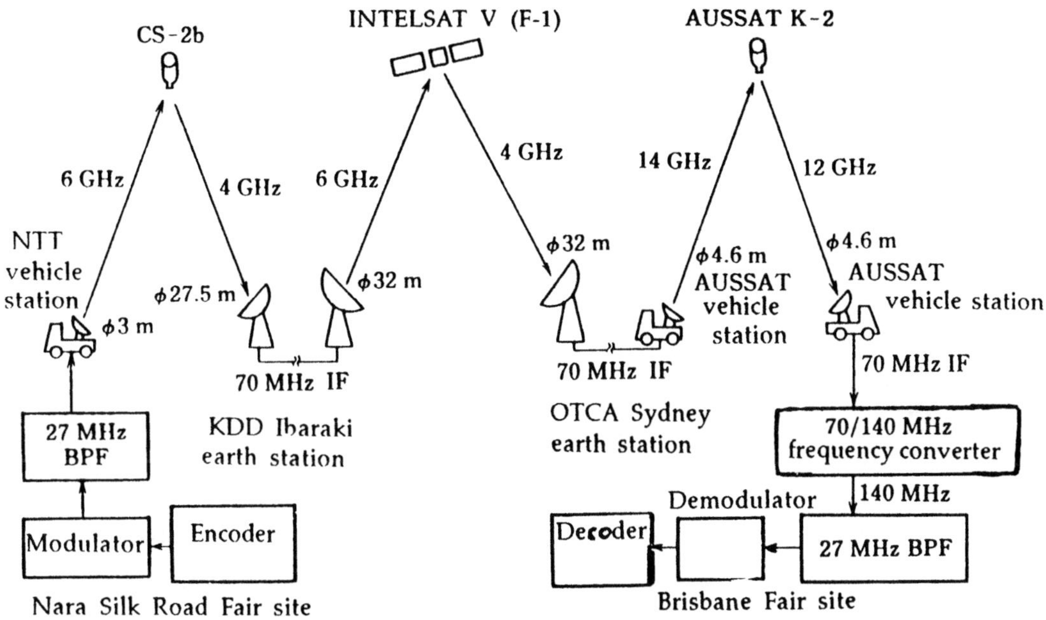

FIGURE 3.43. Transmission system connecting Nara and Brisbane.

nals were possible even from locations with less than favorable conditions. This possibility will continue to be pursued in the future.

3.4 PROGRAM TRANSMISSION

Hi-Vision wireless relay transmission, whether using an FPU (Field Pickup Unit), helicopter and other mobile units, or satellite, is the same as for conventional television. However, the high image quality transmission of Hi-Vision differs in its requirement for high performance equipment, and the unavoidable use of high frequency bands to carry the broad transmission band. For these reasons, the propagation distance is shorter, making improvements necessary to secure the circuit margin.

3.4.1 Radio Frequency Bands and Propagation Characteristics

Because transmission of Hi-Vision program materials requires a broad frequency band, the frequency bands assigned for conventional television relays cannot be used. Although the assignment of new frequencies for Hi-Vision program relays is yet to come, it is expected that the frequency band will be considerably higher than for conventional television relays. Possible downlink frequency bands for SNG (Satellite News Gathering) with satellites are 12.5 to 12.75 GHz for broadcast satellites and 12.2 to 12.75 GHz for Ku band communications satellites. Uplinks could be allocated the band from 14 to 14.5 GHz. For terrestrial FPU lines, the possible frequency bands are 22.5 to 23 GHz and 40.5 to 42.5 GHz.

The 22 GHz and 40 GHz bands are not capable of relaying over long spans because of their large attenuation due to rainfall. Figure 3.44 shows the estimated attenuation caused by rainfall in Tokyo.[18] Because the dB level of attenuation due to rainfall increases with distance, lengthening the relay span causes a sharp increase in the required transmission power, or else greatly decreases the circuit reliability. Thus in setting up a relay, a multi-stage relay with short spans has a larger rainfall attenuation margin and greater circuit reliability than one with fewer stages and longer spans.

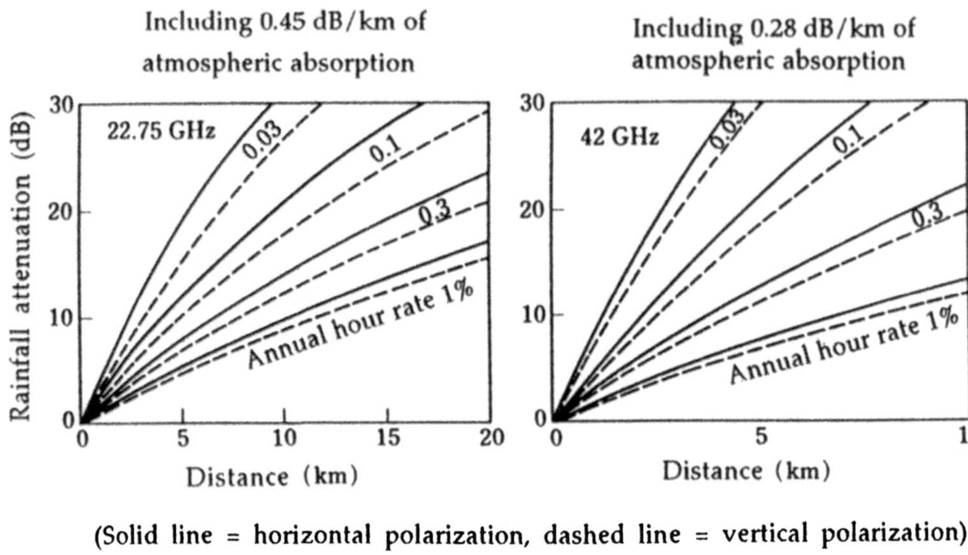

FIGURE 3.44. Dependence of rainfall attenuation on distance.

The difference between the horizontal and vertical polarization in Figure 3.44 is explained by the fact that raindrops fall in a flat disk shape. In the case of circular polarization, the attenuation value is between the horizontal and vertical polarization. While the hourly levels of the curves are displayed with the average annual value as a parameter, in some cases it is more useful to use the hourly level for the worst month. In this case, the annual hourly levels of 0.03%, 0.1%, and 0.3%, and 1% would be switched to the hourly levels for the worst month of 0.14%, 0.4%, 1.0%, and 2.9%.

3.4.2 Radio Relay System

As indicated in Figure 3.45, the radio relay system for Hi-Vision program materials is basically the same as for conventional television. However, as discussed above, because of the higher band frequency and the wideband high quality transmission, the scale and format of Hi-Vision

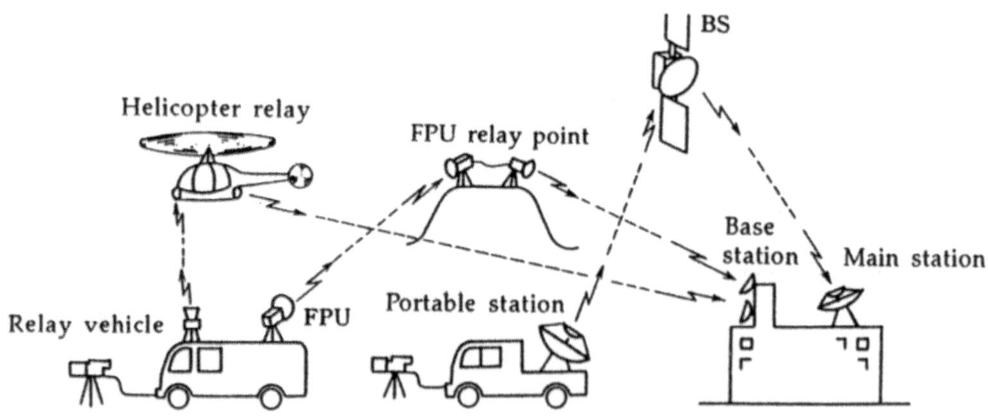

FIGURE 3.45. A wireless relay scheme.

radio relays are different from conventional television.

First, FPU equipment will inevitably tend toward multistage relays. In the case of FM transmission, multistage relays using small scale units will be widely adopted. Although digital transmission is still in the experimental stage, it has an advantage because it is able to prevent signal degradation in multistage relays by using a regenerative system. The large rainfall attenuation requires a wide AGC range to handle fluctuations in the electrical field. However, widening the AGC range has an effect that goes back all the way to the first relay stage and causes a deterioration in the NF of receivers.

In the case of a mobile relay, antenna tracking becomes difficult because of the narrow beam width. However, antennae with an electronic tracking function are under development and are expected to be put into use in the future. In the mobile relay, circular polarization is used because it eliminates the need for polarization tracking and is also convenient for suppressing multipath propagation waves.

While program transmission via satellite is expected to increase for conventional television, the wide transmission band of Hi-Vision is problematic. The 27 MHz transponder bandwidth of Ku band commercial communications satellites requires a transmission system with several channels for the TCI and MUSE-T signals. Since MUSE-E signals can be transmitted on one channel, it can be used for news and other programs that place a secondary priority on image quality.

The broadband channel on the BS-3 satellite, with a bandwidth of 60 MHz, is suitable for the transmission of Hi-Vision programs.

While a small transmission station is desirable for uplinking a remote program to a satellite, the wide band and high CN ratio required for Hi-Vision necessitate transmission equipment several times larger in size than for conventional television transmission.

The output power and antenna size for the earth station transmitter can be calculated from the required CN ratio (C/N) using the equation below.

$$C/N\ (\mathrm{dB}) = P_t + G_t - L_t - L_p - 20 \log \left(\frac{4\pi d}{\lambda}\right) - R + G_r - L_T - N_F - 10 \log (kT_0B) - L_d \tag{3.13}$$

where

P_t: Output power of the earth station transmitter (dBw)

G_r: Gain of satellite receiving antenna (dB)

G_t: Gain of transmission antenna (dB)

L_r: Loss in satellite reception feeder (dB)

L_t: Loss in transmission feeder

N_F: Satellite receiver NF (dB)

L_p: Loss due to transmission beam direction error (dB)

k: 1.3807×10^{-23} (W/Hz K)

d: Distance from earth station transmitter to satellite

T_o: 290°K

l: Wavelength

B: Transmission bandwidth (Hz)

R: Attenuation due to rainfall (dB)

L_d: CN ratio deterioration in the downlink (dB).

For example, suppose the following values apply:

$$L_t + L_p = 2\ \mathrm{dB}$$

$$G_r - L_r = 40\ \mathrm{dB}$$

$$d = 37{,}900{,}000\ \mathrm{m}\ (\text{BS–Tokyo})$$

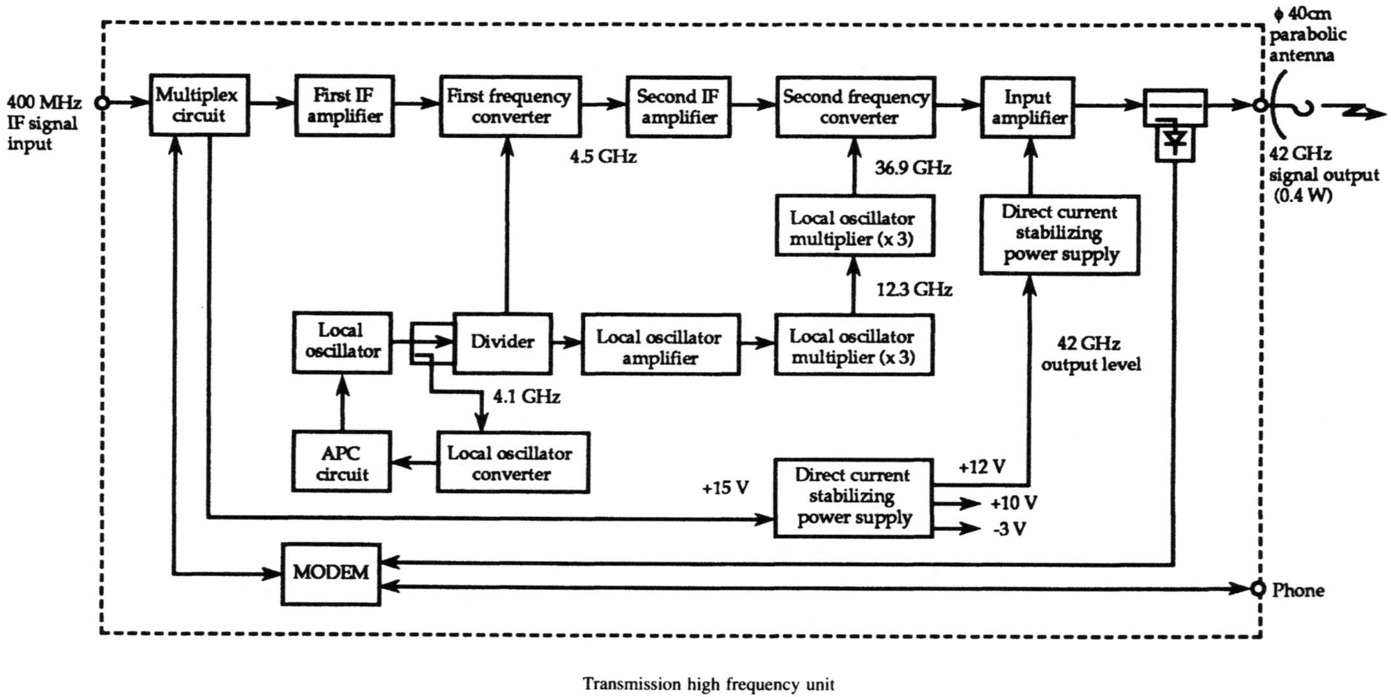

FIGURE 3.46. Configuration of 42 GHz band Hi-Vision FPU.

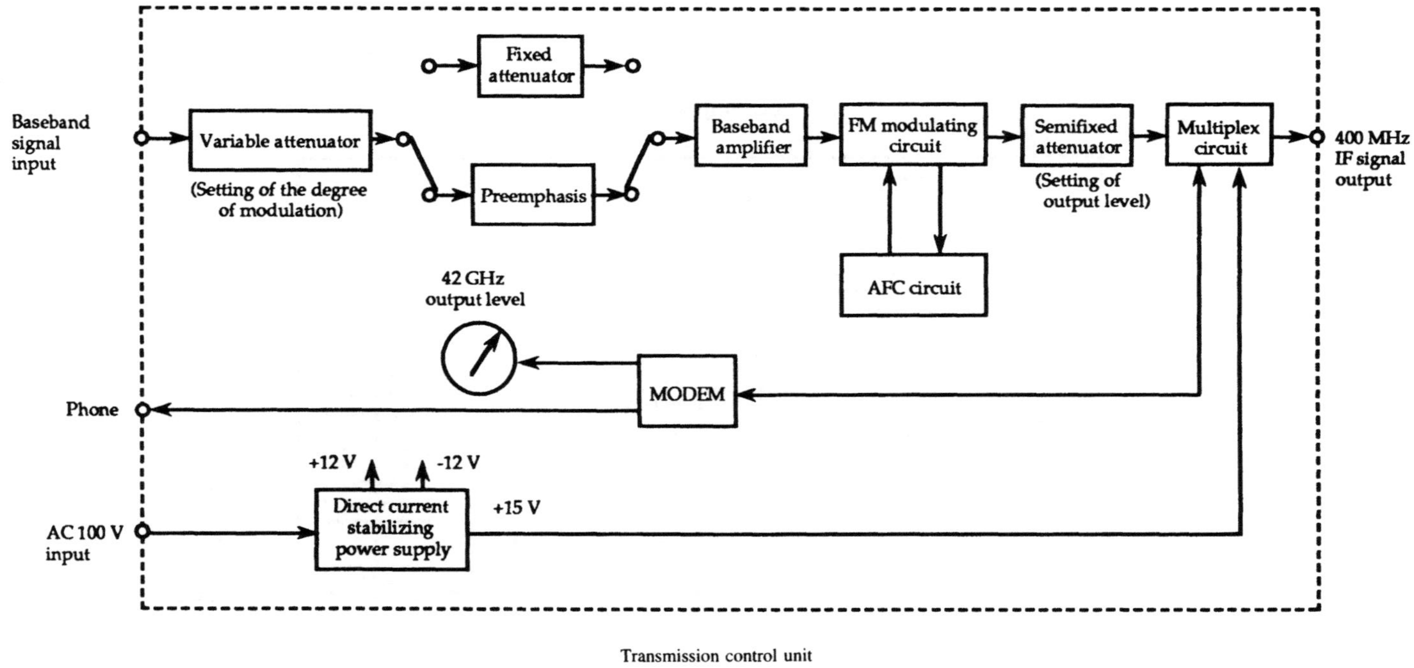

(a) Transmitter

FIGURE 3.46. (continued)

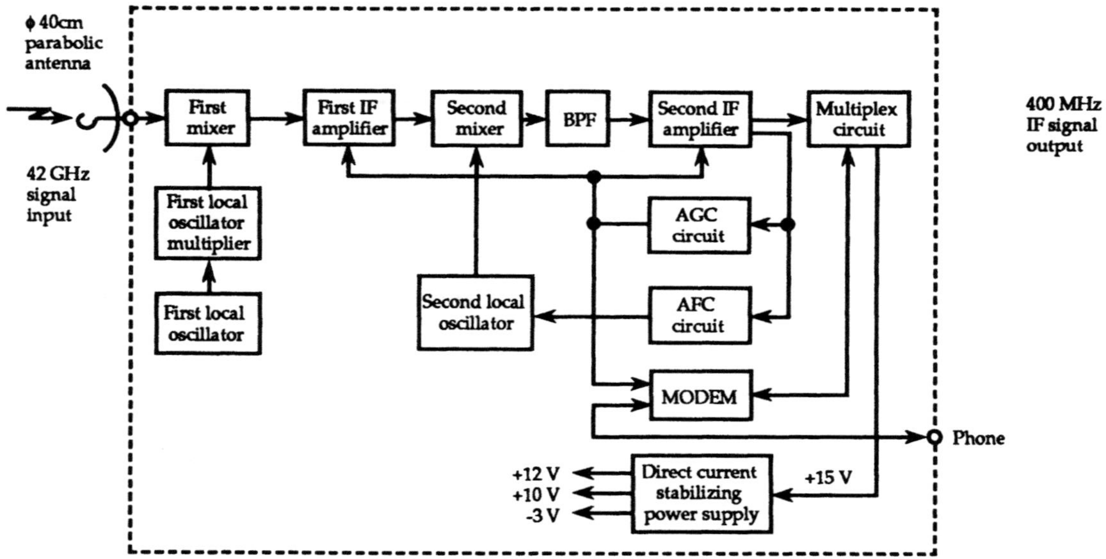

Reception high frequency unit

FIGURE 3.46. (continued)

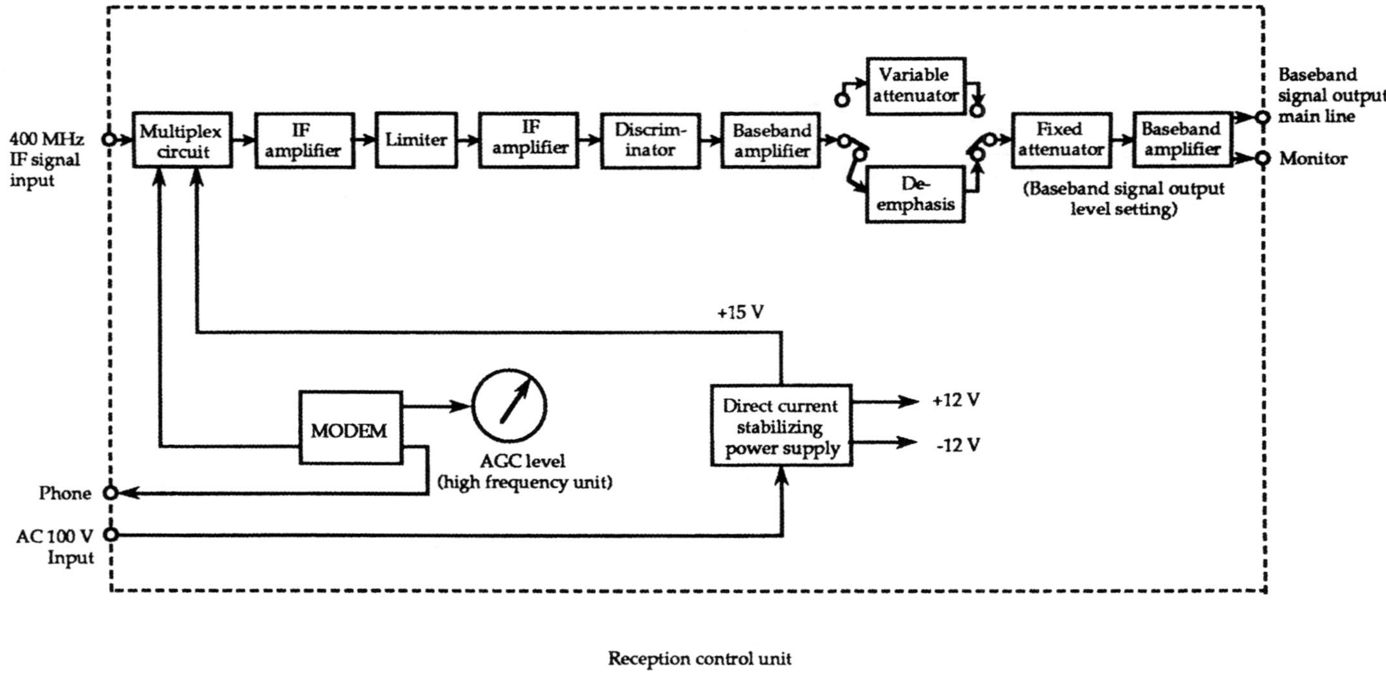

(b) Receiver

FIGURE 3.46. (continued)

N_F = 6 dB (calculated at the antenna)

λ = 0.21 m (14 GHz band)

B = 60 MHz

R = 4.2 dB (0.1% per year for 14 GHz band)

L_d = 0.5 dB

Then we have $C/N = P_t + G_t - 53.6$. Transmitting with a 400 W transmitter and an antenna with a diameter of two meters will result in C/N = 19.7 dB, while transmitting with a 200W transmitter and a 1.5 meter antenna yields a C/N ratio of 14.2 dB.

3.4.3. FPU

Since FPU (Field Pickup Unit) equipment used in Hi-Vision program transmission requires a baseband bandwidth of 16.2 MHz for MUSE-T and about 30 MHz for TCI signal transmission, the FM transmission band must be at least 50 MHz and if possible, 100 MHz. In an FPU, the transmitter and receiver are each divided into a control unit and a high frequency unit directly connected to the antenna. At a multistage relay point, the control unit, that is, the modulation and demodulation unit, becomes unnecessary because the high frequency units are connected with each other. While the delivery IF (intermediate frequency) between the high frequency and control units ideally should be as low as possible to decrease cable loss, a low IF increases the specific bandwidth and so requires a frequency characteristic correction based on the cable length.

Figure 3.46 shows a 42 GHz band FPU[19] that takes the factors described above into consideration. It is a super wideband general purpose FPU that can handle different signal formats, and its delivery IF is 400 MHz.

Regarding the linearity of modulation, which

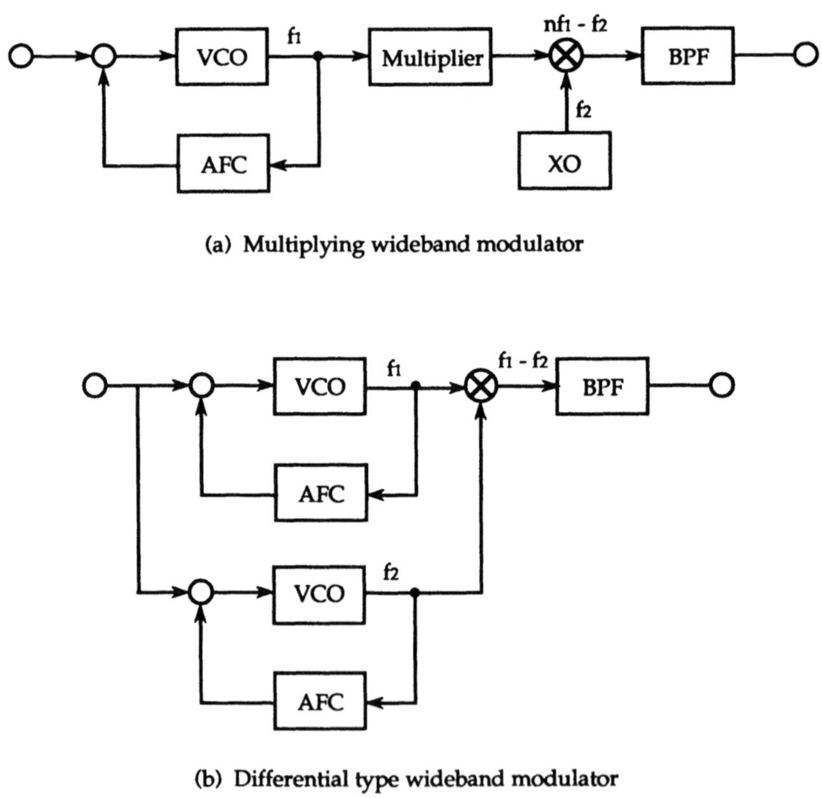

FIGURE 3.47. Wideband FM modulation signal.

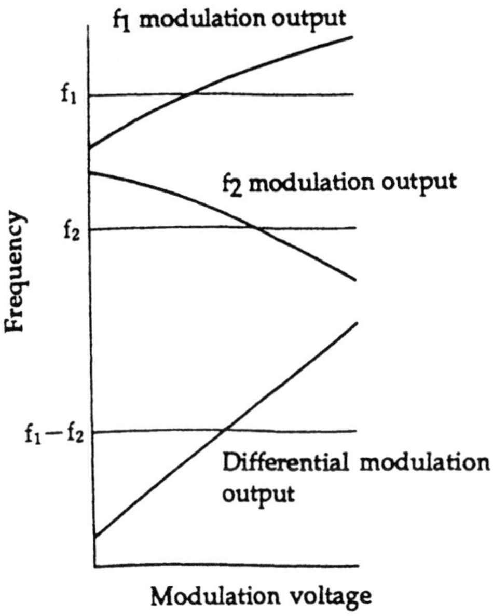

FIGURE 3.48. Operation of differential modulation.

is often a problem with wide band FM, there are two solutions. One method, shown in Figure 3.47, broadens the frequency deviation using FM frequency doubling. In the other method, two varactor modulators are modulated with inverse characteristics so that the difference cancels out the nonlinearity. This latter method of differential modulation operation is shown in Figure 3.48.

3.4.4 TCI and MUSE-T Transmission Systems

The transmission signal, which includes a video signal consisting of luminance signal and color signals and an audio signal, and be multiplexed in one of two ways—frequency multiplexing or time-division multiplexing. While the current NTSC standard broadcasting system uses frequency multiplexing, MUSE, as described in Section 3.1, uses time division multiplexing.

Japan's satellite broadcasting as well as domestic and international television transmission lines are mostly modulated using FM. The FM noise spectrum is a triangular spectrum, and noise is distributed more toward the higher frequency region. Because human vision is more sensitive to noise in the lower frequency region than in the higher frequency region, FM modulation is especially well suited to television transmission. However, with frequency multiplexing, because the color subcarrier is in the high frequency region, when the CN ratio deteriorates, the color noise becomes more conspicuous than the luminance signal. This means that in frequency multiplexing with FM transmission, the power balance between the luminance and color signals is poor. Therefore, low power FM transmission is not suitable for frequency multiplexing unless a sufficient CN ratio is secured.

In time-division multiplexing, because the color subcarrier is not multiplexed into in the high frequency region, the signal's spectrum can be regarded as monochromatic. It is therefore possible to optimally design the power distribution between the luminance and chrominance signal. Another feature of time-division multiplexing is that it is not affected by the triangular noise spectrum of FM transmission. This means that compared to frequency multiplexing, time-division multiplexing can be received at a relatively low CN ratio. The absence of the color subcarrier prevents cross color and cross luminance disturbances from occurring. Furthermore, the system is resistant to nonlinear dis-

TABLE 3.8. Comparison of signal bandwidths.

Method	Signal bandwidth
MUSE	8.1 MHz
MUSE-T	16.2 MHz
TCI	30 MHz
Studio	60 MHz (Y : 30 MHz, C : 15 MHz x 2)

tortions such as DG and DP, so a strong nonlinear emphasis can be applied to obtain a large emphasis gain. Therefore, for low power FM transmission, time-division multiplexing is superior to frequency multiplexing.

(1) TCI

TCI[20] (Time-Compressed Integration) is a time-division multiplexing system. MAC[21] (Multiplexed Analog Component), which has been proposed for the European satellite broadcasting system, is also a time-division multiplexing system. Time-Division Multiplexing is sometimes called by its acronym, TDM.

While there is a TCI signal format that time-compresses one luminance signal and two chrominance signals into each line, in general,

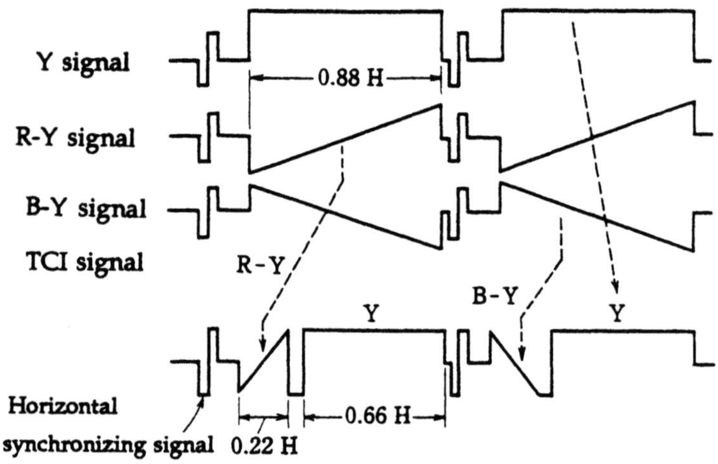

Y and C signals are time-compressed to 3/4 and 1/4, respectively, before they are multiplexed on one scanning line.

(a) Video signal multiplex system

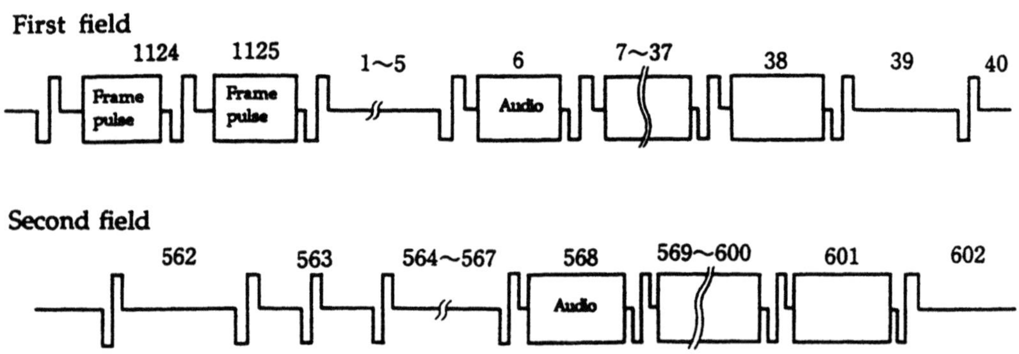

Audio signal is multiplexed during the video signal vertical blanking period. Numbers indicate line numbers.

(b) Audio signal multiplexing format (during the vertical blanking period of TCI signals)

FIGURE 3.49. TCI transmission signal multiplexing.

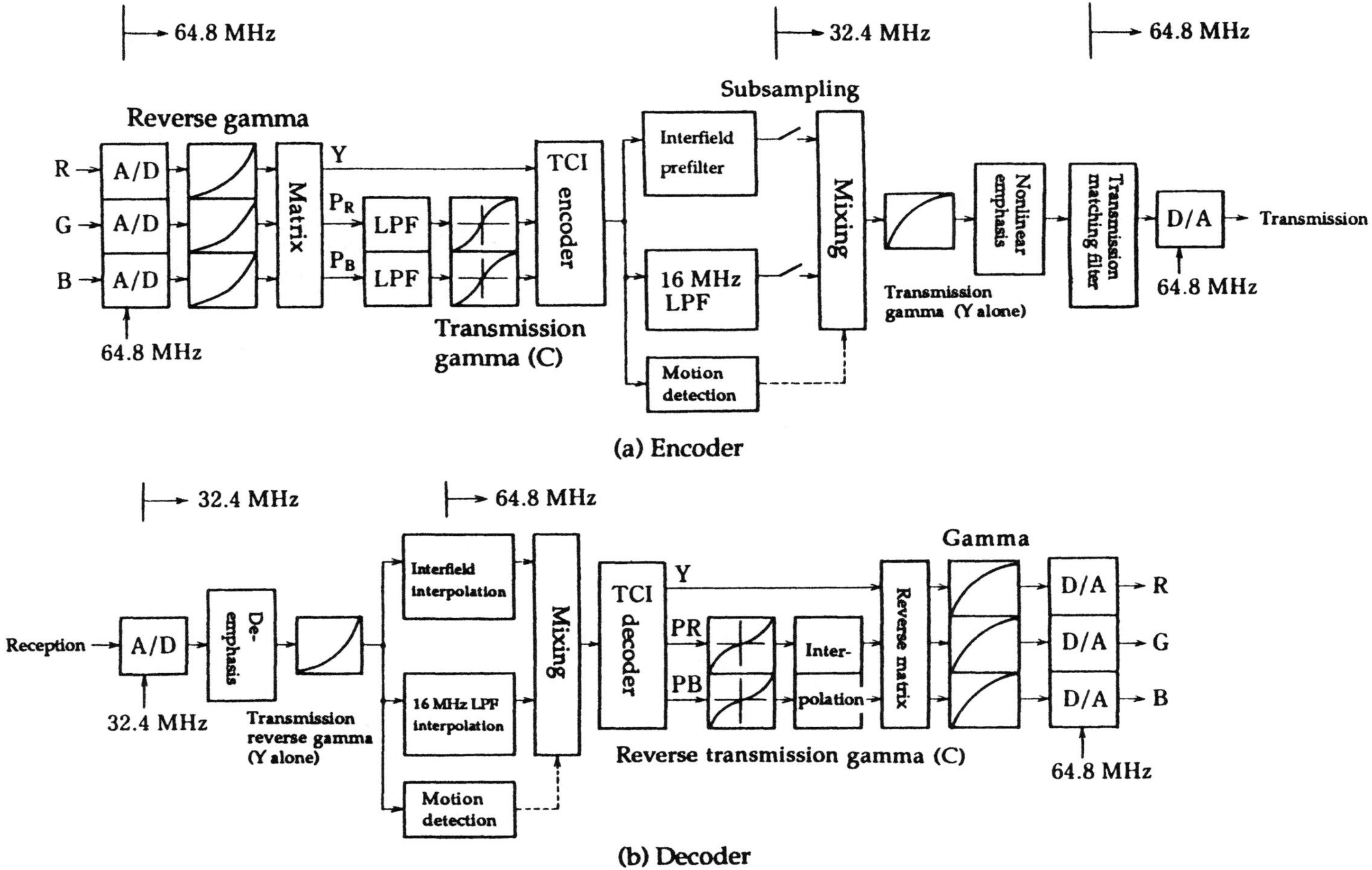

FIGURE 3.50. Example of MUSE-T hardware configuration.

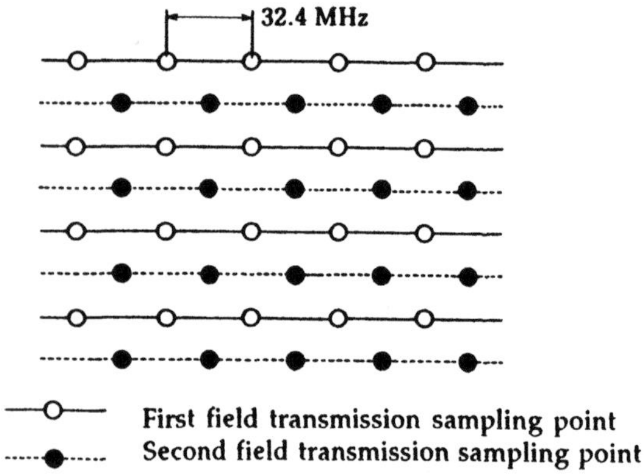

FIGURE 3.51. MUSE-T subsampling pattern.

because chrominance signals have less bandwidth than luminance signals, it is possible to multiplex these two signals into a line in a line sequential manner. In this method, called line sequential multiplexing, the vertical resolution of the chrominance signals is one-half that of the luminance signal. However, because the horizontal resolution of the chrominance signal is also less than one-half of the luminance signal, the balance between the vertical and horizontal resolution is maintained.

There are two methods for station-to-station transmission of Hi-Vision programs: FPU type TCI transmission on the 42 GHz band, and MUSE-T (wideband MUSE transmission).[22] These transmission signal bands are compared to the studio standard and MUSE in Table 3.8.

In TCI transmission, the luminance signal is time-compressed to three-fourths size and the chrominance signal to one-fourth size, and then time-division multiplexed into one scan line period as shown in Figure 3.49(a). The chrominance signals are line-sequentially transmitted. This signal format can transmit a luminance signal with a bandwidth of 22.5 MHz and a chrominance signal with a bandwidth 7.5 MHz. The baseband bandwidth for TCI signals is 30 MHz. The audio signal is time-compressed in the manner shown in Figure 3.49(b), and multiplexed with digital signals into the video signal's vertical blanking period. Being simple, compact, and lightweight, this equipment is well suited for FPU relays of Hi-Vision programs, and has been used for experimental relays of baseball games.

(2) MUSE-T

MUSE-T, which stands for MUSE-Transmission, lies in between TCI and MUSE, as shown in Table 3.8.

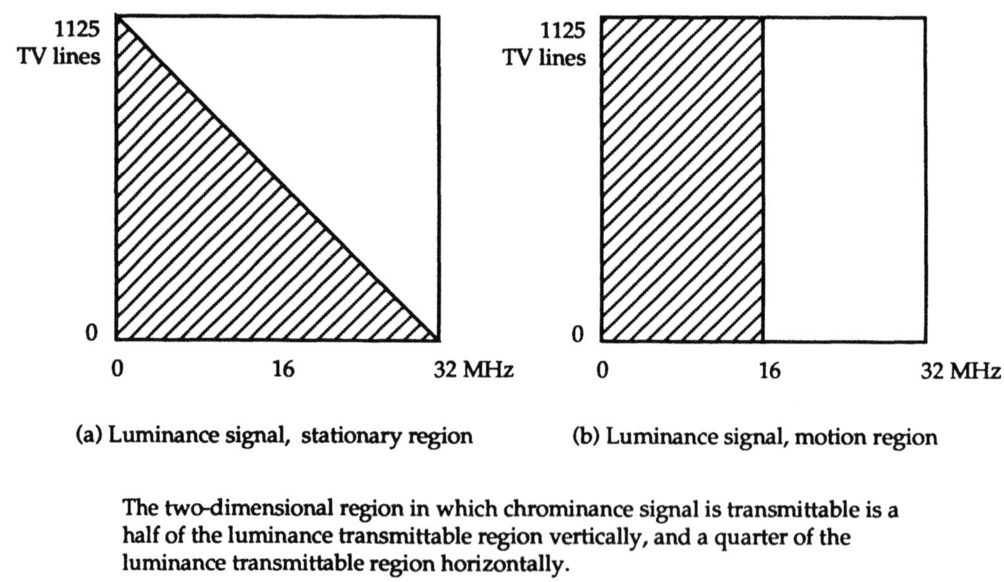

FIGURE 3.52. Region transmittable by MUSE-T, expressed by two-dimensional spatio-temporal frequency (hatched region).

Figure 3.50 shows the hardware configuration for MUSE-T, and Figure 3.51 the subsampling pattern. MUSE-T only uses field offset subsampling, in which the phase is inverted every field. This method reduces the transmission signal bandwidth to half that of the original signal. The transmittable regions for stationary and moving images are expressed by a two-dimensional spatial frequency in Figure 3.52. In the stationary region, the diagonal band is halved, while in the region with motion, the horizontal band is halved. Despite these omissions, resolution loss is not noticeable due to the low diagonal resolving power of the human eye with regard to stationary images,[23] and to the capacitance effect of the camera with regard to moving images. The horizontal bandwidths that can be transmitted in practice for the stationary areas are 28 MHz for the luminance signal and 7 MHz for the chrominance signal.

The transmission signal bandwidth for MUSE-T of 16.2 MHz is exactly twice that of the MUSE system described in Section 3.1. The synchronizing signal format and audio and independent data multiplexing method are exactly the same as in the MUSE system. The emphasis, deemphasis, and transmission line equalizing filters are also the same except that their frequency characteristics are shifted upward by twice the frequency. The matrix and pseudo constant luminance transmission are also similar to those of the MUSE system. However, motion vector detection and correction are not used.

3.5 CABLE TRANSMISSION

3.5.1 Optical Fiber Transmission

(1) Optical Fiber Transmission for Broadcasting

The remarkable progress in long distance optical transmission of digital signals has given the impression that it is now a practical alternative in communications. However, the application of fiber optics in video signal (wideband data) transmission for broadcasting systems has been rather limited. The wide bandwidth of optical fiber transmission has attracted interest in its applicability to Hi-Vision transmission, which has about five times the bandwidth of standard television.

Figure 3.53 shows examples of how optical fiber transmission technology is used in various segments of a broadcasting system such as news gathering, intrastation transmission, interstation transmission, the broadcasting network, and subscriber systems for CATV. For broadcasting purposes, development will be critical in areas such as intermediate distance transmission requiring multiplexing, two-way transmission for news gathering, switching for transmissions with a TV station, short distance transmission using branching methods, and low cost intermediate distance transmission for subscriber systems requiring branching and multiplexing. At present, analog optical fiber transmission is playing the main role in intermediate distance Hi-Vision transmission. The development of multiplexing and branching technologies for this application is considered to be critical.

(2) Optical Fiber Transmission Technology

Because it has a broader transmission band and less attenuation than a coaxial cable, optical fiber is well suited for transmitting wide band signals such as Hi-Vision signals. Furthermore, the fiber is small in diameter, lightweight, and immune to electromagnetic induction. On the other hand, its disadvantage is that fiber bundles are difficult to connect or branch, and optical connectors are difficult to handle.

The optical fiber used in communications is divided into two types by the thickness of the core (the light conducting section)—single mode optical fiber (about 10 μm in core diameter), and multimode optical fiber (about 50 μm). The clad diameter of both fibers is about 125 μm. Since single mode optical fiber is capable of a wider bandwidth transmission than multimode optical fiber, and since its production costs have fallen and are comparable to multimode optical fiber, it is expected to predominate in the future in most applications, including subscription services.

Light wavelengths are categorized into two types—short wavelength (0.8 to 1 μm) and long wavelength (1.2 to1.6 μm). The 1.3 μm band in the long wavelength category has the widest optical fiber transmission band, while the 1.55

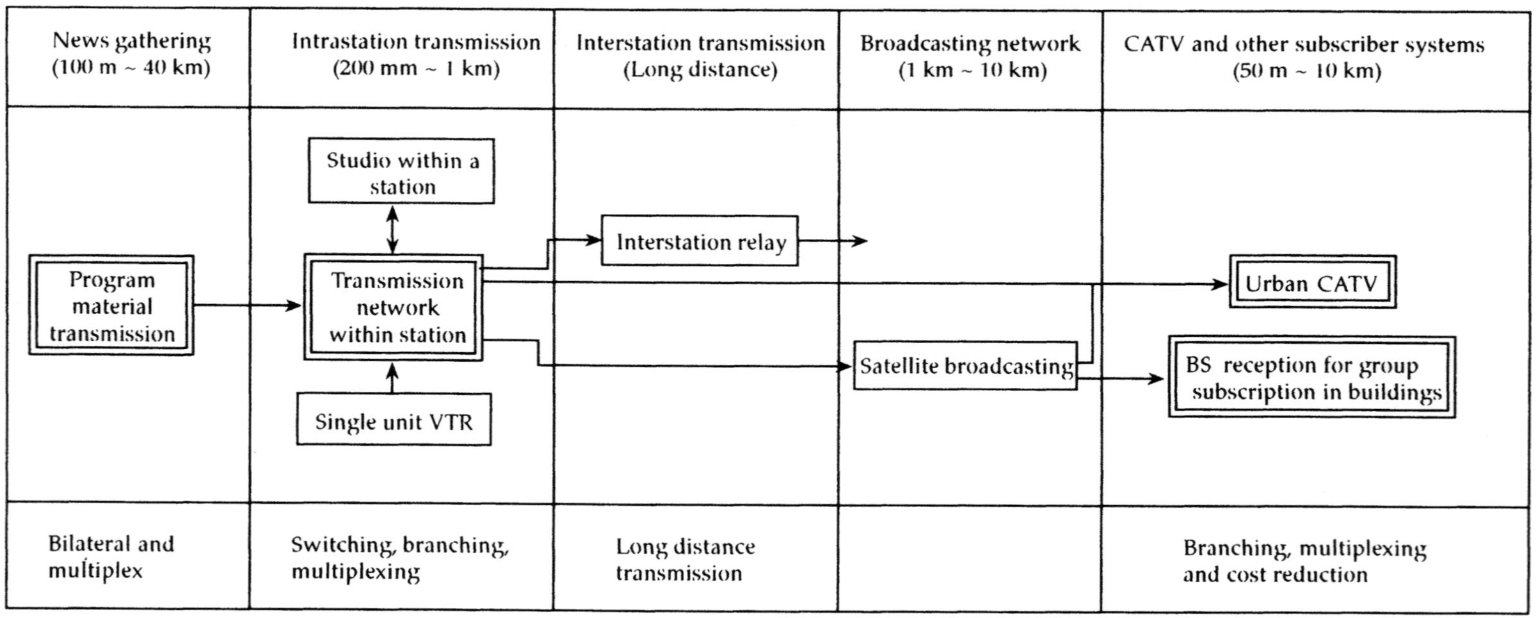

FIGURE 3.53. Broadcasting and optical fiber transmission technology.

μm band has the lowest transmission loss. The optical fiber loss is about 3 dB/km in the 0.8 μm band, 0.5 dB/km in the 1.3 μm band, and 0.2 dB/km in the 1.55 μm band. The short wavelength band was developed earlier and the optical component costs associated with it have fallen over time. However, because of its high transmission loss, optical fiber applications in the future will expand mainly in the longer wavelength band.

Light Emitting Diodes (LED) and Laser Diodes (LD) are light emitting devices. LED light output is about 100 mW. The maximum modulatable frequency is usually about 30 MHz (400 MHz has been reported in experiments). LEDs combined with multimode optical fibers are mainly used in short distance (about 5 km or shorter) baseband transmission. For longer distance transmission, or for transmission at a high modulating frequency, LDs with an optical output of about 1 mW and a maximum modulatable frequency of 1.5 GHz are used.

3.5.2 Optical Transmission of Television Signals

(1) Optical Modulation

Optical fiber transmission methods for television signals can be classified into the following four methods.

1. Direct intensity modulation—the light source intensity is modulated directly by the video baseband signal.
2. AM or FM premodulation—the light source intensity is modulated by the electrical signal obtained from the amplitude or frequency modulation of video signals.
3. Pulse analog premodulation—the light source intensity is modulated by signals obtained from the pulse modulation of video signals.
4. Digital transmission—PCM signals obtained by the A/D conversion of video signals are transmitted.

Method 1 has the simplest equipment configuration. However, unless the wavelength multiplexing (to be explained later) is employed, this method can only transmit one signal over an optical fiber. In addition, it is significantly affected by the nonlinearity of the light-emitting devices. With method 2, frequency multiplexing is possible if the carrier wave is modulated by video signals. FM modulation allows the optical transmission line's SN ratio to be improved relative to the CN ratio with FM improvement. In method 3, because the video signals are converted into a row of pulses with a constant amplitude, the effect of the transmission's nonlinearity is reduced and the SN ratio improved. Method 4, which modulates the light source intensity with PCM signals, makes possible time-division multiplexing.

In the case of Hi-Vision, if the sampling frequencies of the luminance signal and two color difference signals are set to 74.25 MHz and 37.125 MHz respectively, and the quantization bit count is set to 8 bits, the transmission rate will be 1188 Mbit/s.

(2) Multiplexing for Optical Transmission

There are three techniques for multiplexing a large quantity of information onto a single optical fiber: (1) Time-Division Multiplexing (TDM), (2) Frequency-Division Multiplexing (FDM), and (3) Wavelength-Division Multiplexing (WDM). TDM, which performs multiplexing in the time axis, is used for multiplexing digital signals. FDM multiplexes amplitude-modulated waves or frequency-modulated waves by arranging them on the frequency axis at suitable intervals. WDM, a multiplexing technique specific to optical transmission, uses several light sources of different wavelengths to transmit several optical signals through a single optical fiber. This method makes two-way transmission possible.

3.5.3 Optical Transmission of Analog Hi-Vision Signals

Optical transmissions of Hi-Vision analog signals have been reported using direct baseband intensity modulation of TCI signals and FM premodulation of YC separated component signals.

Figure 3.54 is a block diagram of Hi-Vision transmission unit in an FM multiplexing optical

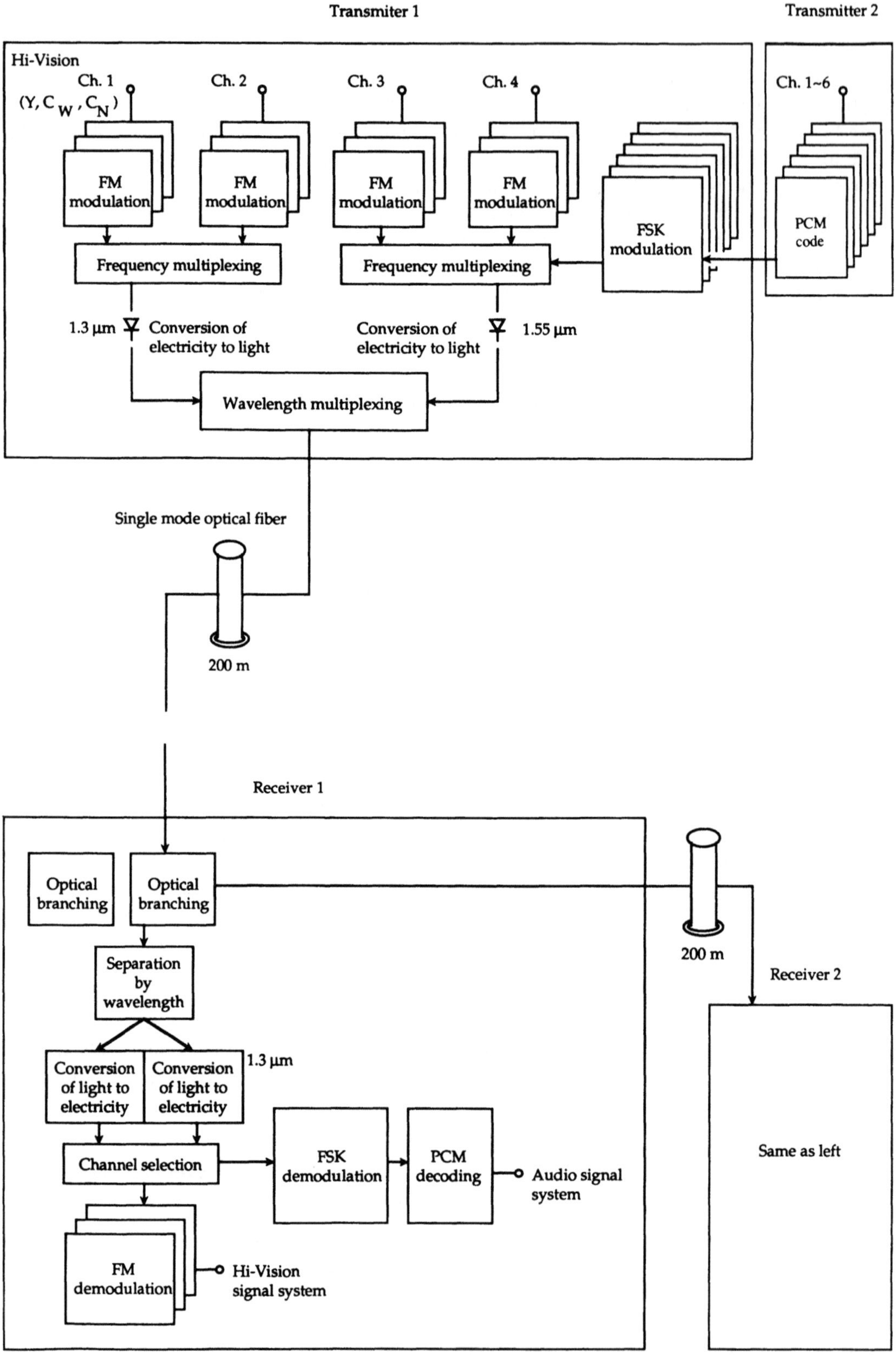

FIGURE 3.54. Configuration of Hi-Vision signal for FM multiplex optical transmission equipment.

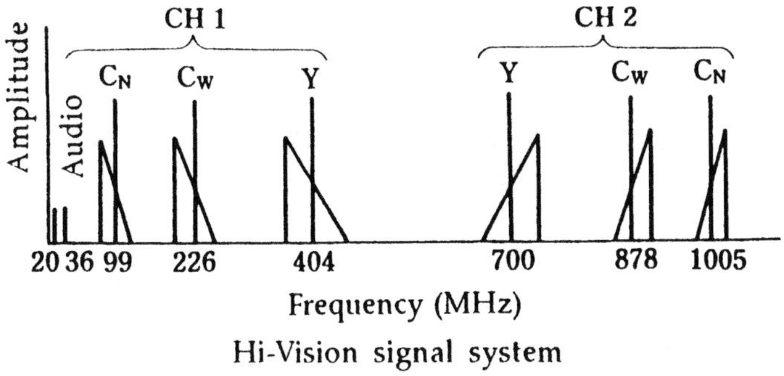

FIGURE 3.55. Frequency arrangement of Hi-Vision signals for FM multiplex optical transmission equipment.

transmission apparatus. It was developed by NHK to distribute Hi-Vision as well as NTSC television signals in a broadcast station. On two Hi-Vision channels of transmitter number 1, two each of the luminance signal (Y), wide band chrominance signal (C_w), and narrow band chrominance signal (C_N), a total of six signals, are placed in a carrier frequency arrangement as shown in Figure 3.55 so as not to be disturbed by secondary distortions.[24] The signals are then frequency-modulated using the FM parameters shown in Table 3.9 and frequency-complexed. Next, the FSK-modulated audio PCM signal from transmitter number 2 is frequency-multiplexed to modulate an LD with a 1.3 μm wavelength. The other two Hi-Vision channels are similarly multiplexed and converted into an optical signal with a 1.55 μm wavelength. The two optical signals are then combined with an optical connector and transmitted on a single mode optical fiber for 200 meters. The combination of FDM and WDM methods makes possible the transmission of four channels of Hi-Vision signals.

The LD module has a built-in optical output-stabilizing circuit and optical isolator that corrects the problem of return light.

In receiver number 1, the light is branched into two signals, one of which is separated with an optical coupler into different wavelengths and converted into electrical signals by an APD (Avalanche Photo Diode). Then one of the four channels is selected and subjected to FM demodulation so that a video output can be obtained.

The audio signal is converted into a PCM signal, FSK-modulated, and then multiplexed into the low frequency region of the video signal. The audio signal transmission frequency band is 30 Hz to 20 kHz, and the sampling frequency is 48 kHz. A 16-bit linear encoding is performed with no error correction. Receiver number 1 is equipped with an FSK demodulator to accommodate one channel and a PCM decoder. Channel selection is performed in coordination with the video processing to obtain stereo sound signals.

TABLE 3.9. FM modulation parameters for Hi-Vision signals in FM multiplexing optical transmission equipment.

Signal	Y	C_W	C_N
Highest video frequency (MHz)	20	7	5.5
Frequency deviation (MHz)	40	14	11
Modulation signal bandwidth (MHz)	80	28	22
FM improvement (dB)	16.8	16.8	16.8
Emphasis improvement (dB)	3	3	3

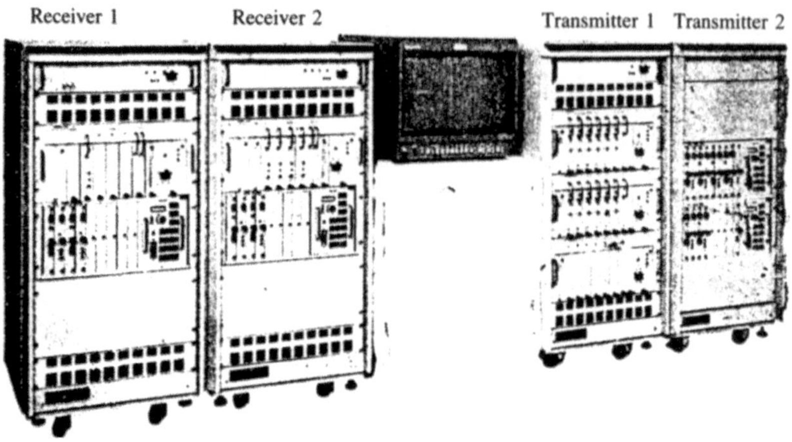

FIGURE 3.56. Reception and transmission equipment for Hi-Vision optical transmission.

A 200-meter transmission over a single mode optical fiber with a branch into two lines at the end yielded good results: the CN ratio for the optical fiber video signal transmission system is at least 38 dB; the unevaluated SN ratio for reception is at least 57 dB; the linearity of the various signals is less than DG 1% and DP 1°; and the SN ratio of the audio signal exceeds 78 dB. This equipment, employed for the transmission of four channels of YC-separated type Hi-Vision signals is capable of a transmission of about 20 km without relay, or of 10 distributions assuming that the unevaluated SN ratio be 50 dB at the receiver terminal. Figure 3.56 shows the outside view of the equipment.

3.5.4 Digital Optical Transmission of Hi-Vision Signals

In an experimental transmission, NTT (Nippon Telegraph and Telephone Corp.) digitized Hi-Vision TCM signals on a single mode optical fiber transmission line (1.3 μm wavelength band) at 400 Mbit/s simulating 20 relay stages and a distance of 525 km.[25] During the Science and Technology Fair at Tsukuba, they transmitted a TCM signal from Tsukuba to Tokyo, and then to Osaka on an F-400M (400 Mbit/s) line.

In preparation for the future digitization of studio equipment, a Hi-Vision digital optical transmission system has been developed to interconnect machines. The light-emitting source is an LD with a wavelength of 0.77 μm that is commonly used in compact disk players. The light-receiving unit is a Si-APD. The luminance signal (Y) and two color difference signals (R-Y, B-Y) are sampled at the respective sampling frequencies of 74.25 MHz and 327.125 MHz. The component encoding is performed with 8-bit quantization. In the transmitter and receiver, component digital signals undergo parallel-to-serial and serial-to-parallel conversions. The transmission speed of the signals over the optical fiber is 1188 Mbit/s. A bit error rate of 10^{-11} has been reported for this system for a 1 km transmission over a graded index multimode optical fiber.[26]

3.5.5 Optical CATV Transmission of Hi-Vision MUSE-FM

In this section, as an example of the optical CATV transmission of Hi-Vision signals, the results of an indoor experiment with MUSE signal transmission conducted at the NHK Science

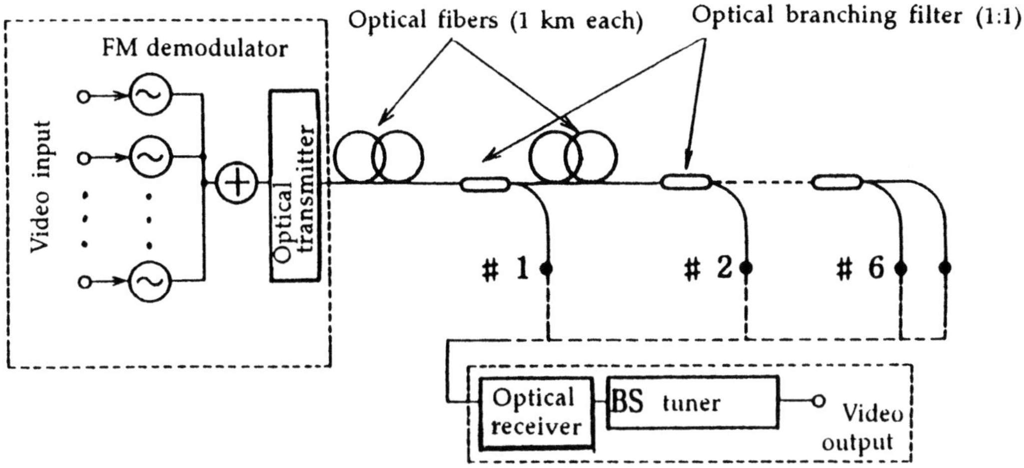

FIGURE 3.57. Simulated optical CATV transmission line.

& Technical Research Laboratories will be described.[27]

The experiment was conducted with a simulated optical transmission line composed of a distributed feedback (DFB) laser, APD, fused optical coupler (excess loss 0.1 dB) and a single mode optical fiber, as shown in Figure 3.57. Figure 3.58 shows the modulation frequency characteristic of the optical transmission system. In transmissions over a distance of about 50 km, the characteristic hardly changes.

Figure 3.59 shows the optical transmission loss of the MUSE-FM signal when transmitted on the BS-IF band (1 to 1.3 GHz) as a function of the unevaluated SN ratio and CN ratio. The results indicate that for 8-wave multplexing, a transmission distance of about 40 km, or 100 branches, is possible without a relay. For 2-wave multiplexing, a transmission distance of

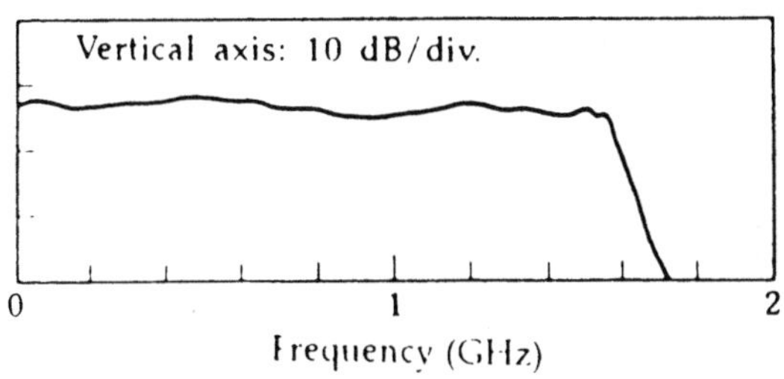

FIGURE 3.58. Modulation frequency characteristic of an optical transmission system.

about 60 km without a relay, or 400 branches, is possible. In super highband (222 to 470 MHz) transmission, equal or better characteristics have been obtained. In addition, for MUSE-FM transmission with an optical relay amplifier, the possibility of 10,000 branches has been confirmed. Incidentally, because the current technique for direct amplification of lights is not yet practical, this optical relay amplifier first converts the light back to electrical signals before emitting a light from the LD.

3.5.6 Transmission of Hi-Vision on Coaxial CATV

Most of the existing CATV systems in Japan are small scale group subscription systems in remote areas with poor television reception and urban areas where buildings obstruct reception. This type of service is mainly limited to signal retransmission. However, as of 1986, CATV had spread to 4.99 million households (about 15% of registered television subscribers). In recent years, large scale urban CATV systems have started to emerge, and their services, including Hi-Vision and other new media, are expected to quickly gain in popularity.

Figure 3.60 shows the frequency allocation for urban CATV systems. The frequency bands expected to be allocated for Hi-Vision and other new services are the midband (108 to 170 MHz),

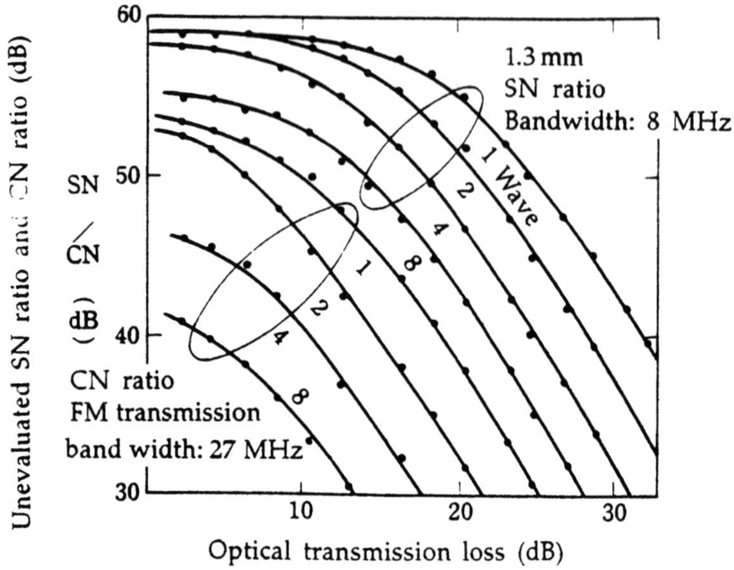

FIGURE 3.59. Optical transmission loss in MUSE FM in BS-IF band transmission versus SN ratio and CN ratio.

excluding frequency bands for existing FM and television broadcasting, and the upper highband (222 to 470 MHz).

(1) Transmission of Hi-Vision MUSE-FM on Coaxial CATV

The main technical issue for the time being is how to retransmit MUSE-FM signals on CATV in correspondence with the NTSC satellite broadcast wave. Figure 3.61 shows a frequency arrangement for transmitting a satellite broadcast FM signal on the super highband. Since a CATV system is composed of a multistage connection of amplifiers, factors that must be taken into consideration include the CN ratio of transmission lines, mutual modulation disturbance ratio, power supply hum modulation, amplitude and phase frequency characteristics, VSWR, radio interference protection ratio, and transmission signal level.

Results of a MUSE-FM transmission experiment conducted at the CATV facilities of the NHK Science & Technical Research Laboratories are described below.[28]

In the system configuration shown in Figure 3.62, the trunk is composed of a 300 meter cable and 20 amplifiers each of which constitutes a stage. The characteristics of these stages can be found by switching the branch outputs from the trunk line branching amplifiers at the 5th, 10th,

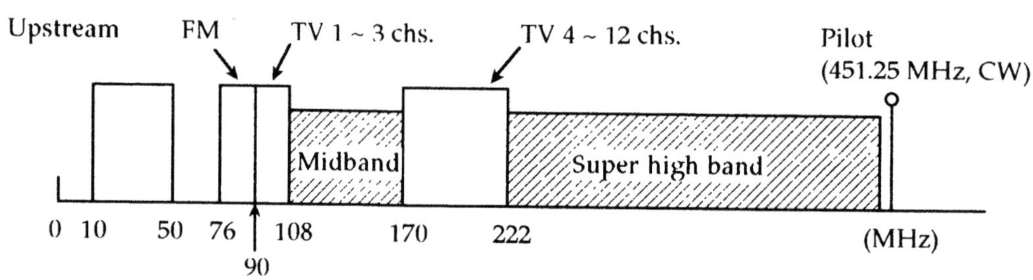

FIGURE 3.60. Frequency allocation for urban CATV.

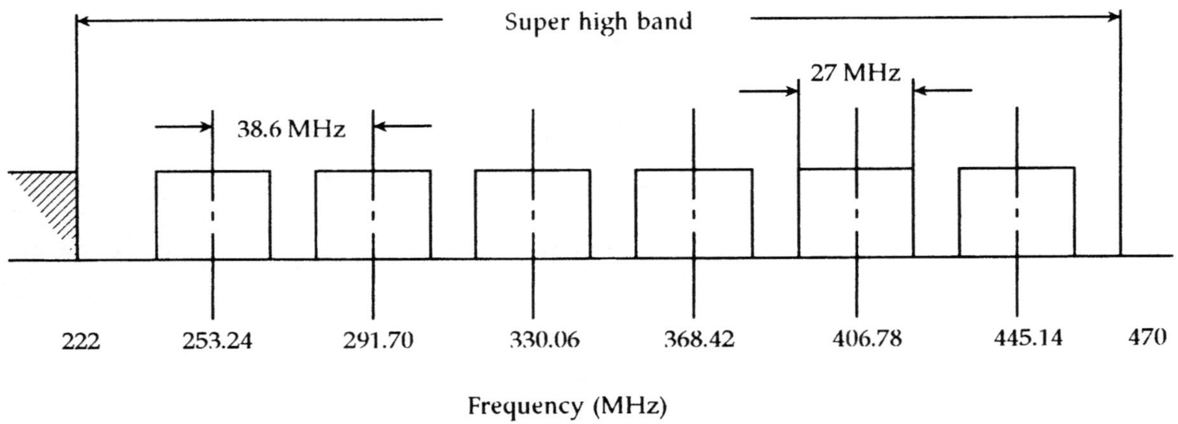

FIGURE 3.61. Frequency allocation of FM signals for satellite broadcasting.

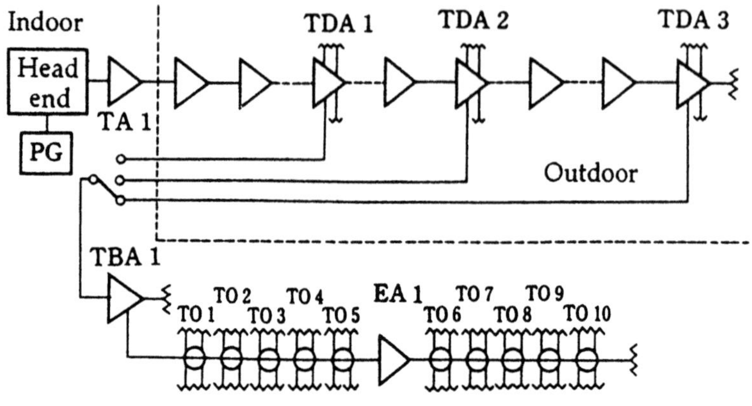

FIGURE 3.62. Configuration of CATV facility employed in experiment.

and 20th stages. There are 10 tapoffs. Downlink and uplink transmissions are performed at 70 to 450 MHz and 10 to 50 MHz, respectively.

In the MUSE signal transmission, measurements with an impulse response method at the last tapoff after the 20 stages in the trunk system and the FM modulator and demodulator showed a flat amplitude characteristic up to 8.1 MHz and a group delay deviation of 16 ns.

Figure 3.63 shows a spectrum of transmission signal that originated as a MUSE-FM signal transmitted from BS channel 11, then was received with a 1.2-meter parabolic antenna, merged with four FM BS standard signals in the BS-IF band, frequency-converted to the super highband frequency, merged with 20 NTSC-AM signals, and finally transmitted. After the transmission, the signal was frequency-converted to the BS-IF band and demodulated. The deterioration of the CN ratio in the 20-stage transmission—22 dB at the input unit versus 21.7dB at the last tapoff—was very small. Neither the resolution and SN ratio showed any deterioration from the transmission, and a high quality Hi-Vision image was obtained.

(2) Transmission of Hi-Vision MUSE-AM Signals on Coaxial CATV

The study of AM transmission of MUSE signals is also important to narrow the required bandwidth and increase the number of channels. In the United States, where the diffusion rate of

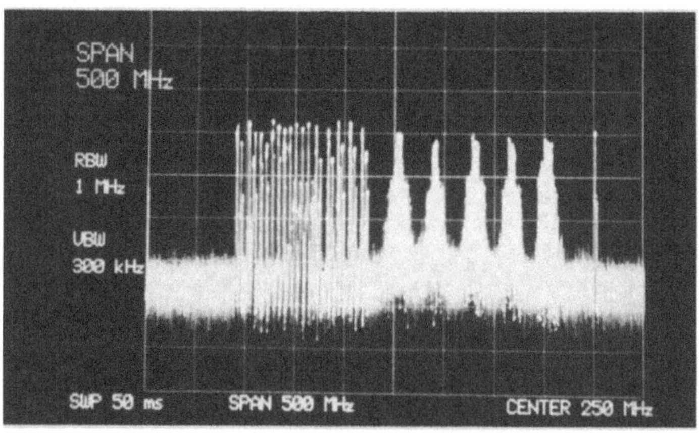

FIGURE 3.63. Spectrum of transmission signals using super highband.

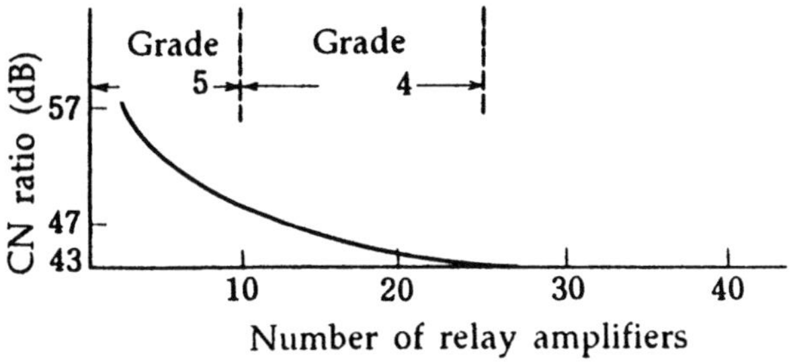

FIGURE 3.64 Relationship between reception CN ratio and the number of CATV relay amplifiers in MUSE-VSB AM transmission.

CATV is 50%, strong demand is expected for decreasing the bandwidth for Hi-Vision and other types of new media so that CATV frequencies can be efficiently used.

From the point of view of increasing the number of channels in Hi-Vision CATV transmission, AM (VSB·AM) transmission of MUSE signals on residual sidebands with a bandwidth of 9 to 12 MHz is a promising technique.

Regarding the CATV transmission of VSB·AM signals of MUSE, the relationship between the number of relay amplifiers and the reception CN ratio is shown in Figure 3.64.[29] To realize an image quality corresponding to grade 4 in the 5-grade evaluation, a CN ratio of about 45 dB (for 8 MHz bandwidth) is required, in which case 20 relay stages are possible.

Compared to FM transmission, AM transmission requires a somewhat stricter reflection characteristic, that is, VSWR. The VSWR of CATV trunk lines is usually satisfactory and poses hardly any problems, but it will be important to secure a satisfactory reflection characteristic between the tapoff and the household protector.

REFERENCES

1. Ninomiya, Ohtsuka, Izumi, Goshi, and Iwadate, "Development of the MUSE System," *NHK Technical Journal*, Vol. 39, No. 2 (Ser. No. 172), 1987.
2. Lash, *Basics of Communications Systems and Data Transmission*, McGraw-Hill, Kogakusha, Chapter 5.
3. Ninomiya and Ohtsuka, "Motion Correction Prediction for Interframe Encoding," *Proceedings of the Institute of Electronics, Information and Communication Engineers, Japan (IEICEJ)*, Vol. J64-B, No. 1, 1981.
4. Institute of Television Engineers, ed., *Handbook of Television Image Engineering*, Ohmsha, 1980, p. 265.
5. Izumi, Ohtsuka, Gohshi, and Ninomiya, "MUSE to 525-Line Converter," *IEICEJ Proceedings*, Spring 1988, D-169.
6. Watanabe and Takegahara, "Near-Instantaneous Compression and Expansion of DPCM for Baseband Transmission of Hi-Vision Audio Signals," *IEICEJ Proceedings*, March 1987.
7. Spot Research News, "High Efficiency Encoding of Acoustic Signals," *NHK Technical Report*, Vol. 30, No. 10, 1987.
8. Ikezawa and Miyasaka, "Study of DPCM Quantization Noise Sensing from the Viewpoint of Auditory Masking," *Journal of the Society of Acoustics (JAS) Proceedings*, March 1988, 1-4-7.
9. Hoshino and Takegahara, "The Optimum Leakage Coefficients in HDTV Audio DPCM Transmission," *IEICEJ Transactions*, B460, March 1988.
10. Tanabe and Takegahara, "Auditory Evaluation of Near-Instantaneous Compression and Expansion of DPCM Signals," *JAS Proceedings*, March 1987, 1-6-16.
11. Tanabe and Takegahara, "Evaluation of Near-Instantaneous Compression and Expansion of DPCM Signals Based on Distance," *JAS Proceedings*, October 1987, 1-6-16.
12. Ninomiya et al., "Development of the MUSE System," *NHK Technical Journal*, Vol. 39, No. 2, 1980, pp. 18–53.
13. Kato et al., "Study of Hi-Vision Satellite Broadcasting Modulation Parameters," *IEICEJ Proceedings*, Fall 1988, B106.
14. Schumid, "How to Determine the 2T Pulse K

Factor of TV Facilities Without the Aid of an Oscilloscope Graticule," *IEEE Transactions*, BC, BC-15, No. 1, 1969, pp. 12–14.
15. Shibuya et al., "Study of Ternary FM Signal Modulation Parameters in 12 GHz Satellite Broadcasting," *Proceedings of the Institute of Television Engineers of Japan (ITEJ)*, 1989.
16. Iwadate et al., "Reception CN ratio vs. Image Quality and Sound Quality in Hi-Vision Satellite Broadcasting with MUSE," *ITEJ Technical Report*, RE 88-32.
17. Kubota et al., "MUSE Hi-Vision Satellite Transmission Experiment," *NHK Laboratories R&D*, No. 2, 1988, pp. 19–27.
18. Ito, "Method for Estimating the Rainfall Attenuation of 12 GHz Band Satellite Broadcasting," *NHK Technical Report*, Vol. 27, No. 5, 1984, pp. 156–161.
19. Takano, "A 42 GHz Band Hi-Vision Experimental Relay Station," *NHK Technical Report*, Vol. 28, No. 12, 1984, pp. 468–473.
20. Ishida, Nishizawa, and Kubota, "High-Definition Television Broadcasting by Satellite," *IEEE Transaction on Broadcasting*, Vol. BC-28, 1982.
21. Windram, Morcom, and Herley, "Extended Definition MAC," *IBA Technical Review*, Vol. 21, 1983.
22. Ninomiya, Ohtsuka, Izumi, Gohshi, and Iwadate, "MUSE: An HDTV Broadcasting and Transmission System," *Third International Colloquium on Advanced Television Systems*, Session 4, 1987.
23. Sakayo, "Frequency Characteristics of the Way Television Noise Appears—Gray Scale and Chromaticity and Sweep Measurement Method," *ITEJ Journal*, Vol. 36, No. 10, 1980.
24. Maeda, Oyamada, Utsumi, and Morii, "High Quality Intrastation FM Multiplexing Optical Transmission," *Transactions of the IEICEJ*, Vol. J69-B, No. 9, 1986, pp. 904–913.
25. Tsuboi, Asatani, and Miki, "Fiber-Optic HDTV Transmission Technology," *SMPTE Journal*, 1985-1, 1985, pp. 4–10.
26. Komuro, Tsujino, Komatsu, and Otobe: "1.2 Gbps High quality Signal Optical Fiber Transmission System," *ITEJ Proceedings*, 17-7, 1987.
27. Oyamada, Maeda, and Utsumi, "BS-IF 8-Channel Transmission with Optical Fiber," *ITEJ Proceedings*, 17-3, 1987.
28. Suzaki, Honda, Sato, and Yokota, "Experimental CATV transmission of BS Hi-Vision MUSE Signals," *ITEJ Proceedings*, 1987.
29. Honda, Suzaki, Tokumoto, Yamazaki, and Miyazaki, "Hi-Vision MUSE Transmission on Coaxial CATV: Part 1," *ITEJ Technical Report*, RE 88-27, Vol. 12, No. 31, 1988.

4

Reception and Display

Takashi Iwamoto, Masaru Kanazawa, Kiichi Kobayashi,
Hiroshi Murakami

4.1 DIRECT-VIEW CRT DISPLAY

Because Hi-Vision signals are ultimately viewed on a display, it is no overstatement to say that the image quality of these signals depends on the display. Thus a Hi-Vision display should meet the following specifications in size and image quality:

Size: The screen should be at least 30 inches in diagonal (100 inches or larger for video theaters).
Image quality: At least 1,000 lines of horizontal resolution and 150 cd/m^2 of luminance.

A Hi-Vision display for home use should also satisfy the following conditions:

Size: Less than twice as large as the display being used for conventional television.
Price: Less than twice that of a conventional television display.
Power consumption: About the same as that of a conventional television display.
Ease of use: Comparable to a conventional television display.

None of the Hi-Vision displays developed so far has met all of these conditions. However, some CRT and projection displays have come close. Direct-view CRT displays, which have excellent image quality and are the most widely used, have been made as large as 40 inches in diagonal.

4.1.1 Color CRT Displays[1]

Figure 4.1 shows the structure of a color CRT display. A shadow mask or aperture grill is used to accurately direct the electrons from the three electron guns to the phosphorous screen. The electron guns are arranged in-line for an aperture grill and in a delta structure for a shadow mask.

(1) Electron Guns
In a CRT picture tube, the three electron guns for R, G, and B colors are arranged in either an in-line or delta formation. To achieve high resolution, the electron beam spot must be reduced across the entire surface of the display. This is done by making the neck of the picture tube as thick as 36.5mm and using a large diameter electron lens, and by applying dynamic focusing over the entire screen surface.

An example of an electron gun is shown in Figure 4.2.[2] Figure 4.3 shows the relationship between resolution and the size of the electron beam spot on the phosphor screen. CRT displays 40 inches in size have achieved a spot diameter

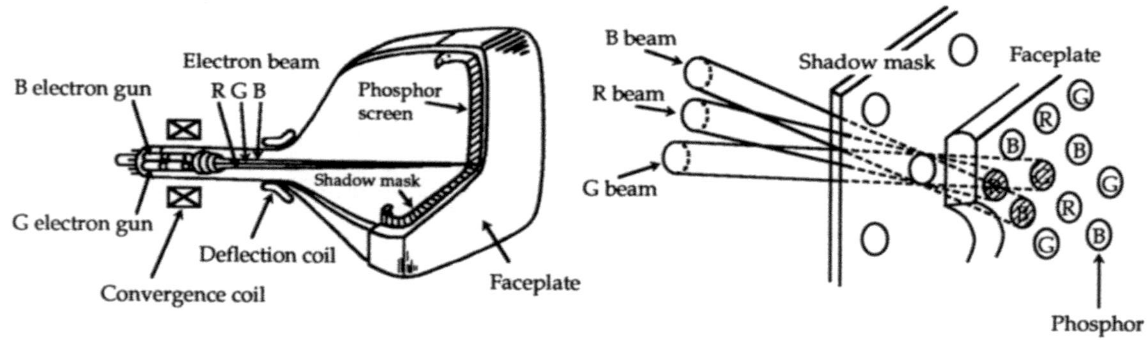

(a) Basic configuration of a color CRT (shadow mask type)

(b) How a shadow mask works

FIGURE 4.1 Configuration and operation of a color CRT.

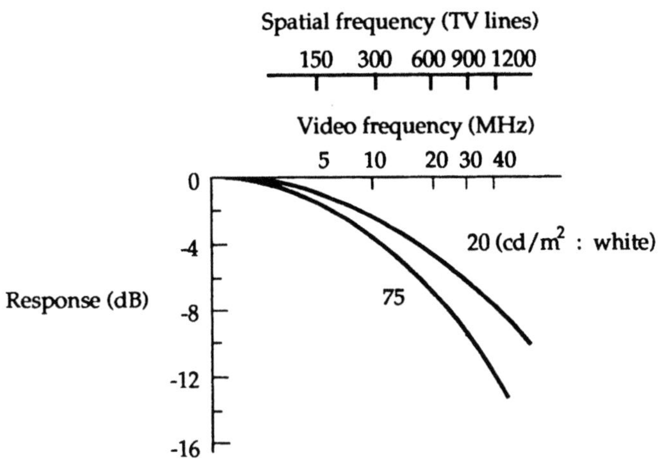

FIGURE 4.2. Electron beam resolution for a 40-inch CRT.

of 1.1 mm (at 5% of peak intensity) and a resolution of over 1,000 TV lines.[3]

To increase luminance, a large beam current is necessary. For example, the beam current for a 40-inch CRT is 0.5 mA or higher on average. Because a regular oxide cathode cannot carry this high a current density, attempting to increase the beam current will result in a larger electron beam spot and lower resolution. By using an impregnated cathode (with a current density at least three times higher than an oxide cathode), a small spot size and high resolution can be achieved with a large current.

(2) Shadow Mask and Aperture Grill

As the electron beams from electron guns are sampled at the shadow mask or aperture grill, the display's resolution deteriorates if these masks have a large pitch. In addition, the display's resolution depends on the bandwidth of the video circuit and the diameter of the spot formed on the phosphorous screen by the electron beam.

Because the input signals are sampled at the mask, the resolution of the mask (MTF) depends on the phase relationship between the mask and input signal, and has a value between Y_{max} and Y_{min}.[3]

$$Y_{max} = \left(\frac{\pi}{2}\,\theta\,\frac{pt}{2}\,\frac{1}{V}\right)$$

$$Y_{min} = Y_{max}\cos\left(\frac{\pi}{2}\,(2 - |\,\theta\,|)\,\frac{pt}{2}\,\frac{1}{V}\right) \qquad (4.1)$$

$$\text{where}\quad \theta = 2\left(\frac{V}{pt} - \left[\frac{V}{pt} + 0.5\right]\right)$$

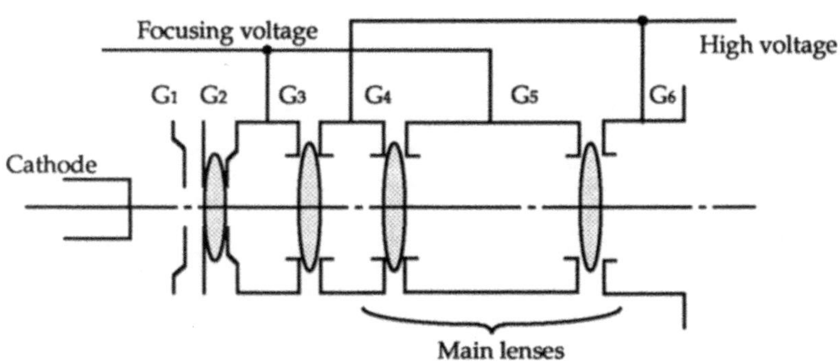

FIGURE 4.3. An electron gun for a 40-inch CRT.

[x] is a Gaussian symbol
that represents the largest integer that does not exceed x.

V: Effective vertical screen height

p: Horizontal or vertical pitch of the mask aperture

t: Resolution in TV lines.

Because a Hi-Vision display must have a resolution of at least 1000 TV lines for any phase relationship, the mask aperture pitch needs to be smaller than for CRT displays currently being used in broadcasting. The difference between a Hi-Vision shadow mask and the shadow mask for conventional television is shown in Figure 4.4.

To ensure that the electron beams hit the phosphor targets after passing through a mask with a narrower aperture pitch, the mask must be accurately positioned relative to the phosphor target.

In manufacturing a shadow mask, the thickness of the mask should be less than roughly half the aperture pitch. However, a thin mask is not just mechanically unstable; because over 70% of the electron beam is absorbed by the mask (the transmission rate for a shadow mask is about 20% and 27% for an aperture grill), the thermal expansion causes the relative position of the phosphors to shift, resulting in reduced color purity (by doming). Thus the mask is made as thick as processing limits will allow, and put under high tension. In addition, materials with a low coefficient of thermal expansion such as invar alloy (which has a coefficient of thermal expansion about one-tenth of iron, the conventional material for shadow masks) are being used.

With a shadow mask, because a moiré is created by the mask and scanning lines, the mask pitch must be set to minimize this effect. In the case of a 40-inch CRT, a 450μm pitch will satisfy the requirements for both resolution and moiré reduction. The mask hole diameter and mask thickness are 220μm and 200μm respectively. The resolution characteristics of this mask are shown in Figure 4.5.

(3) Driver Circuit

The driver circuit for a CRT uses a video circuit with a bandwidth of at least 30 MHz and output power of at least 50 V_{p-p}.

Because convergence error causes resolution to deteriorate, a Hi-Vision display must perform convergence error correction adequately. The correction coils in the neck of a color CRT pic-

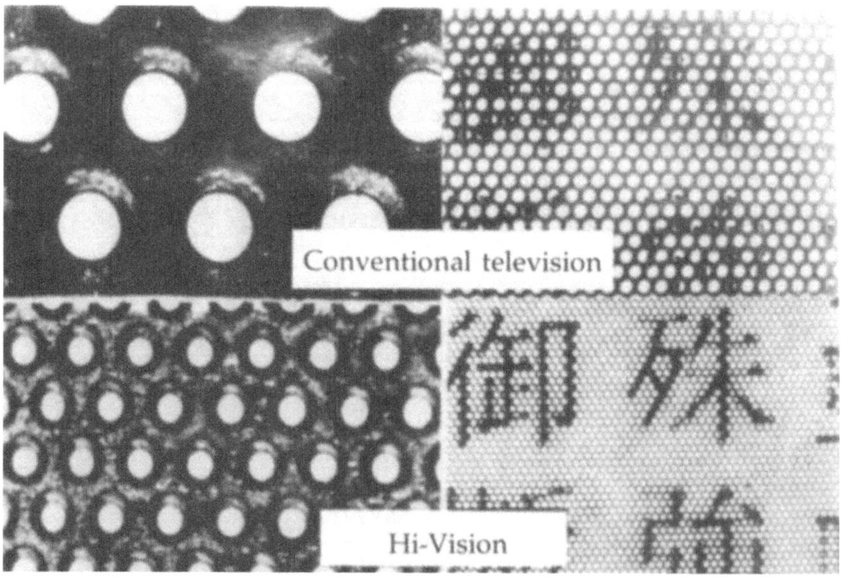

FIGURE 4.4. Comparison of shadow mask pitches.

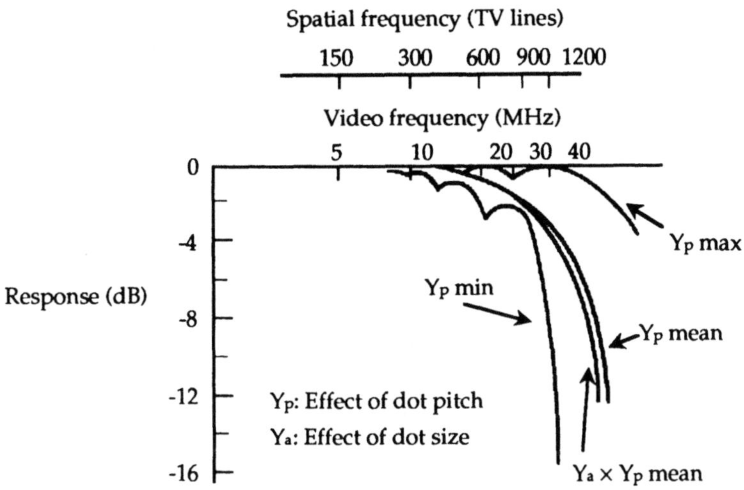

FIGURE 4.5. Shadow mask response of a 40-inch CRT.

ture tube perform convergence corrections over the entire screen by varying the correction current in correlation with the position of the electron beam. Many Hi-Vision displays have a digital convergence correction circuit with about one hundred adjustment points on a screen. The correction values for these adjustment points are inputted with a keyboard and stored in the digital memory beforehand, and the correction data is read as the electron beam scans the screen. A block diagram of a digital convergence circuit is shown in Figure 4.6. Because noticeable levels of irregular interpolation occur when using digital convergence circuits, they are often used in combination with an analog correction circuit.

4.1.2 CRT Display Performance

(1) Resolution and Luminance

Figure 4.7 shows the overall resolution as a combination of such factors as video circuits, electron beam spot size, and the MTF of the mask.

Adequate luminance can be obtained while maintaining resolution by applying at least 30

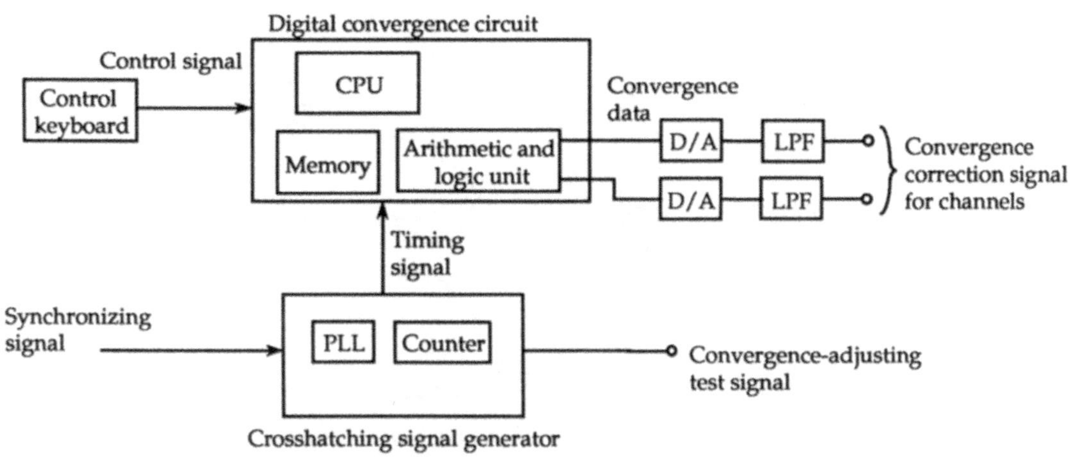

FIGURE 4.6. Digital convergence circuitry.

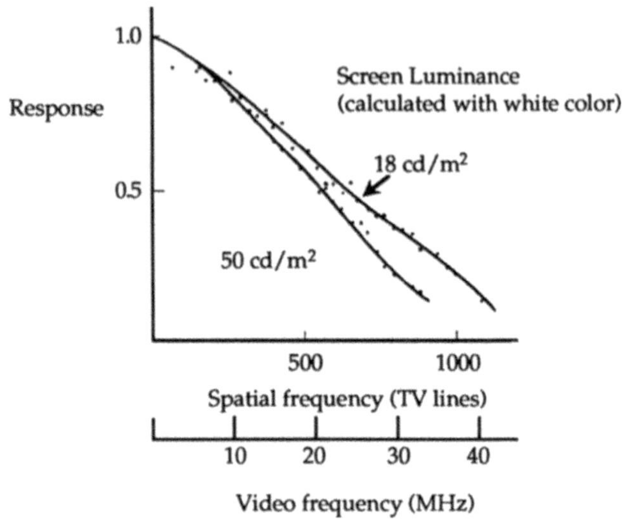

FIGURE 4.7. MTF of a 40-inch CRT.

kV to the cathode. Luminance levels of 150 cd/m^2 (white peak) for a 40-inch CRT and 200 cd/m^2 (average) for a 32-inch CRT have been achieved. These levels are adequate for practical use.

Using the definition for white windows described in Section 4.2 (0.1 screen width × 0.1 screen height), a contrast ratio of over 40:1 has been obtained. This corresponds to a a contrast ratio of at least 100:1 for a normal image.

(2) Uniformity of Screen Image

As a CRT increases in size, the electrons must travel a longer distance. Because this increases the influence of geomagnetism, color purity and other properties tend to deteriorate. To eliminate the effects of geomagnetism, correcting magnets are installed at the neck of the CRT. In addition, the whole display is protected by magnetic shielding, and correcting coils are installed on the edges of the screen.

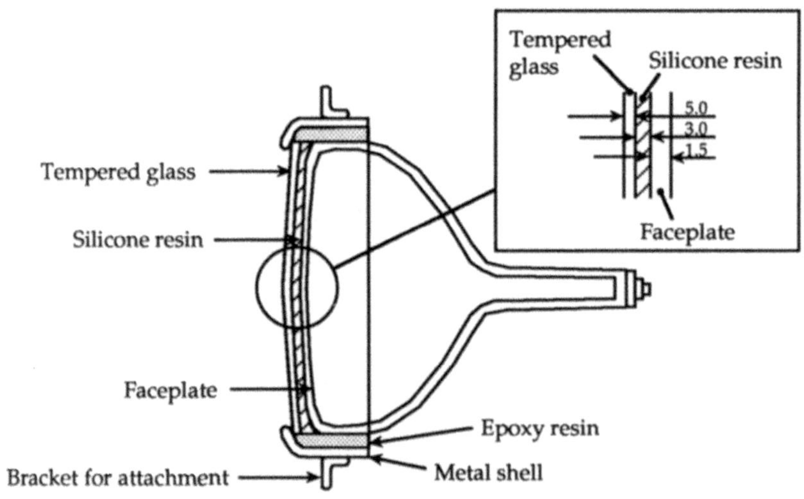

FIGURE 4.8. Structure of a 40-inch CRT faceplate.

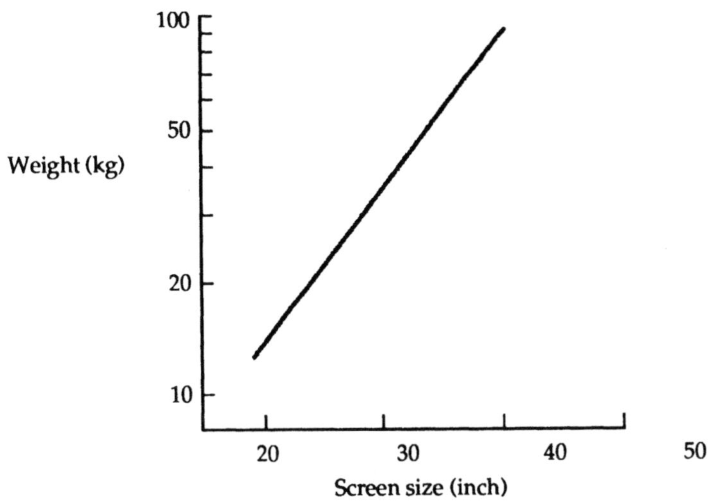

FIGURE 4.9. Weight of the direct view CRT (CRT only).

As stated previously, convergence error is corrected with digital convergence-correcting circuits at a precision of 0.3 scanning lines over the entire screen. Correction for shading, like convergence correction, is also done with many adjustment points on the screen.

(3) Size and Weight

The internal vacuum of a CRT is subjected to increasingly larger atmospheric pressure as the size of the display increases. Because a Hi-Vision CRT has a wide screen, the atmospheric pressure (tensile stress) is especially great near the ends of the longer sides of the screen. These stress values have been brought to the level of conventional CRTs by using finite element computer aided design methods. The thickness of the glass walls has been increased for extra strength. In particular, the face plate is especially strengthened against breakage for safety reasons, as shown in Figure 4.8.

As the size of the CRT increases, so too does its weight. Because the Hi-Vision CRT uses a 90° deflection angle to maintain its high resolution at the edges of the screen, it is heavier than a conventional CRT of the same size. Figure 4.9 shows the relationship between size and weight. A display complete with electrical circuits and case is about twice the weight shown in this graph.

A CRT display weighing more than 100 kg is not only difficult to manufacture but difficult to handle as well. At present, the largest Hi-Vision CRT display is 42 inches. Manufacturing a display larger than this is extremely difficult.

Figure 4.10 is a picture of a 40-inch Hi-Vision color display. Of the various types of displays, the direct-view CRT display has the best display image quality, and is often used in applications such as studio monitors.

4.1.3 Monochrome CRT Display

Monochrome monitors are primarily for studio use, and are usually less than 20 inches in size.

Because the demand for these monitors is not very large, rather than developing displays with a 16:9 aspect ratio, most of these monitors have a 4:3 aspect ratio with the top and bottom portions of the screen masked to display a Hi-Vision image.

The bandwidth of the image circuit is at least 30 MHz, and a large diameter electron lens is used to achieve a resolution exceeding 1000 TV lines. Because of the absence of factors that would reduce image quality such as sampling through a shadow mask, these displays are valuable in monitoring the Hi-Vision signal source and transmission equipment.

FIGURE 4.10. 40-inch CRT displayed.

4.2 PROJECTION DISPLAYS

By using optics to enlarge a small image, projection displays can easily attain screen sizes that are larger than 50 inches, which would be difficult to do with a direct-view CRT. Projection displays today use CRTs (refraction and reflection types), oil film type light valves, lasers, and liquid crystal panels. Further, projection displays are also divided into front and rear projectors.

4.2.1 CRT Front Projection Display

While a front projection display requires a dark room to see the screen, it can easily project an image exceeding 100 inches because it uses a screen similar to a movie screen. Because of this convenience, displays of this type are mainly used for commercial applications.

(1) Projectors

The projector consists of either one set or two sets of R, G, and B CRTs. The optical system is categorized into two types, depending on whether reflection or refraction is used.

A refractive projector consists of 7 to 12-inch CRTs and projection lenses. Because human vision is most sensitive to green light, the green CRT is placed in the center and those for R and B are placed on either side. Because the R and B CRTs project onto the screen obliquely, their images need to be distorted in a trapezoidal fashion. For this reason, a 7-inch CRT (shown in Figure 4.11) has an effective screen size of about 5.1 to 4.4 inches on the phosphor area.

CRTs achieve a beam current of several milliamperes by using high voltages that exceed 30 kV and an impregnated cathode. To attain adequate resolution, CRTs predominantly use electromagnetic convergence.

The optical system has about 10 glass lenses

FIGURE 4.11. 7-inch CRT for projection display.

to ensure adequate resolution. The lenses are usually coupled to the CRT with a liquid medium.

The contrast ratio is usually measured with the pattern shown in Figure 4.12. It is defined as the luminance ratio between white and black observed when this pattern is displayed on screen. Under this definition, the contrast ratio observed on a projection display with no fluid coupling is about 15. This is lower than the contrast ratio for a direct-view CRT display (which exceeds 40 if a black matrix is used), which therefore has a better image quality. The contrast ratio of a projection display is inferior because the light emitted from the phosphor screen must travel through substances with different refractive indexes, such as the CRT face plate, air, and the projection lens. Of these, the main cause of

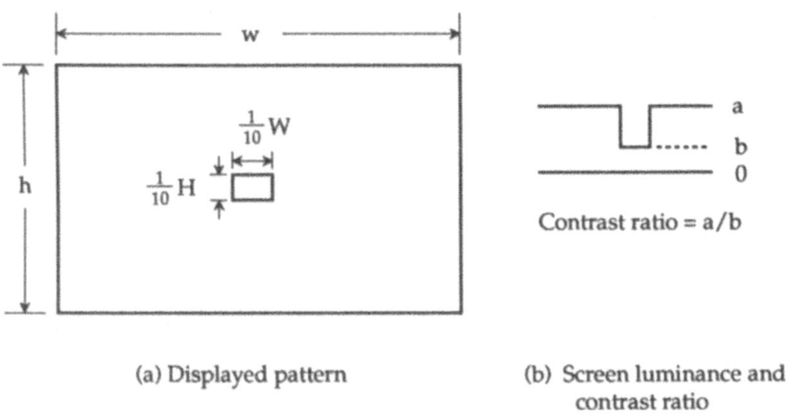

FIGURE 4.12. Definition of contrast ratio.

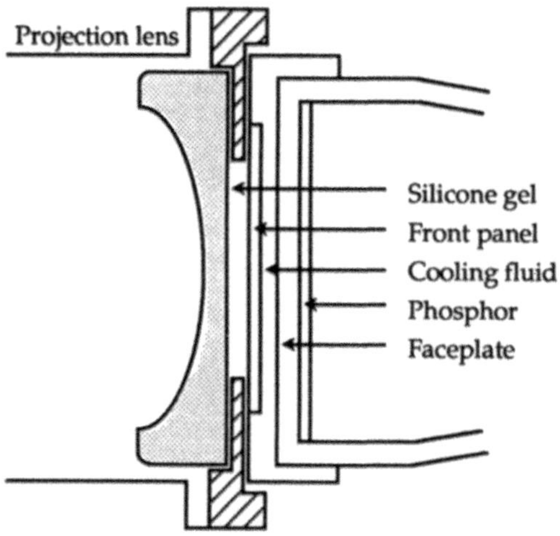

FIGURE 4.13. An improved coupling of the projection lens and CRT.

deterioration is the reflection between the lens and the CRT. By filling the space between the CRT and the lens with a fluid having the same refractive index as glass (such as ethylene glycol), the reflection is eliminated and the contrast improved. Figure 4.13 shows the structure connecting the projection lens to the CRT. A contrast ratio of over 30 can be obtained with this method, resulting in a remarkable improvement in image quality.

Because of the cooling effect of the fluid on the CRT face plate, a large beam current can be used without fear of heat damage while luminance is also improved. The luminance of the CRT is over 70,000 cd/m^2 and the luminous output of the CRT combined with the lens exceeds 200 lm.

The reflection type projector integrates the CRT and the optics into one piece aided by a concave mirror. Figure 4.14 shows the Schmidt projection tube adopted for Hi-Vision. CRTs that are 7 to 10 inches in size have been developed. Because the mirror helps project the light efficiently from the phosphor to the screen, a set of 10-inch CRTs produces a luminous output as high as 400 lm, suitable for a large projection screen.[4] However, because of various limitations in use, including the fixed projection distance which can be changed only by redesigning the CRT, projectors of this type are mainly used for commercial applications.

(2) Convergence Correction Circuit

Accurately combining the images from three (or six) CRTs requires convergence correction. This is performed by both a digital convergence circuit and an analog correction circuit being used together. These circuits also perform the trapezoidal correction of the images on the CRTs.

Since precise convergence correction usually requires a long time to adjust, an automatic method has been proposed.[5] In this method, a low frequency pattern is projected onto the screen, and a TV camera is aimed at the image to detect and correct the convergence error. Because the R, G, and B images are projected separately, the camera can be a monochrome camera. By using a computer to process the signals from the camera, convergence errors can be detected with high precision. An automatic adjustment method such as this will be required in Hi-Vision video theaters.

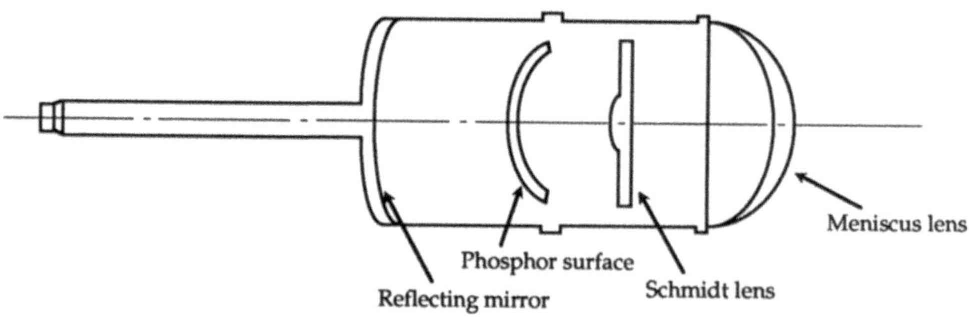

FIGURE 4.14. 10-inch Schmitd projection tube.

(3) Screen

To obtain both high luminance and a sufficiently wide viewing angle, the screen must have adequate directivity. Aluminum is usually used for the screen because of its high reflectivity and good processability. The light can be broadly diffused by matting (roughing) the surface or giving it a lenticular structure (a fine lens array). The ratio of the luminance of a screen to the luminance of a perfect diffuser is called the screen gain. While an increase in the screen's gain improves luminance, it also narrows the screen's directivity. The directivity of the screen, therefore, is determined by increasing the gain while maintaining the necessary view angle. On a flat screen, the gain is held down to four or less because a flat screen will generate a hot spot. With a curved screen, which reflects light more effectively, the gain can be increased to as much as ten.

(4) Performance

By drawing a large current through a small CRT to increase its luminance, a projection display can attain the required resolution for Hi-Vision display of over 1,000 TV lines, although image quality is not as high as on a direct-view CRTs.

The luminance L (cd/m^2) of a projection display is given by the following equation:

$$L = \frac{GK}{M^2} L_0 \qquad (4.2)$$

where

L_0: CRT luminance (Cd/m^2)

K: Utilization rate of light by lens

G: Screen gain

M: Ratio of effective CRT surface area to the screen size.

The luminous output of the projector P (lm) is expressed by

$$P = \pi \cdot S \cdot K_1 \cdot L_0 \qquad (4.3)$$

where S: Effective immage area on CRT (m_2).

One set of CRTs with lenses can emit a luminous output in excess of 200 lm. If this is projected onto a 100-inch screen with a gain of eight, a luminance exceeding 200 cd/m^2 can be obtained. However, due to limitations in the manufacturing of large screens, a gain higher than four is usually unattainable. Because one set of CRTs with lenses will not produce a sufficient luminance on a large screen, several sets of CRTs and lenses are used. In the International Science and Technology Exposition at Tsukuba, four sets of CRTs (twelve altogether) were used to project images onto a 400-inch screen (rear projection was used). A large number of CRTs, however, complicates the problem of convergence correction, and the use of one or two sets of CRTs is more practical. Because of this limitation, 100 to 200-inch screens are more or less the current standard.

The projection distance (the distance from phosphor of the CRT to the viewing screen) is determined by the projection lens, but is usually 2.5 to 3 times the screen height.

4.2.2 CRT Rear Projection Display

Rear projection displays have a transmissive screen with a lower reflectivity on the viewer side. For this reason, the contrast ratio is not decreased very much by ambient lighting. The projector can be made compact by using mirrors, and it does not occupy much floorspace even when the picture being projected is large. With these features, the rear projection display is suitable as a large display for household use.

Rear projectors primarily use a refractive optical system.

A 50-inch display is shown in Figure 4.15. The components of this system are almost the same as those for a front projection CRT display. However, the rear projection display differs in its use of mirrors and in the type of screen.

(1) The CRT and Lens

Because this type of projector is intended for home use, its compactness is important. Those manufactured today use 7 to 9 inch CRTs.

Projection lenses made only of glass tend to be heavy and expensive. To solve these prob-

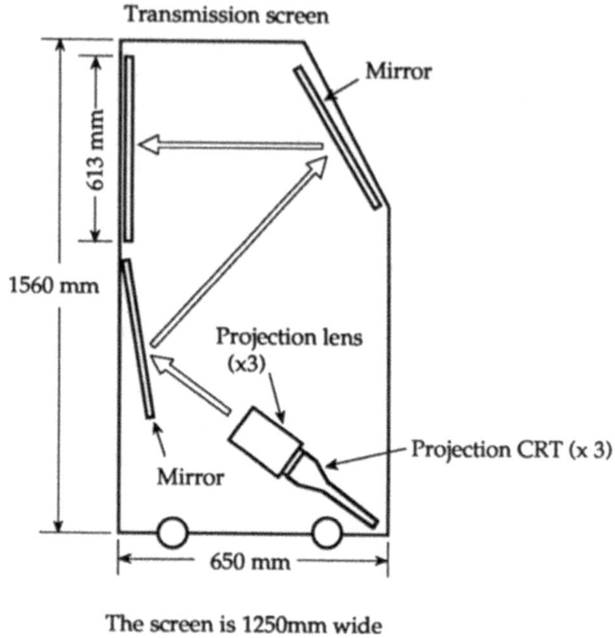

FIGURE 4.15. Structure of a rear projection display.

lems, glass lenses are often combined with plastic lenses, which are light and easily mass produced. Figure 4.16 shows an example of a set of projection lenses. As with front projectors, liquid coupling is used to improve the contrast ratio.

To make the whole system more compact, the light beams from the CRT are reflected off one or two highly reflective surface-reflecting mirrors on their way to the screen.

(2) Screen

A transmissive screen, like a reflective screen, must diffuse light from the projector while

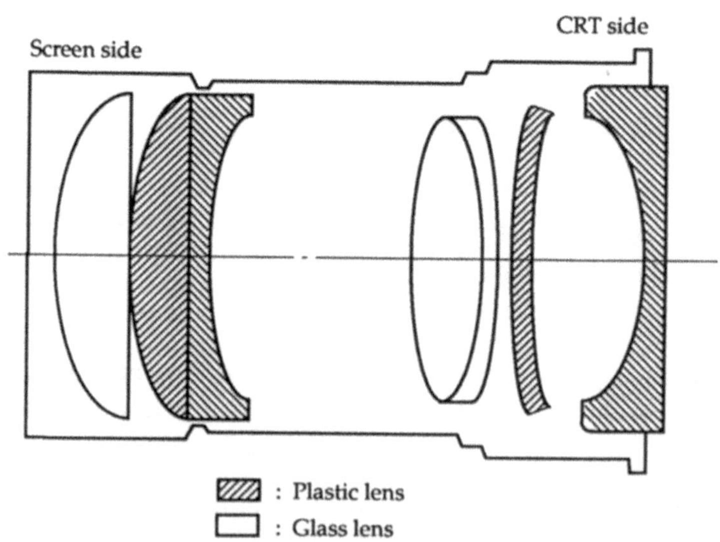

FIGURE 4.16. Configuration of a projection lens.

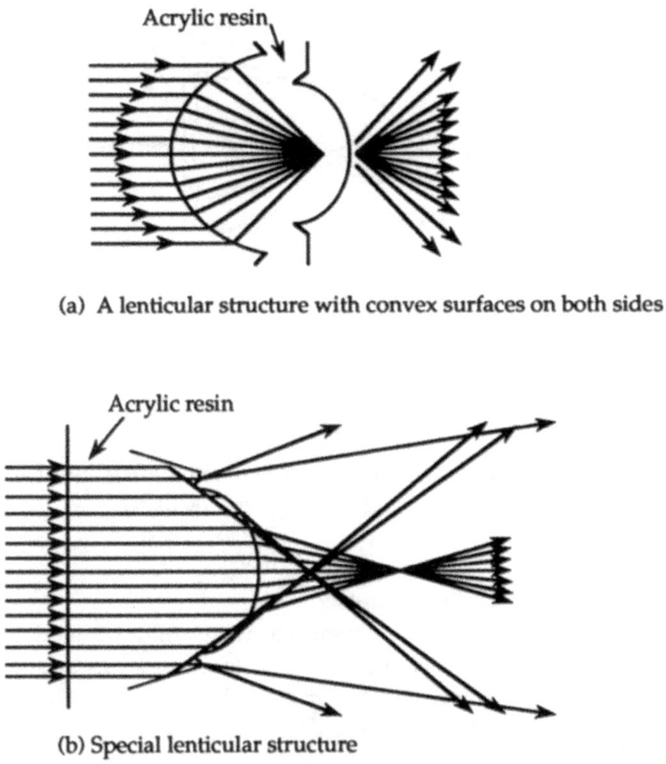

FIGURE 4.17. Lenticular structures.

achieving a suitable degree of directivity. Methacrylic resin (acrylic resin), a highly transmissive and easily processed material, is usually used for the screen.

Light diffusion is achieved by methods such as matting the surface and adding fine particles of impurities. These methods are easily accomplished and achieve the same diffusion characteristic in both horizontal and vertical directions, but are not very effective in expanding the screen's view angle. Another diffusion method uses lenticular structures. This method requires processing to form fine lenticular structures but is suitable for obtaining broad directivity. Shown in Figure 4.17 are lenticular structures that have been commercialized. The structure shown in Figure 4.17 (a) has minute lens structures formed on both the front and back surfaces to obtain a broad directivity. The structure shown in Figure 4.17 (b) has a protruding lens structure so that a broad directivity can be achieved by processing on only one side. Because a broad directivity is more important in the horizontal direction in practical use, diffusion is achieved vertically with impurities and horizontally using lenticular structures. A gain of about four is obtained in this configuration, with a directivity of $\pm 30°$ in the horizontal direction and $\pm 10°$ in the vertical direction.

As with a shadow mask, the lenticular structure samples the video signal. To maintain high resolution, the pitch must be sufficiently small. Thus far, lenticular structures have been developed with a pitch of 0.5mm. On a 50-inch screen, the MTF for a horizontal resolution of 1,000 TV lines exceeds 60%, which poses no practical problem in terms of deterioration in resolution.

Transmissive screens use Fresnel lenses to direct light even from the edge of the screen toward the viewer. As Figure 4.18 shows, the concentrically processed acrylic resin material of the Fresnel lens works as an equivalent lens. In Figure 4.18 (a), the Fresnel lens is on the viewer's side of the screen. Although it effectively redirects the light including light from the edge of the screen, the concentric structure is

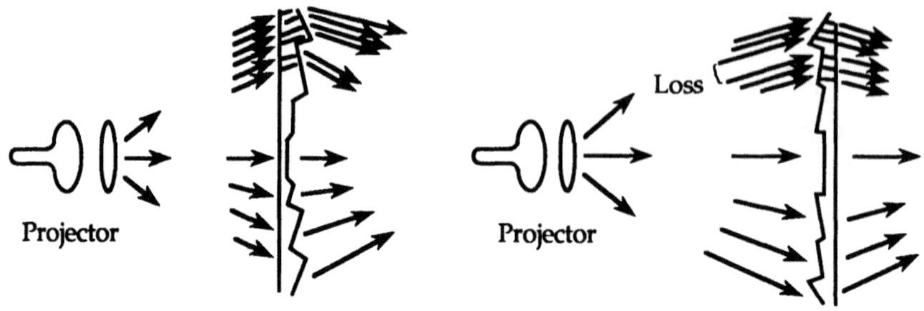

(a) Fresnel lens facing the viewer (b) Fresnel lens facing the projector

FIGURE 4.18. How a Fresnel lens works.

conspicuous and tends to generate moiré patterns due to interference with the scanning line structure. When the Fresnel lens is on the projector side of the screen, the moiré pattern is less conspicuous, but part of the light from the edges cannot be used effectively. In addition, with this structure a lenticular structure can be put on the viewer side, making it possible to form a screen with only one acrylic sheet. Because both structures (a) and (b) have their pros and cons, the choice between them depends on the particular application. The pitch of the concentric grooves of the Fresnel lens needs to be small enough not to interfere with the scanning lines and cause moiré. A pitch of 0.3mm has been adopted for 50-inch screens.

(3) Performance

Display resolution is a product of the resolution of the projection tubes, projection lenses, and screen. Of these factors, the screen contributes relatively little to degradation. Figure 4.19 gives an example of resolution.

Luminance is obtained by multiplying Equation 4.2 by the reflectance ratio of the mirrors. A luminance exceeding 400 cd/m^2 has been obtained for a 50-inch display.

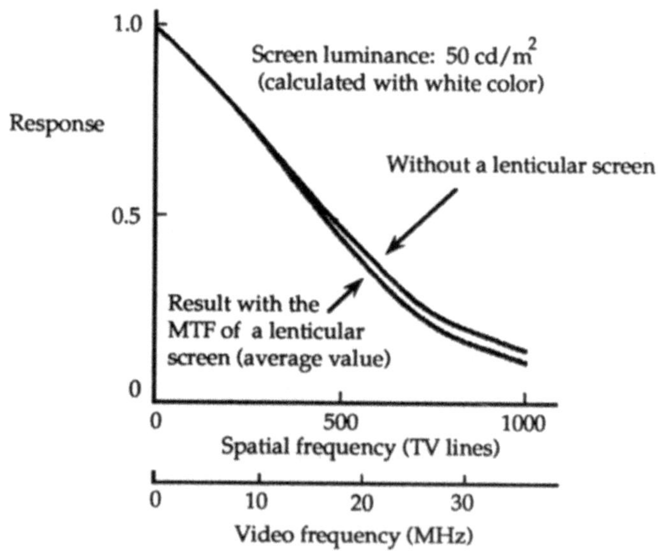

FIGURE 4.19. MTF of a rear projection display.

The fluid coupling of the lens and CRT improves the contrast ratio to a level close to that of a direct-view CRT. Because the reflectance of a transmissive screen is low, image deterioration due to a low contrast ratio is not likely to occur under ordinary indoor light.

Because manufacturing has not begun yet for large Fresnel lenses, the screen size is limited to 70 inches at present. This size, however, is sufficient for home viewing.

A screen can be shaped with lenticular structures in both horizontal and vertical directions (without using a Fresnel lens). In this arrangement, screens over 100 inches in size are possible. The 400-inch screen demonstrated in the International Science and Technology Exposition at Tsukuba had this structure. Recently, a 110-inch screen with finer lenticular pitch has been developed.

Display size is critical for home use. The depth of a rear projector can be reduced by using a projection lens with a short focal distance. Thus various measures including aspherical plastic lenses are being pursued to reduce the focal length of the lens. At present, the depth of the display is about the same as the height of the screen. Automatic convergence correction methods are also being developed to improve the ease of handling and to stabilize the operation of projection displays.

4.2.3 Light Valve Display

In this system, an electron beam modulated by the video signal forms a distortion on an oil film that corresponds to the original image. The light from a high power xenon lamp is shone onto the distortions on the film, and the reflected light is projected onto a screen via a Schlieren optical system. While this system has a more complex structure than the CRT systems, the xenon external light source with a power of over 1 kW produces an image several times brighter than a CRT projector. Two types of projectors, the Eidophor and the Talaria have been developed.[7]

(1) Eidophor Projector

An Eidophor projector has three projection tubes (shown in Figure 4.20), one each for R, G, and B.

The video signal modulates the electron beam's focus to change the spot size on the oil film. In the absence of a video signal (black), the electron beam spot scanning the surface of the oil film is enlarged so that it is connected to the spot of the next scanning line. The oil film in

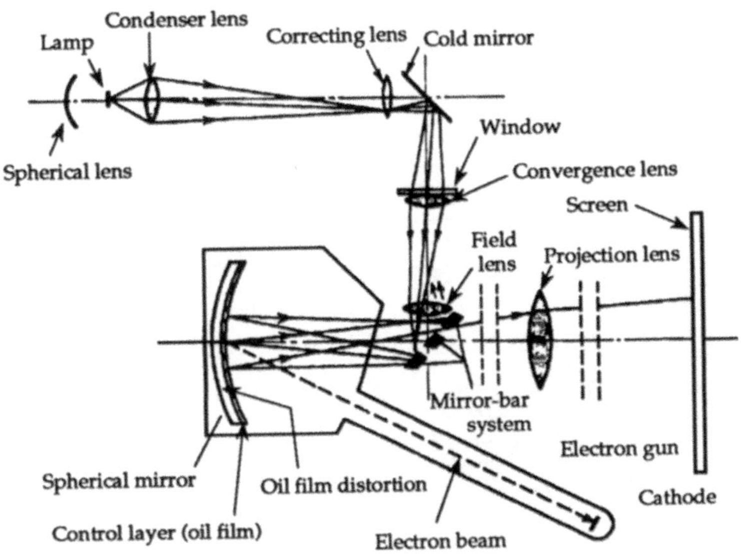

FIGURE 4.20. Configuration of an Eidophor projector.

this case is flat (or has no distortion) because of a uniformly distributed charge. The light from the xenon lamp is reflected by mirror-bars and projected onto the oil film surface. Because the oil film is flat, the light is uniformly reflected and returns to the mirror-bar system rather than being projected onto the screen. When a video signal is present, the electron beam spot area is reduced, causing a distortion on the film surface. In this case, the reflected light goes through mirror-bars and is projected onto the screen. In this manner, variations in the video signal change the quantity of the reflected light, thus modulating the luminance.

An Eidophor projector with a 4.8 kW xenon lamp has a light output of about 7000 lm. When this output is projected onto a 400-inch screen with a screen gain of two, a luminance of 100 cd/m^2 is obtained. At present, the resolution is at least 800 TV lines. There are no high resolution displays that have a higher light output than an Eidophor projector.

(2) Talaria Projector

A Talaria projector is different from an Eidophor projector in that it has two projection tubes, one for G and the other for R and B, and a transmissive Schlieren optical system. Figure 4.21 shows the configuration of the system.

The projection tube for G is monochrome and works like an Eidophor projection tube. The projection tube for R and B, however, produces two colors by wobbling the electron beam vertically with the R video signal, and horizontally with the B video signal, thereby controlling the horizontal and vertical light diffusion independently. The light from the xenon lamp passes the horizontal and vertical input slots before striking the oil film surface. The light from these horizontal and vertical slots is then directed to horizontal and vertical light shielding bars corresponding to the slots before it reaches the projection lens. For example, R light that has passed through the horizontal slot will be shielded by the horizontal bar in the absence of an R signal because the light coming out of the oil film is not diffused in the vertical direction. However, an electron beam modulated by the R signal and having entered the oil film surface has the light diffused in the vertical direction, and is able to pass through the horizontal bar and be projected as R light output onto the viewing screen. In the B light projection process, the horizontal and vertical in the above description are reversed.

A 2-tube Talaria projector has a 0.7 kW projection xenon lamp for G and a 1.3 kW xenon lamp for R and B, and is capable of 2,500 lm or higher luminous output. The resolution is at least 800 horizontal TV lines.

4.2.4 Laser Display

In this method, three lasers are modulated by R, G, and B video signals as they are projected

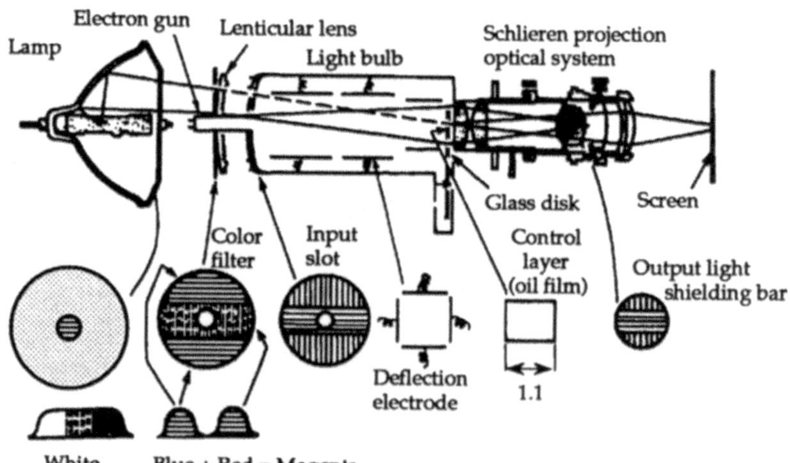

FIGURE 4.21. Configuration of a Talaria projector.

onto a screen. These displays have a very high resolution, and have been developed as large as 100 inches. However, due to low power efficiency, speckle patterns that decrease the image quality, and other problems, this method has not gained broad acceptance.[8]

4.2.5 Liquid Crystal Projection Display

This method uses projection lenses to project images of a small liquid crystal display on a screen. With recent advances in liquid display technology, 100-inch displays have been developed (although for conventional television). However, before this technology can be applied to Hi-Vision, many problems such as achieving higher resolution and contrast need to be overcome.

4.3 PANEL DISPLAYS

To enjoy the telepresence of Hi-Vision images in the home, the best type of display is a large flat panel that can be hung on the wall. Flat panel TV receivers using 4-inch and smaller liquid crystal or flat CRT panels are already on the market. However, large displays for Hi-Vision are still in the developmental stage. This section will describe large, high image quality flat panel displays, focusing mainly on Plasma Display Panels (PDP) and to a lesser extent on liquid crystal displays and flat CRT displays.

4.3.1 Color Plasma Display (PDP)

A PDP display pictures on the screen using light emission caused by an electrical discharge at each pixel. The application of this method to a full color panel is limited to a method that excites the phosphors with far ultraviolet rays generated by discharge. Because the ultraviolet excitation is not as intense as with an electron beam, sufficient luminance cannot be obtained by regular line sequential driving. Thus it is necessary to lengthen the emission time within a field by giving the panel a memory function.

(1) Planar Discharge of the AC PDP

The panel shown in Figure 4.22, with its electrodes covered with dielectric material and concealed from the discharge area, is an AC panel. The AC panel's memory function works with the charges accumulated on the surface of the MgO layer.

Since ordinary AC panels have the simple structure shown in Figure 4.22, large panels can also be made with this structure. A panel that emits orange light and uses neon gas has been developed that is 1067 × 1067 mm^2 (2048 × 2048 dots) in size.[9] However, because of the lack of cell sheets to separate cells from each other, the phosphors for a color display can only be coated in the vicinity of the electrodes. Thus the deterioration of phosphor characteristics due to ion bombardment at the time of discharge is inevitable.

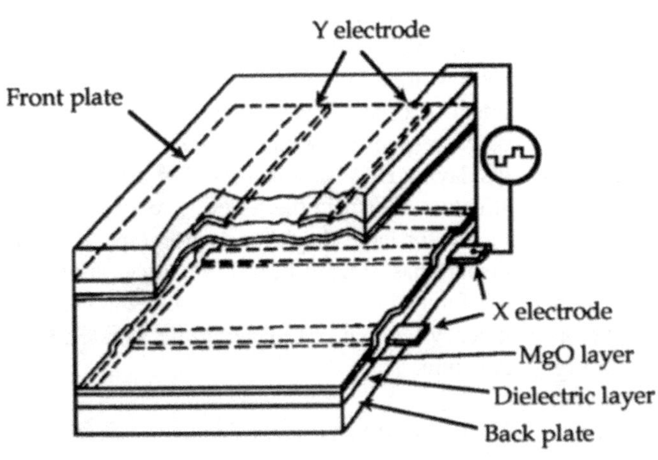

FIGURE 4.22. Structure of an AC PDP with opposing electrodes.

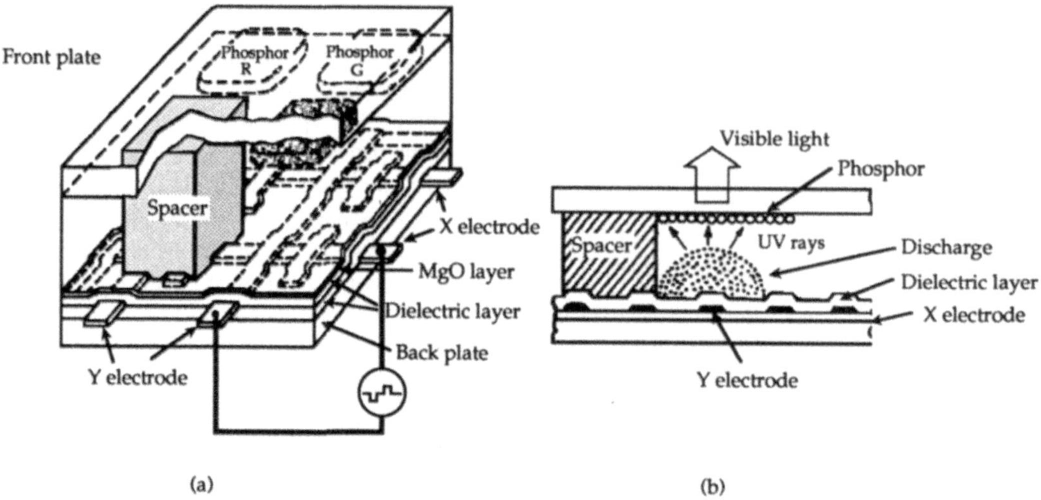

FIGURE 4.23. Surface discharging AC, PDP.

To solve this problem, a planar discharge panel with the structure shown in Figure 4.23 has been developed. In this panel, the X and Y electrodes are formed on the back plate so that they sandwich a dielectric layer. Because the phosphors can be coated onto the inside surface of the front plate, they are not exposed to ion bombardment (see Figure 4.23 (b)). The planar discharge panel in Figure 4.23 has an R, G, and B cell composing a pixel, and a fourth cell position is used for a spacer and auxiliary discharge.[10] The fourth pixel sets the distance between the front and back plates and improves the writing speed. Future issues concerning this structure will be to improve color purity, which deteriorates due to crosstalk between adjacent rolls, ensure the high speed response required for Hi-Vision, and improve emission efficiency.

(2) Pulse Memory PDP

A panel with its electrodes exposed to the discharging area is called a DC panel. Although a DC panel essentially has no memory function, it is possible to add a memory function to it by modifying the driving method. One such modification is the pulse memory method. The operating principle of this method, shown in Figure 4.24, uses the fact that charged particles and metastable particles generated by the discharge gradually decrease with time after the discharge has terminated (Figure 4.24 (c)), and the fact that reignition is likely to occur in the presence of these particles. However, in experiments with simple pulse memories, the time between the application of the writing voltage and the start of the discharge varied widely from a few μs to several ms. This result indicates that a broad write pulse is necessary to cause the discharge without fail. This system, therefore, was not applicable to the display of television images.

To solve this problem, auxiliary cells have been added, as shown in Figure 4.25. The charged particles and metastable particles generated by the auxiliary cells are diffused to the display cells via the priming space so as to facilitate the initiation of the discharge. In addition, by modifying the driving method, a stable discharge start is made possible even with low amplitude, short writing pulses. In a panel driving experiment, the panel has been confirmed to show a quick response rate that should be capable of displaying Hi-Vision images. The simple panel structure shown in the figure is relatively easy to use for large displays. A prototype 20-inch display has already been built.[11] An issue for the future is the improvement of emission efficiency without complicating the simple panel structure.

(3) Townsend Discharge Memory Panel[12]

Figure 4.26 shows the structure of a Townsend discharge memory panel. This memory panel,

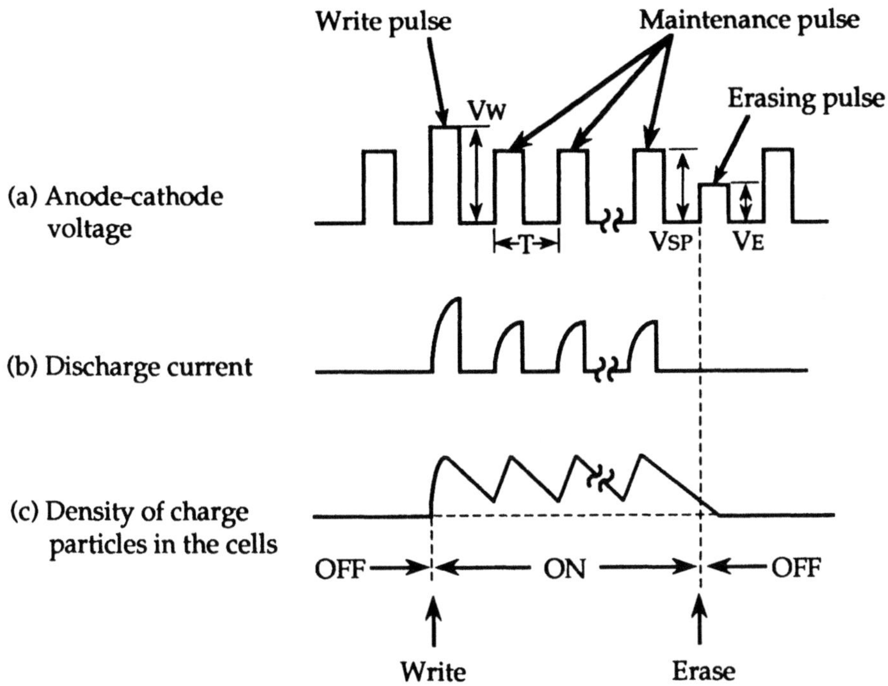

FIGURE 4.24. Principle of the pulse memory method.

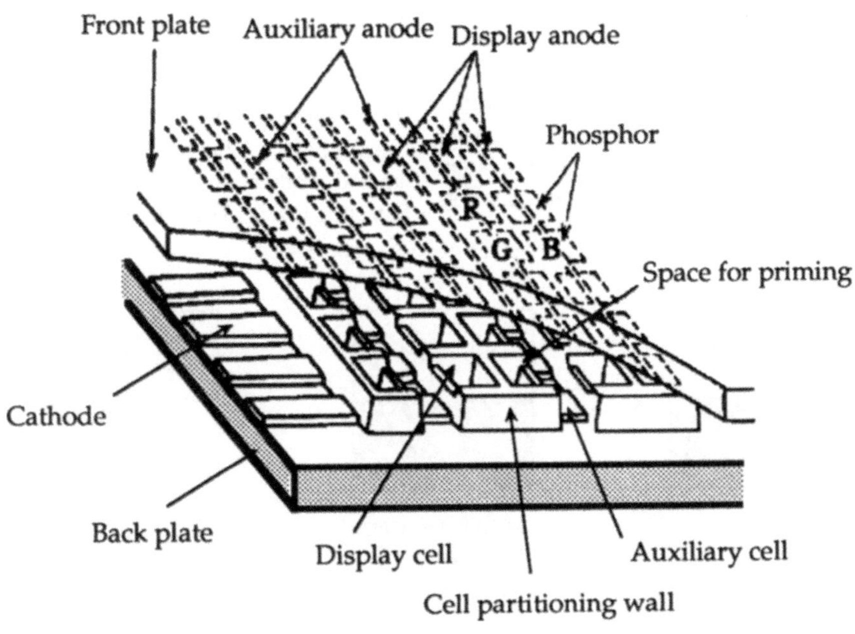

FIGURE 4.25. Structure of a pulse memory flat panel.

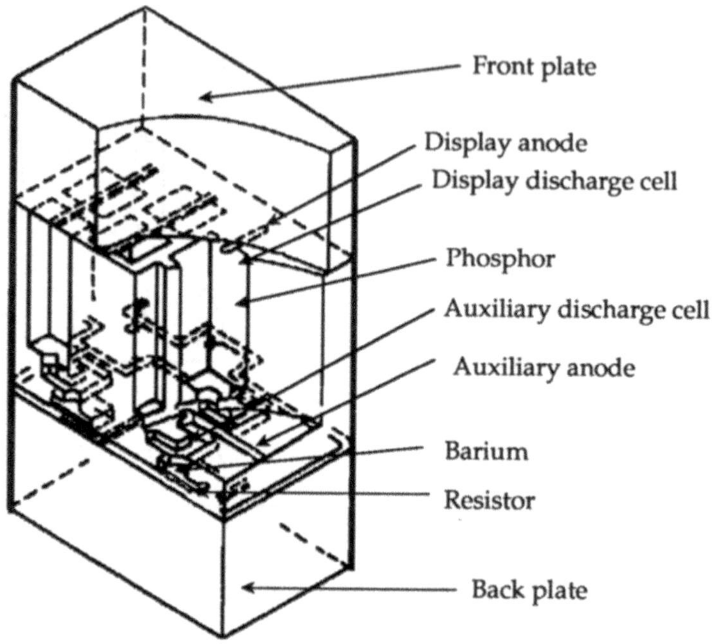

FIGURE 4.26. Structure of the Townsend memory panel.

like the pulse memory method, obtains its memory function by pulse discharge. This panel is characterized by relatively deep (2mm) display cells whose inside walls are coated with phosphor, barium cathodes, a resistor placed in each display cell, and the use of extremely short sustain pulses for driving. The result is a luminous efficiency of 1.6 lm/W and a luminance of 200 fL—levels much higher than on other panels and sufficient for practical use. However, because of the complexity of the panel structure, large panels with high resolution will be an issue for the future.

4.3.2 Color Television Display Systems Based on PDP

Figure 4.27 shows the basic principle in a method for displaying television images on a display panel with memory. The figure shows how a one-field image with four gray levels would be displayed using two bits.

When MSB (Most Significant Bit) and LSB (Least Significant Bit) bit surfaces are displayed consecutively within one field period, they overlap and produce an image that appears to have four gray levels. The brightness ratio of the on-

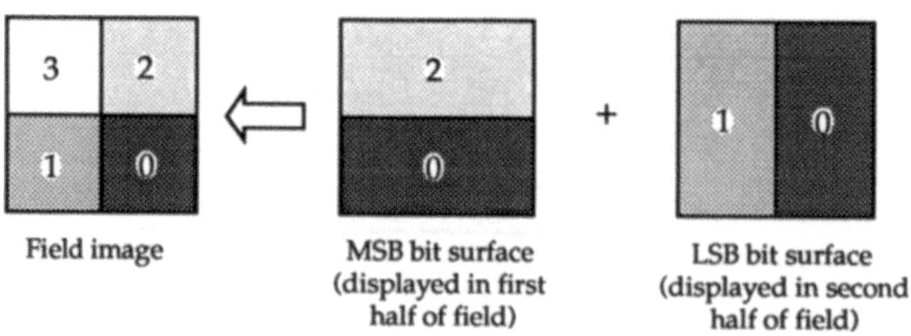

FIGURE 4.27. Television image display principle in a panel with memory.

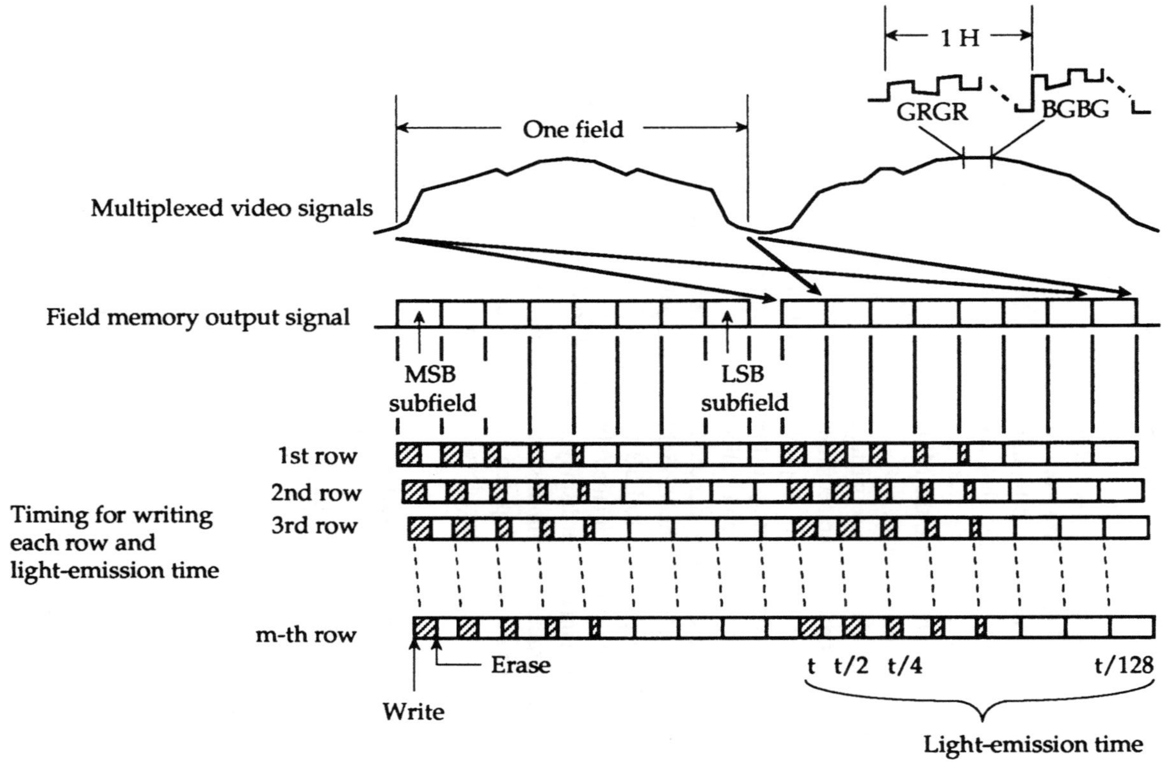

FIGURE 4.28. Time chart for television display.

regions of the MSB and LSB bit surfaces is set at 2:1 by setting the LSB light emission time to be a half that of MSB.

Figure 4.28 shows the timing for writing and erasing row electrodes on an 8-bit, 256-level display. First, the three primary video signals (R, G, and B) are switched and multiplexed to correspond with the phosphor dot array on the panel. These signals are converted to 8-bit digital signals by an A/D converter. After they are stored in the field memory, they are read sequentially at high speed starting from the MSB bit surface. These are called subfields.

In the subfield period, the lines are written in order from top to bottom, then erased after a prescribed time from the top. This displays two-level on/off images (corresponding to the bit surfaces shown in Figure 4.27 and called subfield images here). The discharge durations are set at t, $t/2$, $t/4$, . . . , $t/128$ of the subfield corresponding to the bits from MSB to LSB to adjust the luminance of the subfield images. In this manner, eight subfield images are displayed consecutively with gradually diminishing brightness. These subfields, superimposed in time, appear as an image with 256 gray levels.

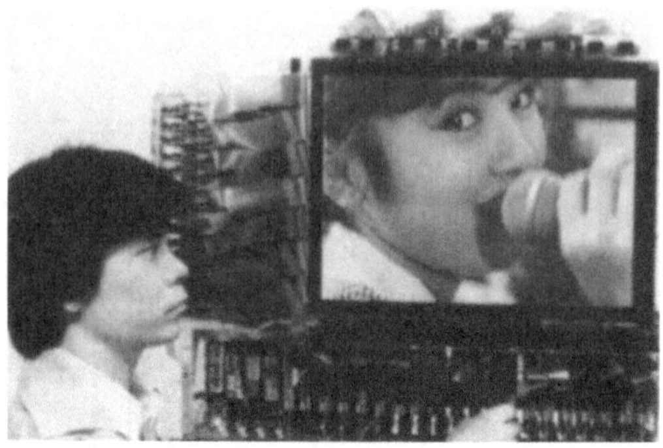

FIGURE 4.29. A 20-inch pulse memory panel displaying a TV signal.

Table 4.1 lists the results of recent color television image display experiments. Figure 4.29 shows a 20-inch panel prototype with the structure shown in Figure 4.25. This panel is driven by the pulse memory method.[11]

4.3.3 Liquid Crystal Display

TN mode liquid crystal displays, the most widely used type of LCD, have a smooth and gradual relationship between applied voltage and light

TABLE 4.1. Experimental PDP color display results.

	Surface discharge AC panel [10]	Pulse memory panel		Townsend memory panel [12]
		Vertical cells [13]	Horizontal cells	
Surface area (mm^2)	50 × 50	160 × 126	103 × 83	160 × 120
Number of display cells	$100 \times 100 \times \frac{3}{4}$	160 × 126	160 × 126	160 × 120
Cell pitch (mm)	0.4	1.0	0.65	1.0
Cell arrangement	R B / ■ G	G R / B G	G R / B G	R G B / B R G
Brightness (white, cd/m^2)	52 (15 fL)	135 (40 fL)	58 (17 fL)	690 (200 fL)
Efficiency (lm/W)	0.2	0.34	0.11	1.6
Contrast	45:1	75 to 100:1	90:1	
Gray levels	64	256	256	128
Access time (μs)	8	2~4	4	9

transmittance and do not have a distinct threshold. For this reason, to show a high contrast image with good color reproduction, it is necessary either to draw a lead from each cell and drive them statically, or else use an active matrix method. The former method is used for super large displays that have many modules consisting of several hundreds cells. However, because a lead is drawn from each cell, the display's pixel pitch becomes quite large. Furthermore, the image quality is limited by the gaps between modules and the variations in their luminance. The latter method attaches either a triode such as a thin film transistor, or a diode or other 2-terminal device to each cell. By switching each cell with one of these devices, an effect similar to a static driving method is obtained. Three- to four-inch liquid crystal televisions with an active matrix LCD have already been commercially produced. A 14-inch prototype display has already been made at the experimental level. However, it is still extremely difficult to fabricate thin film transistors and other nonlinear devices without any defects for a one-meter diagonal Hi-Vision LCD flat panel.

Liquid crystal displays have been developed using STN and SBE mode liquid crystals, which have a steep applied voltage-transmittance curve, and ferroelectric liquid crystals, which have a fast response time. However, before these can be applied to Hi-Vision, problems such as operating speed and display size must be solved.

4.3.4 Flat CRT

CRT displays, which excite the phosphorous screen with high speed electron beams, have a proven track record and advantages such as superior color and high luminous efficiency. Development is under way for flat CRTs that take advantage of these qualities. Small displays have already been produced wherein the electron gun is parallel to the phosphorous screen and the electron beam is bent at a 90° angle. However, it is difficult to apply this method to large panels.

Figure 4.30 shows the structure of a FLAT-SCREEN® display, in which electrons generated by the discharge between the cathode and anode grids are accelerated with 4 kV to excite the phosphor. A 35-inch monochrome display (252 × 352 pixels with a pixel pitch of 2mm) can display television images at a brightness of 35 fL. Another method called MDS has also been developed, in which multiple electron beams

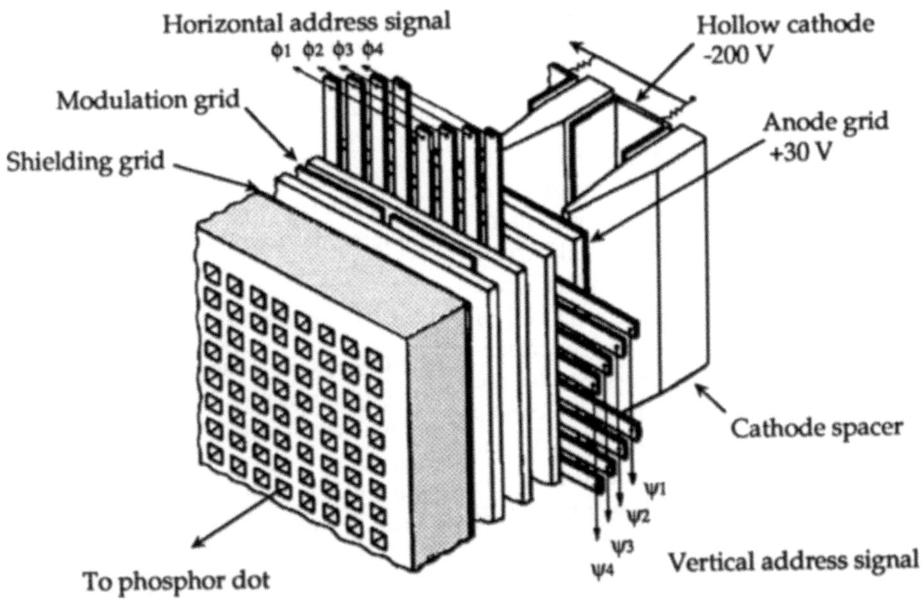

FIGURE 4.30 Structure of flat CRT (FLAT SCREEN®).

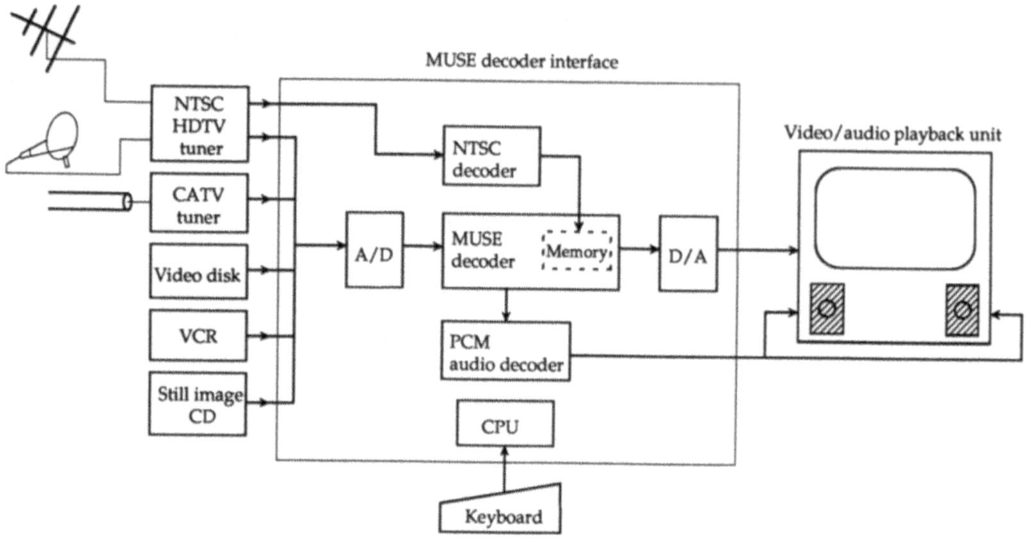

FIGURE 4.31. Block diagram of Hi-Vision receiver.

from linear thermionic cathodes placed over the entire screen are deflected by a matrix of deflecting electrodes, and then accelerated.[14] A 10-inch MDS color television display (RGB trio pitch:0.5 mm) can display TV images at a brightness of 70 fL.

The critical development themes in the future for flat CRTs are the achievement of both high resolution and large panel size, as well as uniform image quality.

4.4 MUSE RECEIVER

A Hi-Vision receiver set is a MUSE receiver in the sense that it receives signals broadcast in the MUSE format. Special characteristics of the MUSE receiver include a substantially larger image display, higher resolution, and an aspect ratio of 16:9, which is wider than the 4:3 of conventional television. These display characteristics create a sense of telepresence for the viewer. In addition, because Hi-Vision carries about five times more visual data than conventional television, images are extremely clear. With these features, households will be able to perform applications unheard of with conventional television receivers.

Conventional television receivers are also used in many ways other than broadcast reception, such as for display terminals for VCRs, video disks, and game machines. Hi-Vision receivers, which have a far superior display capability, are expected to be used in even more ways.

For example, Hi-Vision receivers in the future will be used not only for receiving Hi-Vision and conventional broadcast programs, but also as reception terminals for CATV and other cable media, VCRs, video disks and other packaged media. If a Hi-Vision receiver is combined with the information processing functions of a microcomputer, it can be used as a comprehensive household image information terminal.

Figure 4.31 shows a block diagram of a Hi-Vision receiver used as a comprehensive reception terminal.

4.4.1 MUSE Reception

Hi-Vision broadcasting with MUSE was planned for broadcast satellite BS-3, which was launched in 1990. MUSE signals broadcast from the satellite can be received using a parabolic antenna, BS converter, and a Hi-Vision BS tuner.*

This BS equipment is basically the same as

*Experimental Hi-Vision broadcasting has been underway since November 1991 on a daily 8-hour schedule.

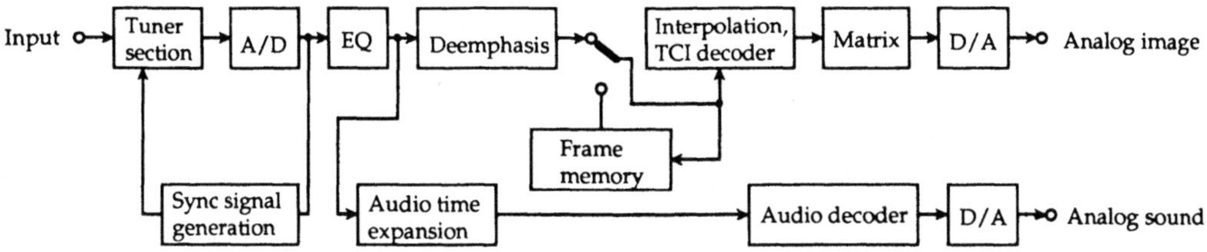

FIGURE 4.32. Block diagram of MUSE decoder.

that which is being used to receive the current satellite broadcasts. A small parabolic antenna with the required CN ratio works well for satellite reception. Recent improvements in low noise amplifying semiconductor devices, especially the introduction of HEMTs (High Electron Mobility Transistor), have decreased the noise level of BS converters (below 1.5 dB) and made even smaller antennae possible. Flat antennae have also improved in efficiency and gain because of improvements in substrate materials and feeders. Lightweight and thin flat antennae which have been commercially produced can easily be mounted on walls. BS tuners will need a broader video bandwidth than is needed for conventional television, to handle Hi-Vision.

A MUSE decoder converts MUSE signals received from satellite broadcasts or packaged media back into Hi-Vision signals. It is the central digital signal processing unit of the Hi-Vision receiver, and consists of large capacity video memory chips and other LSIs.

Expansion of the functions of a Hi-Vision receiver will involve using the approximately 20M of video memory in the MUSE decoder in a wide variety of ways. Figure 4.32 is a block diagram of a MUSE decoder.

Although the high image quality of Hi-Vision broadcast programs is best viewed on a Hi-Vision receiver, these programs can also be seen on a conventional television receiver using the MUSE 525-line down-converter developed for this purpose.

4.4.2 Customizing LSIs for MUSE Receivers

The MUSE system[14] was developed for broadcasting Hi-Vision signals over one broadcast satellite channel. Since it compresses video and audio signal bands with digital technology, the reception of these signals requires large and complex circuits. To reduce the cost, size, and power consumption of the MUSE receiver while improving its reliability, LSI circuits become essential, especially for the main component which is the MUSE decoder.

This section discusses the technological issues involved in making customized LSIs for the Hi-Vision receiver from the viewpoint of LSI and semiconductor technologies.

(1) Considerations in Adopting LSIs

The acceptance of Hi-Vision receivers as a home appliance by the general public rests critically on reducing their cost by minimizing the number of components. This cost reduction is the most important reason for developing customized LSIs.

The normal procedure in developing LSI chips is to divide the system configuration into functional blocks while referring to LSI technological requirements, and then to optimize the functions that can be put on each LSI. However, as the scale of integration increases, technical difficulties that limit the applicability of LSIs tend to increase, resulting in a cost increase.

This is true of LSIs for the receiver. Economies of scale will not apply in the initial marketing stages of the receiver, and so it is crucial to find a way to develop the LSIs at a low cost. Realistically, one should expect that the LSI chip set will be developed gradually as the Hi-Vision market grows. Thus for the first stage it is important to achieve a suitable level of integration that will reduce costs and at the same time improve the ease of use by limiting the number of external components that will need to be attached. In addition, aggressive implementation of low cost development methods is necessary.

LSI technology is undergoing rapid advances in the area of development technologies for application specific ICs (ASICs). Among them, gate array and standard cell methods are especially powerful ASIC development methods that use CAD (Computer Aided Design) technology to drastically reduce the cost and time of LSI development. Recent developments in on-chip memory are raising the level of performance and functions of LSIs. The effective implementation of these ASIC technologies in receivers is a crucial issue for the future.

The second important issue in developing LSIs for receivers is reducing power consumption while increasing reliability. At present, most of the logic ICs that could be used in the receiver are bipolar—TTL (Transistor-Transistor Logic) and ECL (Emitter Coupled Logic) devices. While these are effective in increasing operating and reducing the impedance of circuits, they consume a lot of power and therefore are not suitable for large scale LSI circuitry. On the other hand, CMOS (Complementary MOS) technology is widely used for LSIs with a high level of integration, but it is inferior to the other two technologies in speed and loaded driving power. However, with advances in fine processing leading to improved device performance, the best avenue is to incorporate improvements in circuitry such as pipeline and parallel processing into CMOS technology.

Another characteristic of CMOS LSIs is that the signals have a large logic amplitude and the circuit impedance is high. The latter makes them vulnerable to external noise. To prevent the receiver from making errors, special attention needs to be paid to improve the mounting of LSIs and to keep out external noise having a high frequency component.

(2) Technological Issues in Incorporating LSIs

Figure 4.33 shows a block diagram of LSIs for the MUSE decoder. Although high performance A/D and D/A converters are necessary, all circuits except the input/output sections are stable digital processing circuits.

In developing LSIs for the decoder, the special characteristics of the decoder's circuits raise the following issues.

(a) High Speed Synchronous Operation. In general, LSIs that perform video and audio signal processing differ from those used in computers in that they must simultaneously process signals having 8 bits and 16 or more bits. This complicates the circuit design due to the fact that with the LSI logic gates operating simultaneously, the instantaneous operating current increases in proportion, causing greater potential fluctuations in the power source and signal lines. In the case of a CMOS LSI with low load driving power, the current in the input/output section may comprise nearly half the current for the whole chip, thus requiring ample consideration for the decline in operating speed caused by an increase in chip temperature.

The clock frequencies needed for the video circuits in the MUSE decoder are 16.2 MHz, 24.3 MHz, 32.4 MHz, and 48.6 MHz. To control the clock phases and synchronize the circuits, a master clock has a frequency of 97.2 MHz. For high speed LSIs having several clock inputs, phase control of the clocks is critical to compensate for delay time on the printed circuit board. For example, a new method is needed for handling low speed clock pulses on an equal basis with other video signals, and then performing phase control within the LSI after incorporating them into a high speed clock pulse.

The present CMOS technology for logic LSIs has attained design rules of 1.5 to 1.2μm, with which it is possible to achieve the MUSE decoder's highest operating frequency of 48.6 MHz. However, since the cycle for each operation is about 20 ns, if we consider LSI design factors such as variations in manufacturing processes, fluctuations in power source voltage and temperature while in operation, and a normal margin of 100%, then it is necessary to design for 10ns speed. Thus in multiplications involving extensive logical depth, it is necessary to improvise with methods such as a table lookup system using memory.

(b) Nonlinear Operations. One of the characteristics of MUSE is its use of a pseudo-constant luminance principle. For this reason, nonlinear logic operations are used not only for nonlinear deemphasis, but also for noise coring,

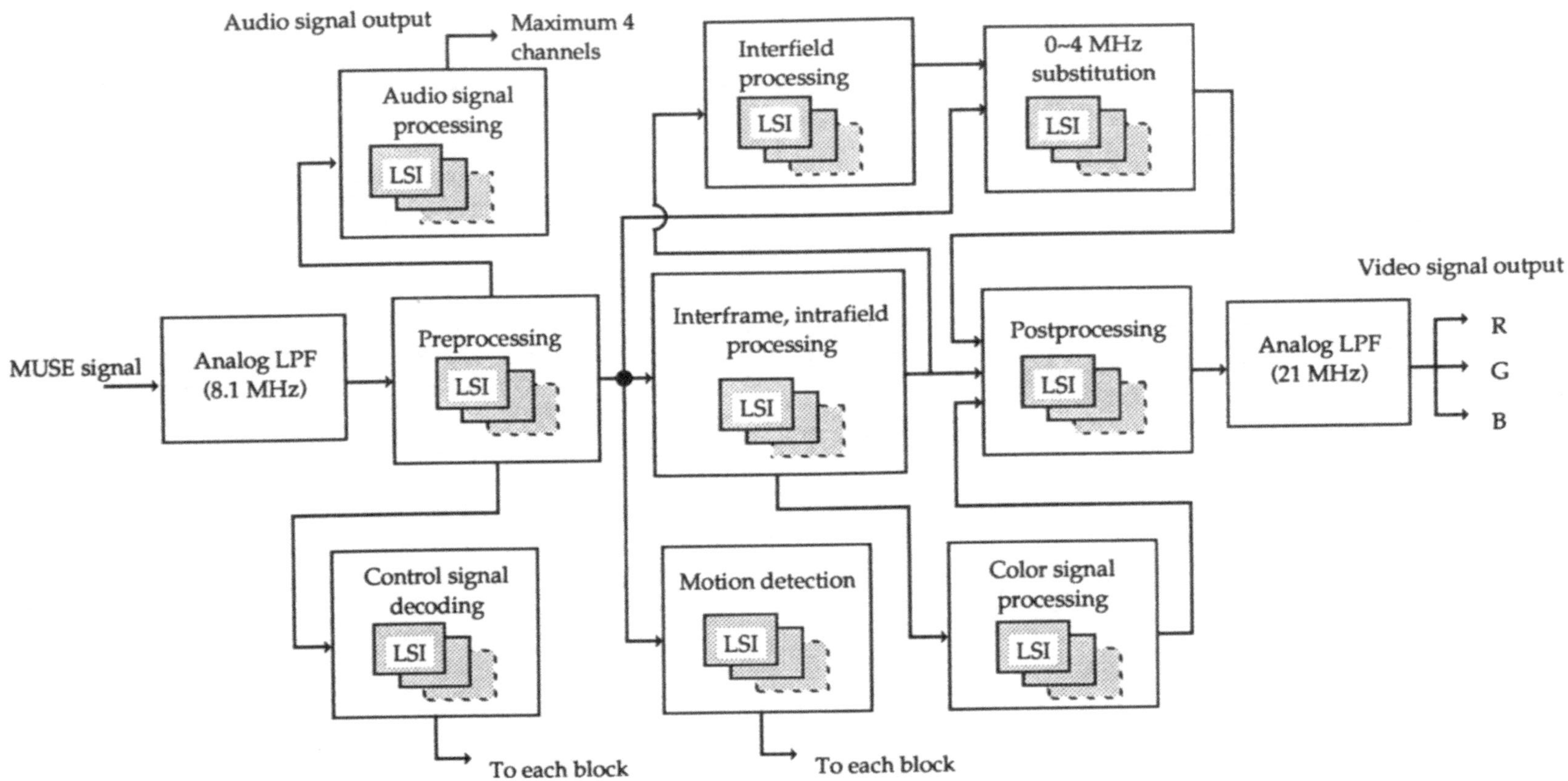

FIGURE 4.33. LSI block diagram for the MUSE decoder.

luminance signal enhancement, motion quantity detection, and color signal gamma correction.

Nonlinear logic operations normally use a ROM memory. The input signal is led to the world line, and the results of the nonlinear multiplications stored beforehand in the memory are read out from the bit line. However, due to differences in the manufacturing processes, fabricating memory and logic devices on the same chip is very difficult with currently available semiconductor process technology, especially when high speed memory is required. Thus in some cases it is necessary to substitute nonlinear characteristics with a piecewise linear approximation.

(c) Digital Filter. In the MUSE transmission system, band compression is performed by the offset subsampling of video signals in the spatio-temporal region. Thus the high quality reproduction of an image requires stable and constant band characteristics (frequency and phase characteristics) throughout the transmission and reception systems. To provide these conditions, digital filters are frequently used in a variety of ways not only to limit the signal bandwidth, but also for various interpolation processes (symmetric filter) and the conversion of sampling points from 32 MHz to 48 MHz (asymmetric filter).

High speed sum of products operations are necessary with digital filters (Figure 4.34). When the coefficient values can be approximated by powers of two, multiplication can be performed by bit-shift addition, and large scale integration is easy. Otherwise, as in the case of nonlinear operations, a high speed ROM is necessary and the use of LSIs becomes complicated.

(d) High Pin Count. Because the LSIs in the receiver have numerous input and output signals, and because the design must assure the stability of high speed operations, high pin counts are inevitable. However, a high pin count presents a major obstacle to realizing low cost LSIs and thus a low cost receiver. Packaging costs could even account for over half of the cost of the LSI. Furthermore, a large number of pins create LSI design difficulties and increase power consumption as described before. They also pose problems in high density mounting on the circuit print board. In the use of LSIs for building a receiver, we have to study the functions of LSIs thoroughly to find the ways to reduce the number of pins.

Generally speaking, a high speed, highly integrated LSI needs to have several power source and ground pins for its stable operation. Further, control pins may need to be added to enable LSI functions to be set externally, again increasing the number of pins. For instance, since video signals in the MUSE transmission system are controlled in accordance with the quantity of the motion and motion vectors, these controls may also require additional pins. An effective but slow method for limiting the pin count is to supply the control data from an external processor to the LSI via a serial bus. This makes the standardization of the LSI control bus a critical issue. At present, an interim standard shown in Figure 4.35 is in force.[15]

4.4.3 Additional Functions of the MUSE Receiver

Extra functions that can be added to the Hi-Vision receiver include the reception of various broadcasting services described below, as well

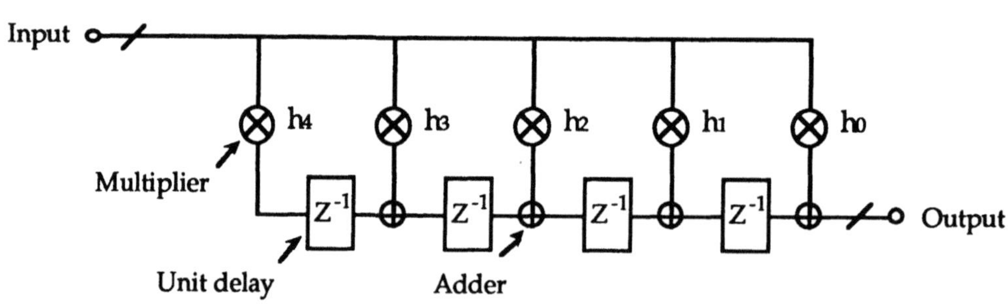

FIGURE 4.34. Configuration of a transversal filter.

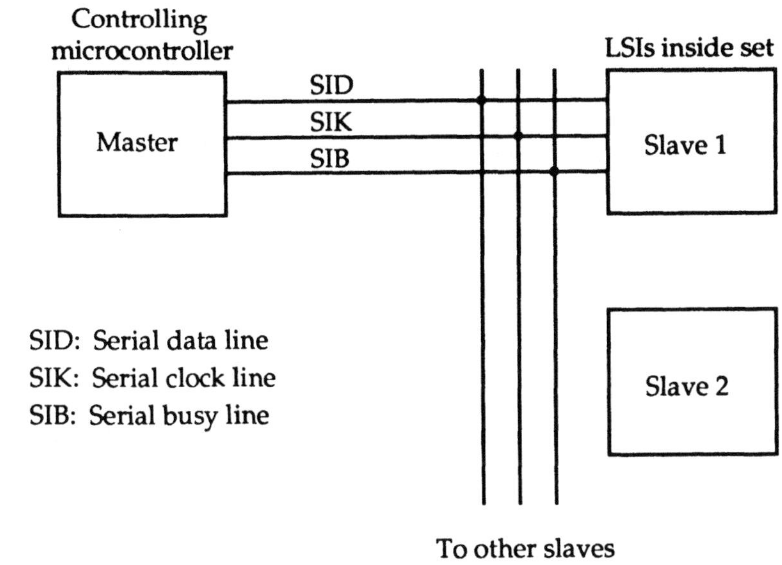

(a) Serial bus connection

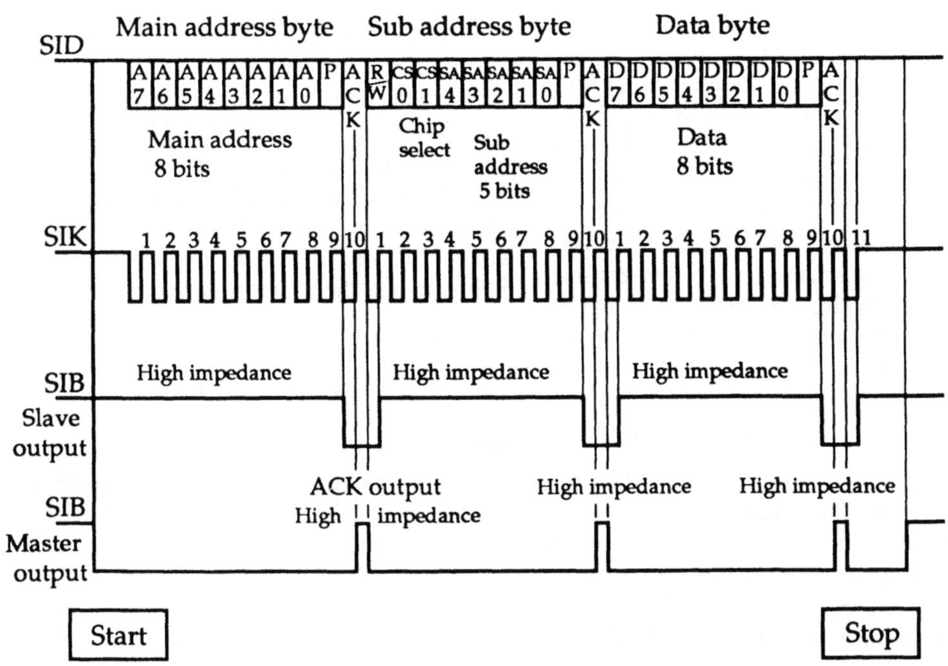

(b) Example of 1 byte data writing

FIGURE 4.35. MUSE serial bus standard (interim).

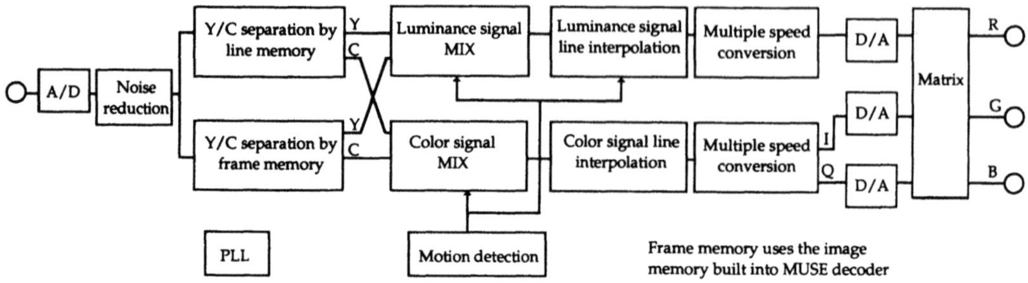

FIGURE 4.36. Block diagram of IDTV.

as interconnections to packaged media devices such as VCRs and video disk players.

(1) Reception of Conventional Broadcasting

A basic function that a Hi-Vision receiver must have is to be able to receive conventional television broadcasting as well as Hi-Vision broadcasting. Because Hi-Vision and conventional television have different aspect ratios, a Hi-Vision receiver receiving conventional broadcasting will have blacked out areas on both sides of the screen. However, these blacked out areas can be used to show other information.

When using a MUSE receiver for conventional television broadcasting, the cross color and cross luminance image deterioration caused by the imperfect YC separation in the conventional television standard can be corrected at no additional cost by using the large memory capacity in the MUSE decoder. In addition, interline flicker can be eliminated by converting the 525-line 2:1 interlacing of the current television standard to either 525-line 1:1 sequential scanning or 1050-line 2:1 interlacing. Figure 4.36 shows a block diagram of an IDTV (Improved Definition Television) system that shares the image memory of the MUSE decoder.

(2) Broadcasting Still Images in Hi-Vision

Besides broadcasting Hi-Vision moving images in MUSE, there are also multichannel services such as audiographics which broadcast MUSE still images accompanied by music. While current television broadcasting has an audiographic service in the form of environmental images and B-mode stereo, Hi-Vision's combination of high definition image and 4-channel stereo makes still-image broadcasting far more realistic and impactful.

As Figure 4.37 shows, the signal format of still-image broadcasting repeats two frames of video signals and two frames of audio signals. One still image is composed of two frames of video signals. The audio data for a four frame interval that includes the two video frames is transmitted for all the channels in the vertical blanking period and the two audio frames. The signal format allows a Hi-Vision receiver without a still-image decoder to receive one channel of still-image broadcasting. When a 6-channel service is performed with this signal format, 24 frames comprise one cycle.

The video and audio signal formats are the same as those of the analog MUSE system. The audio multiplexing method takes an audio signal packet having the same format as that multiplexed into the vertical blanking period of the current MUSE system, and time-division multiplexes it into the frame assigned to audio signals.

(3) Broadcasting on Data Channels

The audio signal of the MUSE system is transmitted in the VBL interval with time-compression multiplexing as a bit stream with an average bit rate of 1.35 Mb/s. Bits that are not used for the audio signal can be used for data transmission in what is called the data channel. Although the capacity of the data channel may vary depending on the audio mode and the number of channels used, it ranges from 112 Kb/s (with B-mode stereo) to a maximum of 912 Kb/s (with A-mode monaural).

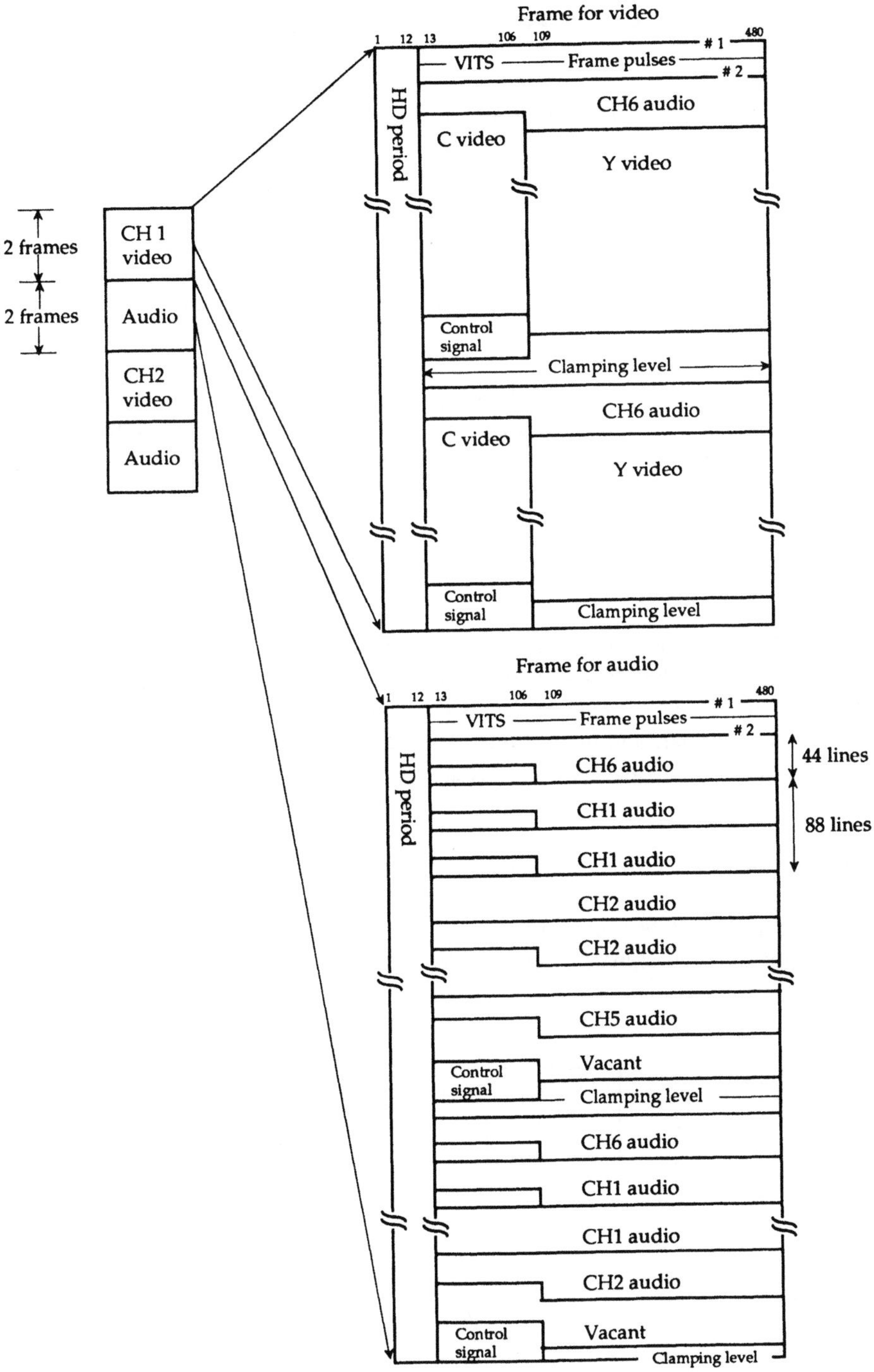

FIGURE 4.37. Signal format for still image broadcasting.

The data channel makes the following types of broadcasting possible:

1. Fascimile broadcasting: Detailed information can be broadcast by printing the information on paper. The information being broadcast may supplement a Hi-Vision program or be independent information.
2. Teletex: The character display is sharper than that of conventional text broadcasting, and the information displayed is accessible at a glance.
3. Telemusic broadcasting: This broadcasting transmits data for controlling musical instruments (such as pitch, duration, timing) such as an automated piano or synthesizer in the home. By combining a musical performance with Hi-Vision images, programs can become lifelike.

To receive these data broadcast programs, a decoder is connected to the Hi-Vision receiver to separate the data signals of each service, distinguish the desired signal, and input it into the personal computer, musical instrument, or other equipment.

(4) Interface with VCR, Video Disk and CATV

A MUSE receiver must be able to interface with and receive MUSE signals from CATV and packaged media such as VCRs and video disks.

Because Hi-Visions's wideband signals have five times the data of conventional broadcasting, a home VCR or video disk player would only be able to make short recordings. For longer recordings of Hi-Vision programs on a VCR or video disk, band compression becomes necessary. The best choice for compression technology is MUSE, because it would then be compatible with Hi-Vision satellite broadcasting. This band compression method compresses the Hi-Vision signal to about 8 MHz, making possible VCR recordings of three to four hours and video disk recordings of about 60 minutes.

(5) Still-Image CD

Hi-Vision still-image CDs record still images and two channels of audio data on a 12cm compact disk (recording capacity: 540 Mbytes). The physical format and error correction method are exactly the same as for CD-ROM. Besides still image programs, the CD can be used as an image file of pictures and photographs that can be searched randomly. Figure 4.38 shows a block diagram of a Hi-Vision still image CD system. Table 4.2 shows the signal formats. The video signals are digitally recorded after compressing the MUSE signal a second time with DPCM to one-half size.

The audio signal is the same as for the MUSE

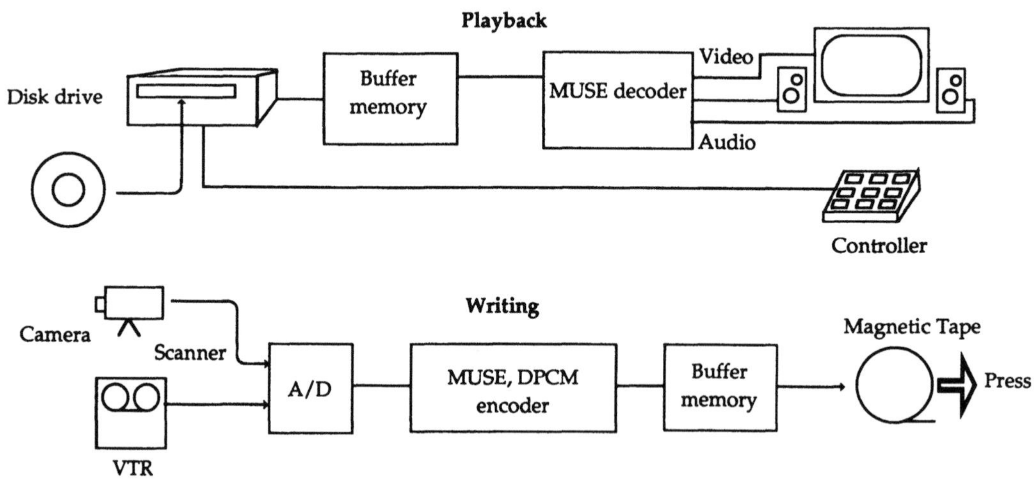

FIGURE 4.38. Block diagram of Hi-Vision still-image disk system.

TABLE 4.2. **Signal format for still-image CD recording.**

	Capacity	Transfer time	Signal format
Video	640 images	About 4.5 seconds	Digitally recorded after compressing MUSE signal with DPCM to one-half size.
Audio	60 minutes Two channels	---	Near instantaneous compression and expansion DPCM 32 kHz sampling (range bits: 3 bits, differential data bits: 8 bits)

A-mode audio signal, and is recorded at a sampling rate of 32 kHz using near instantaneous compression and expansion with DPCM. The CD can record 640 still images and 60 minutes of two-channel audio. The video signal transfer takes about 4.5 seconds because a CD-ROM disk drive for which the data transfer rate from the disk drive to the buffer memory is only 150 Kbytes/second, is used.

In terms of printed matter, a 12cm × 12mm compact disk stores as many pictures as does a 5cm thick picture album, and the compact dish also stores a recorded narration. In addition, while it would take time to search through a picture album for a specific picture, a CD can find it very easily.

(6) Interface with a Personal Computer

Recently, even conventional television receivers have been used not only for broadcast reception but as image data terminals. This trend will increase greatly in the Hi-Vision era.

For home television viewers to be able to use the multifunction Hi-Vision receiver as a comprehensive reception terminal, the Hi-Vision receiver needs to have control functions corresponding to the various application modes not only for the internal units of the receiver, but

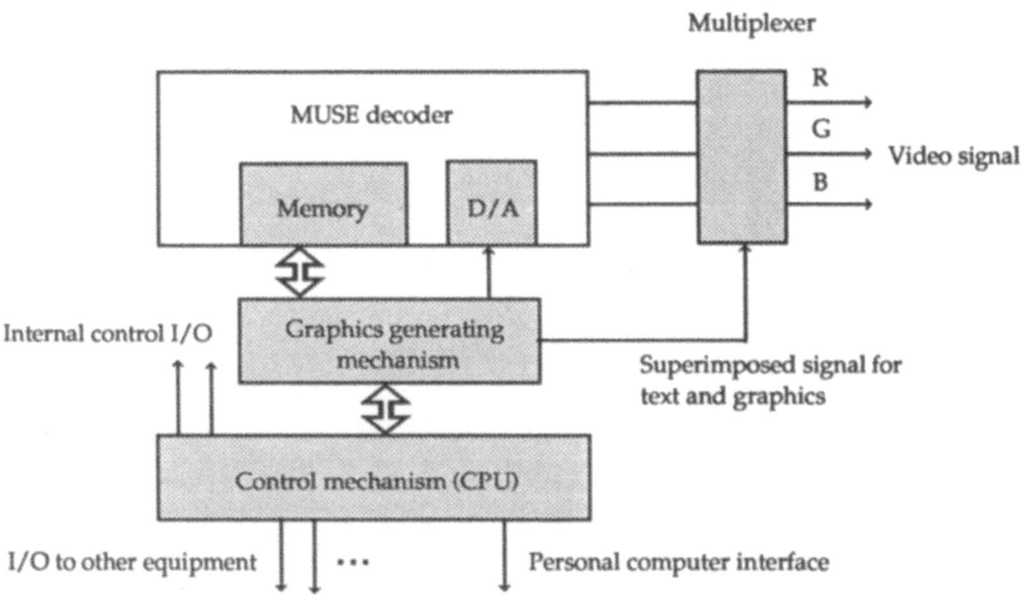

FIGURE 4.39. Block diagram of a personal computer interface.

also for external equipment and the internal operations of display terminals.

With regard to maps and encyclopedias recorded on CDs, to equip the Hi-Vision receiver with the ability to search and display data quickly and to generate graphics, interfacing with the capabilities of a personal computer is crucial.

Figure 4.39 shows a block diagram of an interface between a MUSE decoder and a personal computer.

REFERENCES

1. Taguchi et al., "High Definition Television Equipment: 5. Displays," *Journal of the Institute of Television Engineers of Japan (ITEJ),* Vol. 39, No. 8, pp. 685–689 (1985).
2. Sugimoto et al., "40-Inch High Definition Color Display," *ITEJ Technical Report,* ED 821 (1984).
3. Mitsuda et al., "Color CRTs for a 40-inch High Definition Television," *National Technical Report,* Vol. 33, No. 2, (1987).
4. Kanazawa et al., "Large Screen Projector for the High Definition Television," *ITEJ Proceedings,* 11–4 (1984).
5. Kanazawa et al., "Automatic Convergence Correction Device for Projection Displays," *ITEJ Proceedings,* 12–48 (1987).
6. Kanazawa et al., "A 50-Inch Hi-Vision Rear Projection Display with an Automatic Convergence Correcting Device," *ITEJ Proceedings,* (1988).
7. True, "High Performance Video Projector Using Two Oil Film Light Valves," *1987 SID* (1987).
8. "Image Display," *Image Electronics Lecture 3,* Korona Publishers, pp. 167–187 (1975).
9. Knauer et al., *1987 SID International Symposium, Dig. Tech. Papers,* pp. 84–87 (1987).
10. Yokozawa et al., *ITEJ Technical Report,* Vol. 8, No. 30, ED 824, pp. 13–18 (1984).
11. Murakami, Kuriyama, et al., *ITEJ Technical Report,* Vol. 12, No. 15, ED 88-11, pp. 37–42 (1988).
12. Mikoshiba and Shinada, *1984 SID International Symposium, Dig. Tech. Papers,* pp. 91–94 (1984).
13. Watanabe et al., *1985 SID International Symposium, Dig. Tech. Papers,* pp. 185–186 (1985).
14. Ninomiya et al., "MUSE, A Hi-Vision Satellite Transmission System," *ITEJ Journal,* Vol. 42, No. 5, pp. 60–69.
15. "Standardization of the Serial Bus for a MUSE Hi-Vision Monitor," *NHK Technical Report,* Vol. 32, No. 2, pp. 44–46.

5

RECORDING TECHNOLOGY

Hiromichi Shibaya, Tatsuo Nomura, Hotaka Minakuchi

5.1 ANALOG VTR

Since the Hi-Vision signal bandwidth is five to six times wider compared to the NTSC format, the development of a videotape recorder that can record a signal of this bandwidth without degradation required that we reexamine all components, including the tape head, tape drive mechanism, and signal processing. An analog (FM recording) VTR was first developed to a practical level, and a digital VTR is under development. The analog VTR will be discussed in this section, and the following section will discuss the digital VTR. This section will also discuss the high density wideband recording technology common to both VTRs.

5.1.1 Recording Wideband Video Signals

At first, the analog VTR was developed with the aim of achieving image reproduction that fulfilled the provisional standard of the NHK Hi-Vision transmission signal (see Table 5.1). To record and replay this wideband signal, it is necessary either to expand the usable bandwidth of the tape head, or divide the signal for multichannel recording. Table 5.2 compares these two alternatives.

Let us consider how wide a signal bandwidth can be recorded. The relationship of the frequency, f, which determines the upper limit of the recorded signal, to v, the head-to-tape speed, and the recording wavelength λ is shown in the following equation.

$$f = v/\lambda \tag{5.1}$$

Recording a wideband signal thus becomes an issue of high speed recording and short wavelength recording. Figure 5.1 shows one solution to these issues. The most basic issue here is to increase the coercivity of the magnetic layer of the tape and decrease the recording wavelength. Historically, coercivity has been increased with the development of oxide tape, high coercivity oxide particle tape, and then metal particle tape. Various means have also been devised to enable the core of the recording head to generate a sufficient magnetic field to record on these tapes.

A Hi-Vision VTR must be designed by taking onto consideration a wide variety of issues, including the development of wideband tape heads as described above, signal processing for multichannel recording, and mechanical issues associated with high speeds and narrow tracks such as stable contact between the tape head and tape, and precise tracking.

TABLE 5.1. Provisional standard for Hi-Vision transmission signal.

Number of scanning lines	1125	
Screen aspect ratio	5 : 3	
Fields per second	60 (2:1 interlacing)	
Video signal bandwidth	Y	C
	20 MHz	C_W: 7 MHz
		C_N: 5.5 MHz
Required SN ratio (S/N) (with waiting)	53 dB	
Waiting coefficient (W) *	Y	C
	13.4 dB	9.5 dB
Optimal viewing distance	3H **	

* With respect to triangular noise. ** H: Screen height.

5.1.2 Configuration of a Hi-Vision VTR

Figure 5.2 (a) is a block diagram of a VTR showing the signal inputs and outputs. The figure shows the distribution of functions for analog recording along the top of the units, and for digital recording along the bottom. Figure 5.2 (b) shows the many issues and design parameters raised in developing each of the units. The target values for the items in the left column were set first. This determined the selection of the tape head, and the signal parameters were set to fully exploit the performance characteristics of the tape head. Several VTR prototypes were built using this procedure.

The goal for the image quality of the playback was to satisfy the provisional standard in Table 5.1. We used (Co)γ-Fe_2O_3 oxide particle tape, which at the time was considered a high coercivity tape. We used a modified a 1-inch open reel tape mechanism, which permits high speed tape movement relative to the tape head. The VTR's mechanical parameters are shown in Table 5.3.

In addition to the tape and mechanical mea-

TABLE 5.2. Magnetic recording methods for wideband signals.

	Multichannel low speed recording	High speed recording with relatively few channels
Number of recorded channels that need to be replayed	Many	Few
Head rotation speed	Slow	Fast
Bandwidth per channel	Narrow	Wide
Mechanical issues	Configuration of a multichannel head	Countering centrifugal force; tape head contact
Video head	Narrow track; narrow gap	Wideband
Signal processing	Complex	Simple

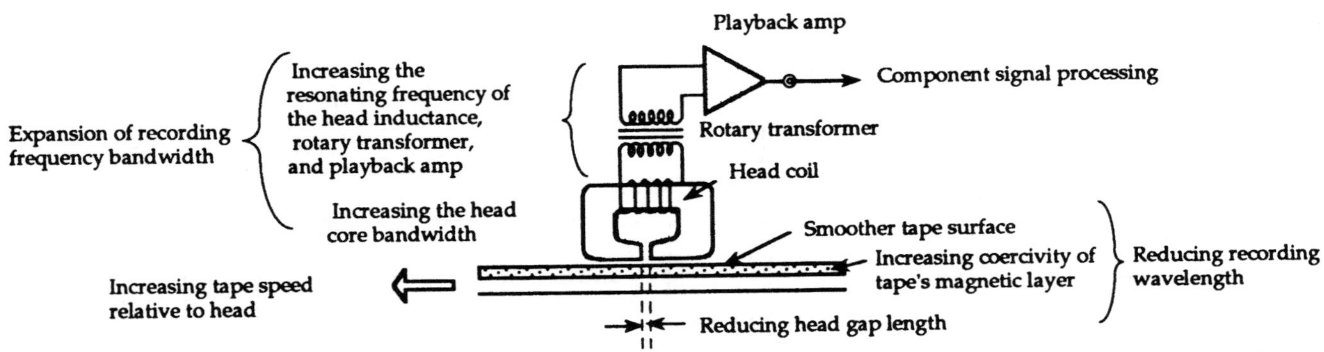

FIGURE 5.1. Wideband magnetic recording technology.

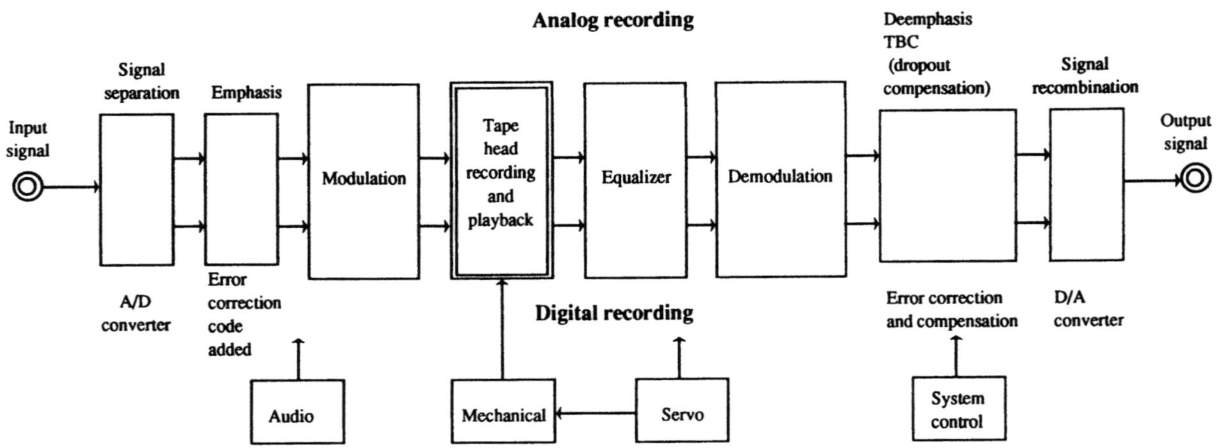

(a) Block diagram of functional units for analog (top) and digital (bottom) recording

FIGURE 5.2. Block diagram of functional units for analog (top) and digital (bottom) recording.

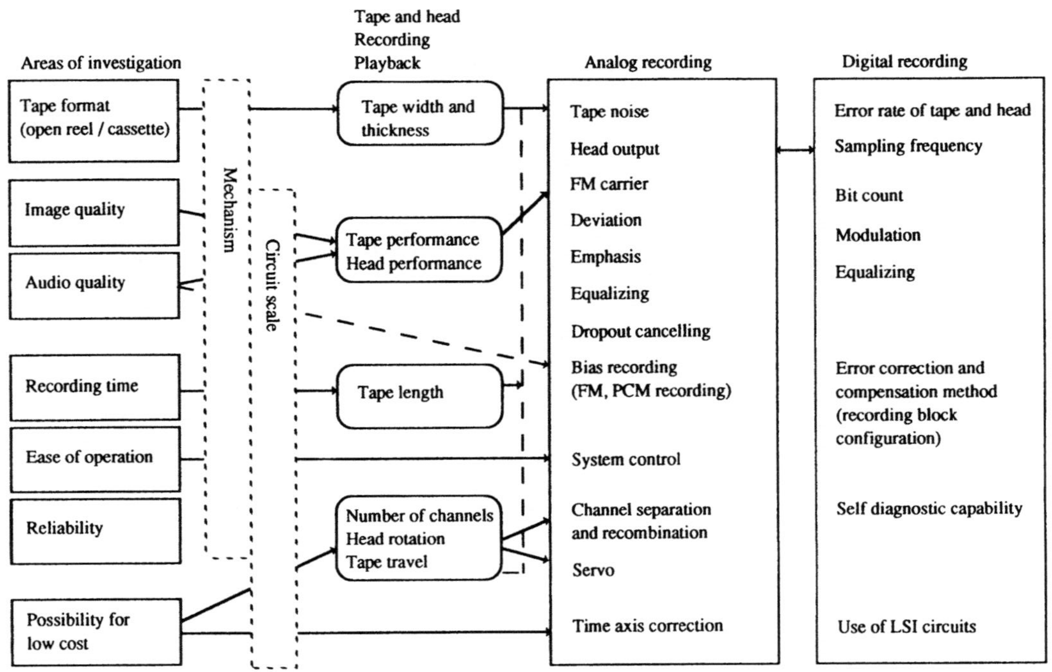

(b) Development issues

FIGURE 5.2. (continued)

TABLE 5.3. Prototype VTR mechanism.

Model	Mechanism	Drum rotation speed	Relative speed	Recording signal
I	1-inch Type-C format	60 rps	26 m/s	b
II	Same as above	120 rps	52 m/s	a, c
III	1-inch Type-B format	380 rps	60 m/s	d

sures described above, by treating the recording signal as a component signal, we eliminated the degradation in the color subcarrier's SN ratio caused by triangular noise during FM demodulation, and used a signal dividing method that conforms to the bandwidth of the tape head. Figure 5.3 shows the signal dividing method for the component recording signal used in the prototype VTR.

5.1.3 Design of a Hi-Vision VTR

Following is a discussion of VTR design for component signal FM recording.

(1) Determining FM Allocation of Component Signals[3]

Previously, the influence of the color subcarrier was the largest factor in determining the frequency allocation for FM recording (such as carrier frequency and deviation). Component signal recording, which has no subcarrier, was designed as described below. Figure 5.4 shows the Y spectral distribution in Hi-Vision composite and component signals.

As the figure indicates, since the component signal has a small amplitude in the high frequency region, even if the FM carrier frequency is set rather low, the unnecessary wave constituents (amount of moire)[4] entering the demodulated video band are minimal.

With regard to the commonly used frequency doubling FM demodulator in Figure 5.5, the amplitude of the frequency doubling FM signal (the lower sideband component U of carrier 4) that is mixed into band B of the modulated signal 5, in other words the amount of moire, can be expressed as a ratio of D to U:

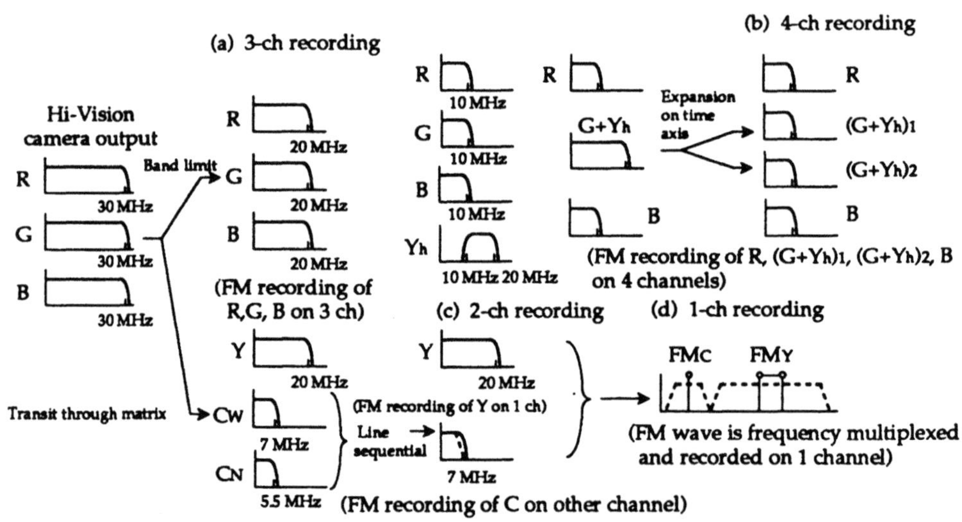

FIGURE 5.3. Recording signals for recording on one to four channels.

$$\frac{D}{U} = \frac{\beta \cdot B}{J_\gamma(2\beta)\,\{2f_c - \gamma B\}} \quad (5.2)$$

where β: Modulation index (FM deviation $\Delta f = \beta \cdot B$)

γ: Order of sideband causing moire interference

$J_\gamma(2\beta)$: First Bessel function of the γ order

f_c: FM carrier frequency

In order to make this DU ratio equal to 40 dB, if the relationship between f_c and B is

$$f_c \geq \left(\frac{3}{2}\right) B \quad (5.3)$$

then the relationship between B and the maximum deviation $2\Delta f_{max}$ will be as shown in Figure 5.6. For various B values, the maximum deviations within the figure's broken line will be allowed. These calculations are based on a f_c of 30 MHz and video emphasis of 8 dB.[3]

Figure 5.7 compares the high band FM allocation used in conventional composite signals having a color subcarrier (a) and in component signal recording (b). This shows that the carrier frequency can be reduced below the previous level while still maintaining a sufficiently large deviation. Since the color subcarrier frequency f_{sc} is close to the value of B, the tape head recording and playback bandwidth for component recording can be conserved. However, even in this case the maximum frequency f_{max} of the FM signal reaches three times that of B. If the f_{max} signal's recording and playback are difficult even if Methods 1 and 3 from Figure 5.1 are used, then it is necessary to reduce f_{max} either by increasing the relative speed as in Method 2, or dividing the signal as in Method 4.

(2) Head-to-Tape Speed and the Recording/Playback Band

The recording and playback band can be expanded by increasing the head playback output for high frequencies. The head playback output e_h for shortwave, long duration recording is defined in the following equation.

$$e_h = (8/\pi) \cdot \eta\, N\, v\, W\, B_r\, L_g\, L_{id}\, L_{sr}\, L_{sp} \quad (5.4)$$

where η = Head playback efficiency

N = Number of turns in head coil

v = Head-to-tape speed

W = Track width

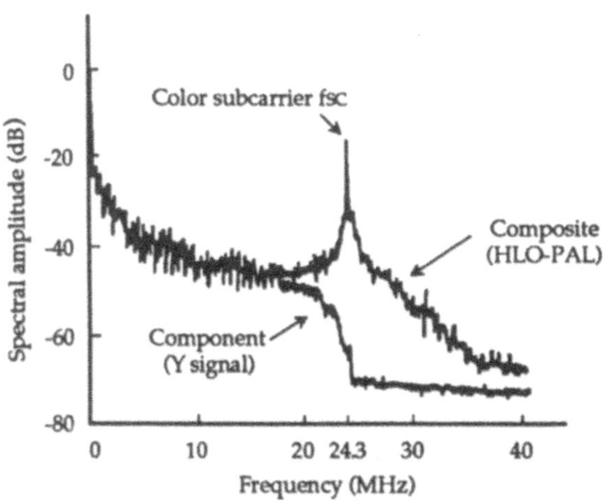

FIGURE 5.4. Spectral distribution of Hi-Vision composite and component signals.

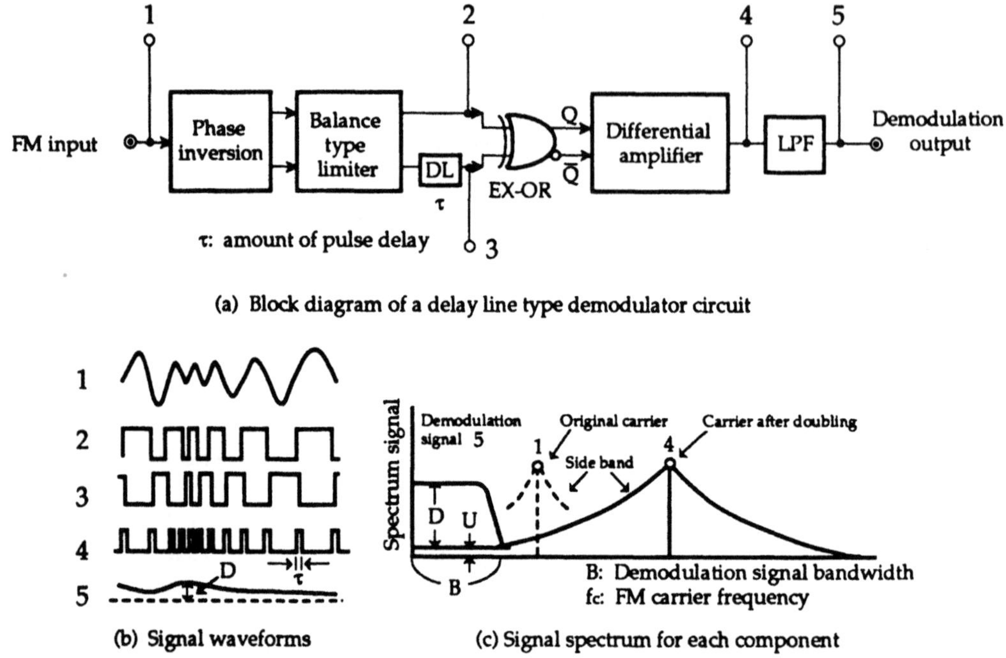

FIGURE 5.5. Operation of a delay line type frequency doubling demodulator.

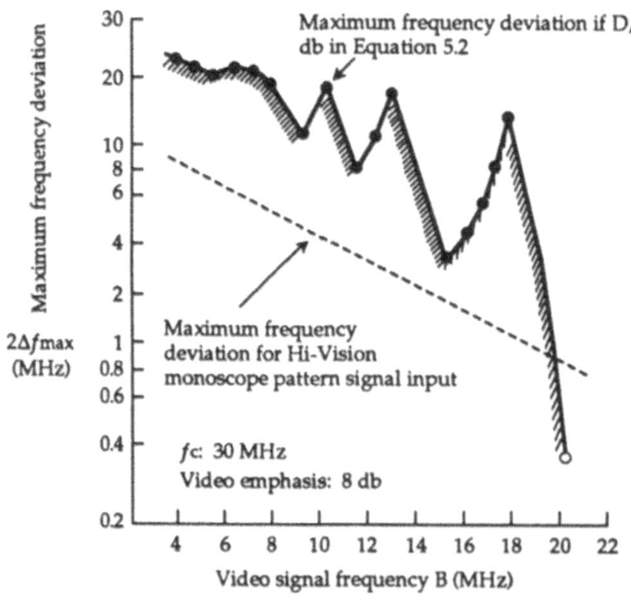

FIGURE 5.6. Video signal bandwidth and maximum frequency deviation.

B_r = Residual magnetic flux density of tape

L_g = Signal reduction caused by loss at head gap

L_{id} = Self-demagnetization loss

L_{sr} = Space loss during recording

L_{sp} = Space loss during playback

Thus to record a high frequency signal, that is, a signal with a short wavelength, Equation 5.4 suggests alternatives such as reducing the various losses while increasing the head-to-tape speed, or increasing the residual magnetic flux density B_r. Increasing the coercivity of the tape's magnetic layer helps to maintain a large B_r and decrease L_{id}, the reduction due to self-demagnetization loss.

The effect of head-to-tape speed and tape characteristics on the recording and playback bandwidth is depicted in Figure 5.8. However, to be able to compare measurements made under

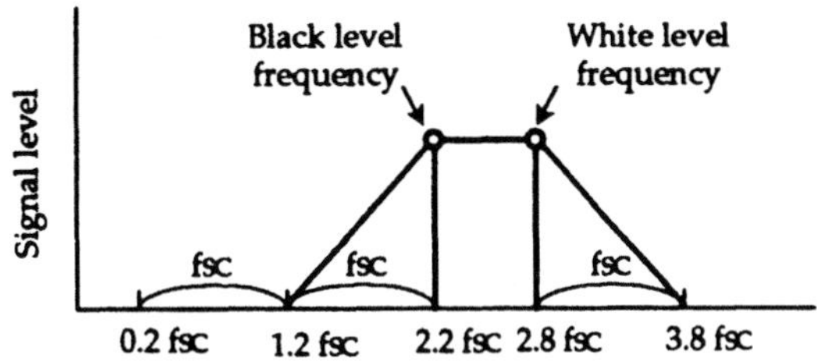

(a) High band for recording NTSC signal

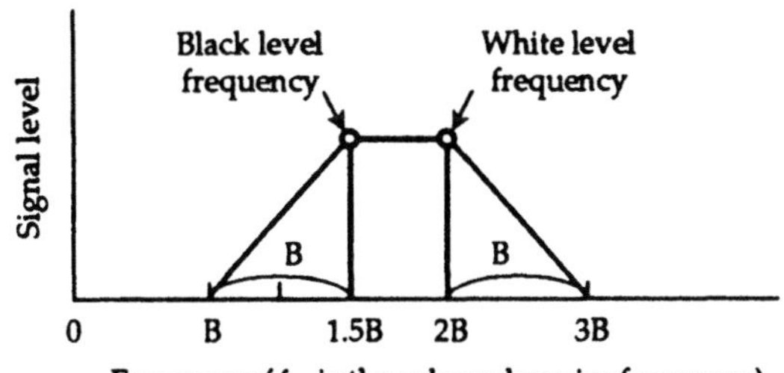

(b) For component signal recording

FIGURE 5.7. FM allocation for composite signal and component signal recording.

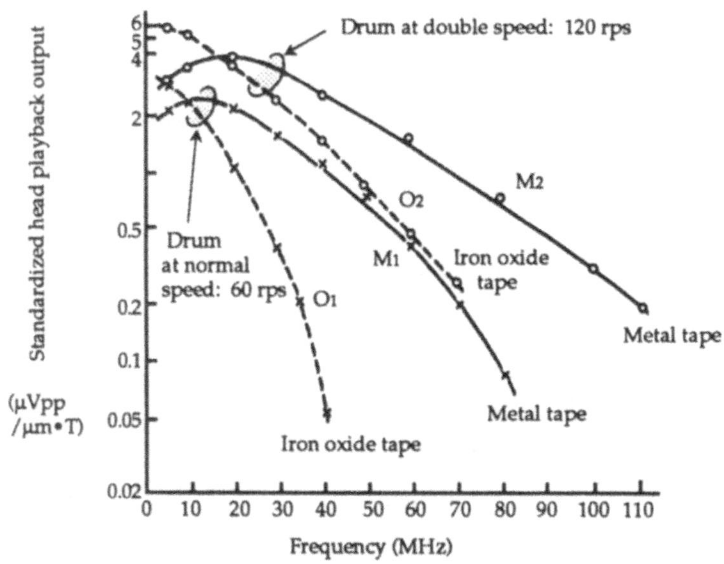

FIGURE 5.8. Relationship between recording and playback signal bandwidth, and relative speed and tape type (measured with 1-inch C format mechanism.

TABLE 5.4. Specifications of 1-inch tapes used in experiment.

Category	Name of tape	Specifications
Metal oxide magnetic particle coating	(Co) γ - Fe_2O_3 tape	H_c: 650 Oersted B_r: 1200 Gauss Rectangular ratio: 0.8 Coat thickness: 5.5μm Base thickness: 20μm Total thickness: 27μm.
Alloy magnetic particle coating	Metal particle tape	Hc: 1500 Oersted B_r: 2500 Gauss B_r / B_s: 0.8 Coat thickness: 4μm Base thickness: 12μm Total thickness: 17μm

H_c: coercivity, B_r: residual magnetic flux density, B_s: saturation flux density.

TABLE 5.5. Specifications of video heads used in experiment.

Name of head	Core material	Specifications	Application
Ferrite head	Mn-Zn * Ferrite single crystal block	W: 40~500μm g: 0.35~0.7 μm N: 4~13 turns B_s: 4700 Gauss H_c: 0.02 Oersted	Metal oxide tape
Metal head	Fe-Si-Al ** Sputtered film laminated structure	W: 5 ~ 60 μm g: 0.15 ~ 0.35 μm N: 12 ~ 40 turns Bs: 11,200 Gauss H_c: 0.12 Oersted	Metal particle tape

* Manganese • Zinc
** Sendust

W: Track width g: Gap length N: Number of turns in coil
B_s: Saturation magnetic flux density H_c: Coercivity

TABLE 5.6. Design implemented for Tsukuba Expo Hi-Vision VTR (extract).

(a) Rating

Video input signal	Y, C_W, C_N (30 MHz bandwidth for each)
Video output signal	Y, C_W, C_N (20, 7 and 7 MHz bandwidths)
Audio signal	3 I/O systems
Recording duration	At least 45 minutes (with 10.5-inch reel)
External interface	For editing and parallel operation
Video signal recording mechanism	
Rotating head mechanism	Similar to 1-inch Type-C format VTR
Number of rotating heads	4 video, 1 erasing
Diameter of head drum	134.6 mm
Rotation speed of head drum	60 rps
Tape format	
Tape width	25.35 mm
Tape forwarding speed	483.1 mm/s
Tape speed relative to head	25.9 m/s
Audio signal track width	Audio 1 and 2: 1.1 mm; audio 3: 0.45 mm
Control signal track width	0.45 mm
Video recording signal	
Channel division method	
Y	2-channel FM
C	C_W, C_N each has direct FM

(b) Performance

Video output signal	
Frequencies	Y: DC ~20 MHz 0~ -3 dB, descends above 20 MHz
	C_W: DC ~7 MHz 0~ -3 dB, descends above 7 MHz
	C_N: DC ~7 MHz 0~ -3 dB, descends above 7 MHz
SN ratio	At least 41 dB (p-p/rms)
Pulse characteristic (2T)	Less than 1
Tilt (horizontal, vertical)	Less than 3%
Linearity in low frequency region	Less than 3%
Moire	Less than −40 dB
Residual timing jitter	Less than 3ns
Time difference between Y and C signals	Less than 3ns

TABLE 5.7. Video recording and playback parameters.

FM allocation for black level ~ white level	16.1 MHz ~ 20.2 MHz	
Emphasis amount	9 dB	
Center frequency	1.4 MHz	
Required recording and playback bandwidth	6 MHz ~ 30 MHz	
Video head	Recording	Playback
Core material	Mn-Zn ferrite	Mn-Zn Ferrite
Track width	80μm	70μm
Gap length	0.7μm	0.35μm
Track pitch	360 μm for R, $(G+Y_h)_1$, $(G+Y_h)_2$, and B combined	

different conditions, the head output on the vertical axis was standardized for track width and number of coils in the head coil. Characteristic O_1 in the figure belongs to Model No. 1 from Table 5.3, and O_2 from Model No. II. Table 5.4 and Table 5.5 list the parameters of the tapes and heads used in these measurements.[5]

5.1.4 VTR Specifications for the Tsukuba Science Exposition

Based on the improvements discussed above in the signal and tape head characteristics, we proved the possibility of a prototype level Hi-Vision VTR that satisfied the standards in Table 5.1

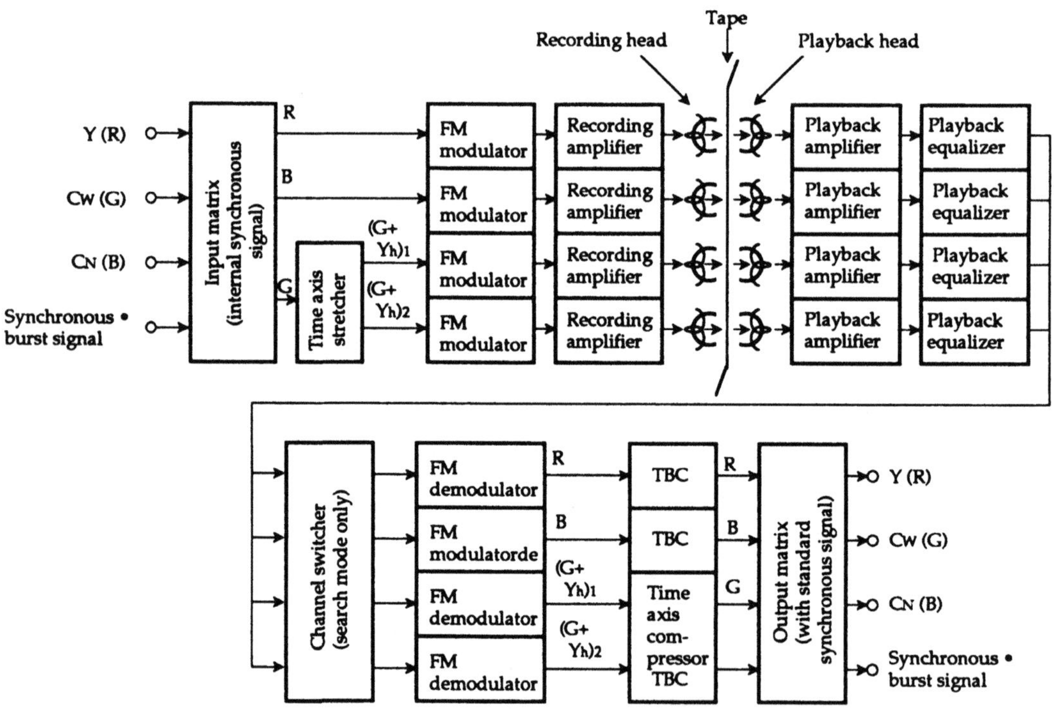

FIGURE 5.9. Video signal processing used in Tsukuba Expo Hi-Vision VTR.

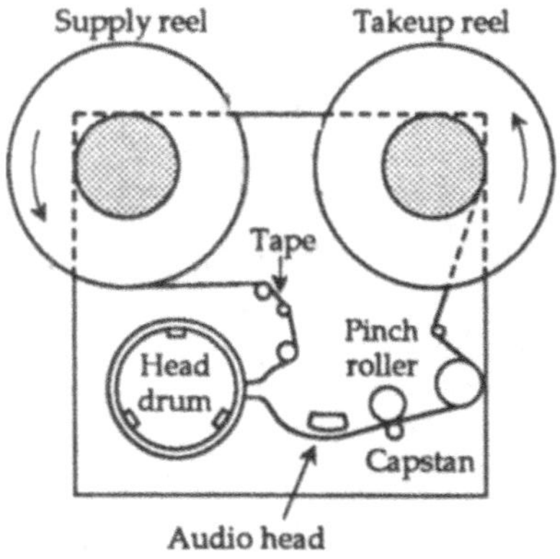

FIGURE 5.10. Tape drive mechanism of Tsukuba Hi-Vision VTR.

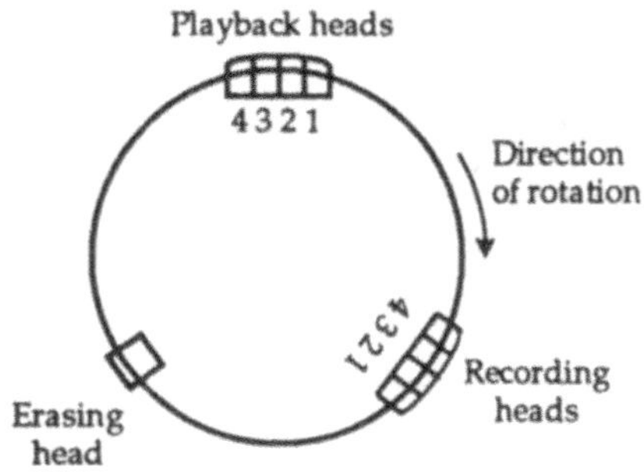

FIGURE 5.11. Positioning of heads on drum of Tsukuba Hi-Vision VTR.

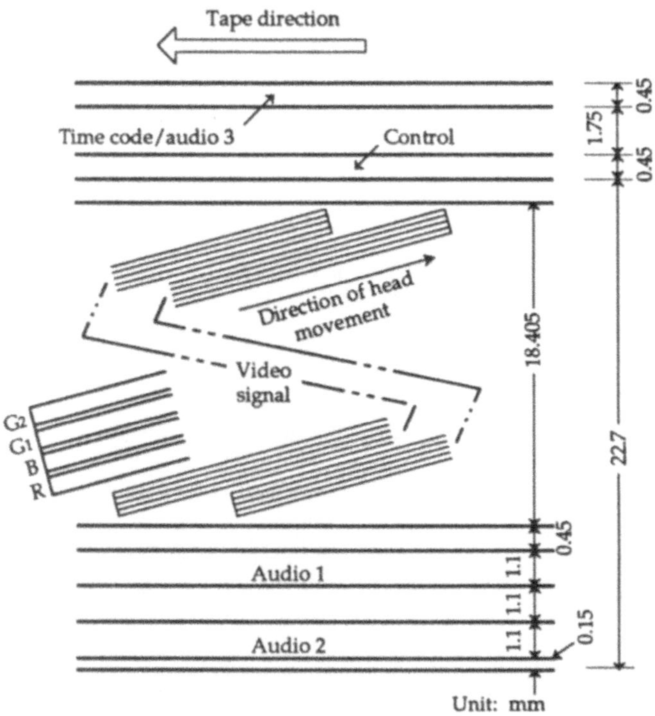

FIGURE 5.12. Recording pattern on tape of Tsukuba Hi-Vision VTR.

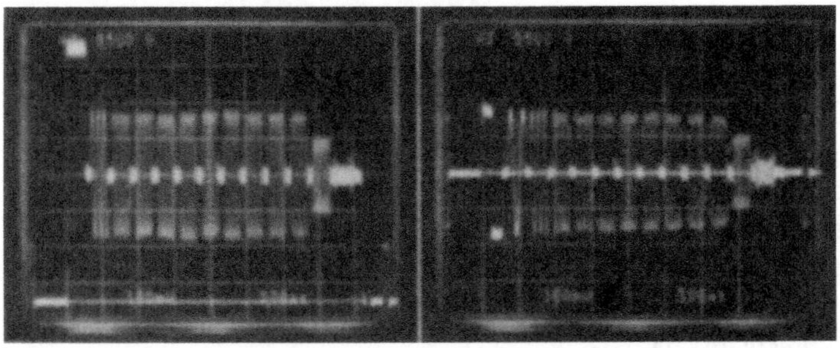

FIGURE 5.13. Playback signal frequency characteristics.

for demodulated signal bandwidth and SN ratio. Furthermore, we were able to record for 48 minutes on a regular 10-5-inch reel tape, with a tape consumption of only twice that of an NTSC VTR.

When the decision was made to set up a Hi-Vision system at the Tsukuba International Science and Technology Exposition in 1985 (referred to as Tsukuba Expo below), NHK used this opportunity to unify a Hi-Vision VTR tape format as part of the Tsukuba Expo specifications and completed development of the first generation of equipment for practical use. The VTR specifications are described in Table 5.6.

As already described in Figure 5.3(b), the recording signal for the VTR built to these specifications takes the three input signals (Y, C_W, and C_N or R,G, and B) and constructs a 4-channel signal consisting of R, $(G + Y_h)_1$, $(G + Y_h)_2$, and B, each having a 10 MHz bandwidth. This signal undergoes FM modulation as shown in Table 5.7 and is recorded on the tape in parallel with four compactly arranged heads.

To be able to magnetize the high coercivity tape adequately with a ferrite head, the dedicated recording heads differ from the playback heads in having a wide gap length. Since both the track width and recording heads are wider, there is more leeway in tracking during playback. Furthermore, the first stage amp of the playback heads are built into the rotating drum and the playback resonating frequency has been increased.

Since a Type-C format mechanism is used with a drum rotation speed of 60 rps, the head blanking interval, in which the head is not in contact with the tape, matches the vertical blanking interval. Thus a sync head for vertical sync recording is not needed.

With regard to the Tsukuba Expo VTR, the video signal processing block diagram is in Figure 5.9, tape path in Figure 5.10, head arrangement on the head drum in Figure 5.11, and recording pattern on the tape in Figure 5.12. Figure 5.13 shows an example of the frequency characteristics of the Y and R and B playback signals.

While the VTR's operability is practically the same as the Type-C format NTSC VTR, the

FIGURE 5.14. Two Tsukuba Expo VTR machines.

absence of a dynamic tracking function prevents slow and still image playback. However, a channel switching function allows images to be captured for monitoring while fast forwarding or rewinding.

VTRs with Tsukuba Expo specifications are currently being manufactured by two companies using a compatible tape. Several dozen of these machines are already being used around the world not only for the production and exchange of Hi-Vision programs but for movie production as well. The VTRs from Sony and Toshiba are shown in Figure 5.14.

5.2 DIGITAL VTR

Open reel analog VTRs with 1-inch tape that have been developed and commercialized need further improvement in reducing the degradation in the playback image quality when programs are rerecorded during production and editing.

The digital VTR responds to this problem. Several prototypes have already been announced, and product development conforming the NHK's guidelines is currently under way.

In this section we will discuss the design parameters involved in constructing a digital VTR.

5.2.1 Required Bit Rate and the Tape Head System

The most important design parameter is sampling frequency. Sampling frequencies we have studied for the luminance signal (Y) are shown in Table 5.8. In the initial stage of digital VTR development, a sampling rate of 46 MHz was used that barely satisfied the tentative standard shown in Table 5.1 of Section 5.1.[6] However, in August 1987 the BTA (Broadcast Technology Association) decided on a studio standard of 74.25 MHz, which is the rate that has been used more recently.[7]

If the samples are quantized at 8 bits, the bit rates will have very high speeds as shown in Table 5.8. Further, the digital signal bandwidth will be about three times wider than that required for FM recording (3B, see Figure 5.7), which is a considerable increase over analog VTRs that makes the need for a shorter wavelength obvious.

These considerations favor a metal tape with high coercivity. By combining the use of this tape with a high head-to-tape speed, the M_2 wide band characteristics described in Figure 5.8 of Section 5.1 can be obtained. Even with M_2 characteristics, the maximum frequency that can be obtained with the required CN ratio of 25 to 30 dB (0-/rms) is 70 to 80 MHz, for a bit rate of 140 to 160 Mb/s per channel.

On the other hand, if the sampling frequency for the chrominance signals is 37.125 MHz and each sampling point signal is quantized at 8 bits, the total bit rate is

$$(74.25 + 2 \times 37.125) \times 8 = 1188 \text{ (Mb/s)} \quad (5.6)$$

This total is distributed over eight channels.

TABLE 5.8. Sampling frequencies that have been studied to date. (However, B for the Y signal indicates baseband bandwidth.)

Sampling frequency	Bit rate (8 bits /sample)	Digital signal bandwidth	FM signal bandwidth (3B)	Remarks
46.0 MHz	368 Mb/s	184 MHz	60 MHz	
54.0 MHz	432 Mb/s	216 MHz	66 MHz	CCIR Recommendation 601; Four times 13.5 MHz
64.8 MHz	518.4 Mb/s	259.2 MHz	75 MHz	MUSE processor frequency
74.25 MHz	594.0 Mb/s	297 MHz	90 MHz	BTA S-001 standard

5.2.2 Parallel Signal Processing

Since digital signals are amenable to serial-parallel conversion, the ultra high bit rates used in Hi-Vision digital VTR signals can be distributed over several low speed channels for recording and playback. Figure 5.15 is one example of this.[7] The Y, R-Y, and B-Y signals sampled at 74.25 MHz are divided into four parts on the screen, and one part (Part A) is time-expanded fourfold using memory. This time-expanded image is Image A in the figure. As a result, the sampling frequency is reduced to one-fourth, or 18.5625 MHz. At this stage, the Y, R-Y, and B-Y images still exist as three separate images. Next, to conserve the number of channels, the sampling points of both the R-Y and B-Y signals are thinned out to reduce the bandwidth by one-half, and recombined with the original Y signal using time division to produce the A′ and A″ images. The A′ and A″ images are each recorded and played back on a separate channel using a 2-channel tape head. Thus far we have described the processing for one-fourth of the original screen. For the full screen, at two channels each for the four parts, a total of eight channels is necessary.

The bit rate in this case is 148.5 Mb/s, and the bandwidth to be processed by one channel matches the tape head characteristic described in Section 5.1.3.

5.2.3 Coding Format

(1) Block Construction and Attached Data

The digital VTR recording signal is not a direct and continuous signal but rather has a block construction, and the signal is recorded onto tape in blocks. As Figure 5.16 shows, in addition to video data encoding, the block construction includes sync encoding, ID encoding, and vertical and horizontal error correcting code (ECC) as described in Table 5.9. These addition codes have allotted to them the horizontal and vertical blanking intervals of the input signal, and the total bit rate is not to exceed Equation 5.6. The additional codes need the space between the margin for head switching and audio encoding.

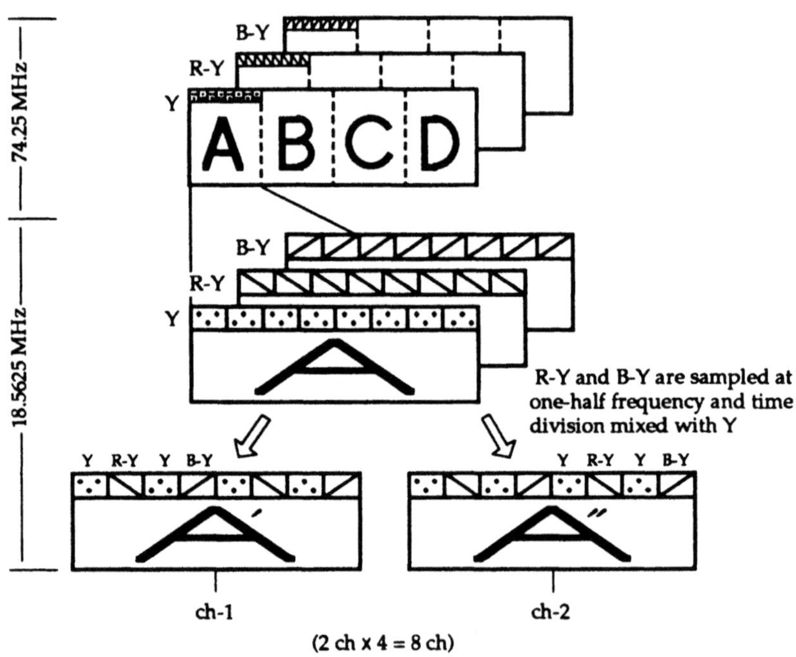

FIGURE 5.15. Parallel signal processing for prototype digital VTR.

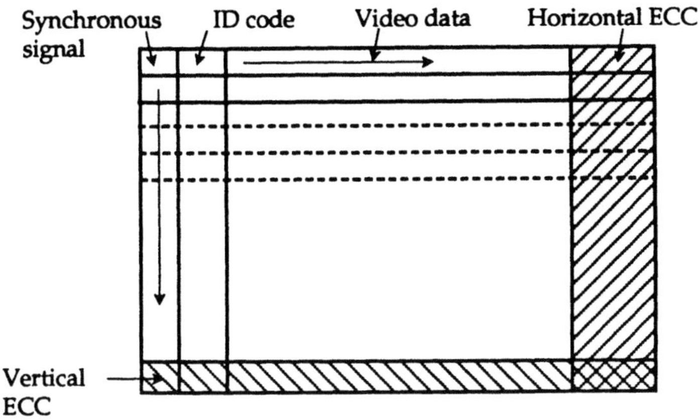

FIGURE 5.16. Block structure of digital VTR recording signal.

(2) Error (Coding Error) Correction and Concealment

(a) Error Correction and Tape Consumption. While a digital VTR uses a wide signal bandwidth as indicated in Table 5.8, tape consumption can be reduced to a level comparable to analog VTRs by implementing error correction and compensation technology and using frequency regions with a low CN ratio. Figure 5.17 shows how much tape consumption can be reduced in digital VTRs by introducing error correction.[8]

The figure shows that when a high SN ratio is needed in playback, a digital VTR will consume less tape than an analog VTR.

The reason for this is as follows. The SN ratio for FM recording on an analog VTR is

$$[S/N]_{FM} \propto A^{0.5\text{-}1} \tag{5.7}$$

where A is the tape area that is used. In comparison, the SN ratio for a digital VTR improves by 6 dB for each 1-bit increase in the bit count such that

$$[S/N]_{PCM} \propto 2^{A/A_O} \tag{5.8}$$

where A_0 is the tape area required to record a 1-bit signal.

(b) Burst Errors and Interleaving. Of the random and consecutive or burst errors that occur in encoding, the majority of errors in tape playback are burst errors, as shown in Figure 5.18.

A method was therefore devised to simplify burst error correction by distributing them. Called interleaving, this method rearranges the data so that burst errors approximate random errors, as shown in Figure 5.19.

TABLE 5.9. Additional codes and their bit rates.

Code	Application	Bit count
Sync code	Determines beginning of block	16~24
ID code	Identifies recording track, segment, field, and position on screen (for special playback features such as slow motion and search)	16~32
Horizontal ECC	ECC for horizontal row of block structure	
Vertical ECC	ECC for several vertical rows in block structure	

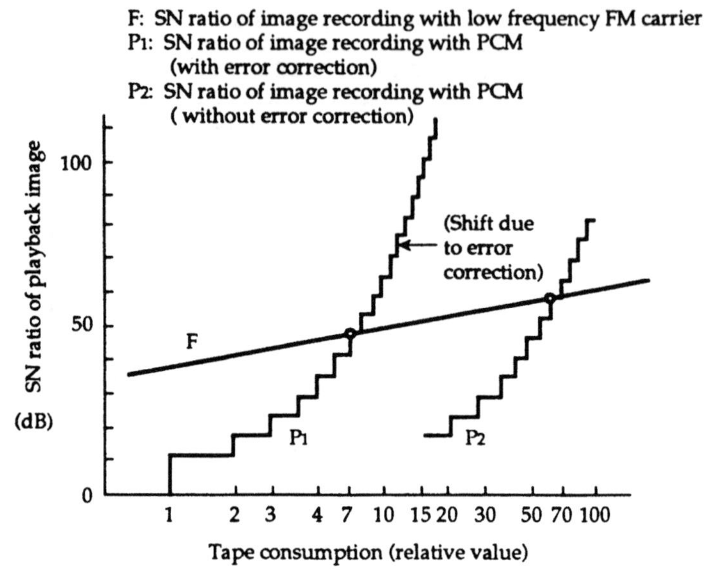

FIGURE 5.17. Decrease in digital VTR tape consumption resulting from error correction technology.

(c) Multiple Construction of Error Correction Code. This method increases error correction capability by composing correction coding in multiplex. Figure 5.20 shows the dual composition of internal and external codes. The first step is the correction of internal errors with internal code C_1. Then in the second step the external code C_2 corrects what was left out by the internal code. In the block structure depicted in Figure 5.16, C_1 corresponds to the horizontal code row and C_2 to the vertical row of code.

(d) Error Correction and Concealment. The error correction code for the Hi-Vision digital VTR prototype uses adjacent code[9] and Reed-Solomon product code.[9] Error concealment is done with an adaptive concealment method that

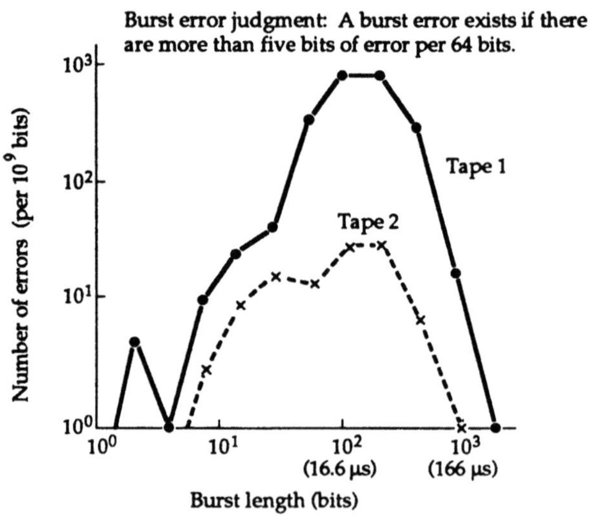

FIGURE 5.18. An example of error distribution.

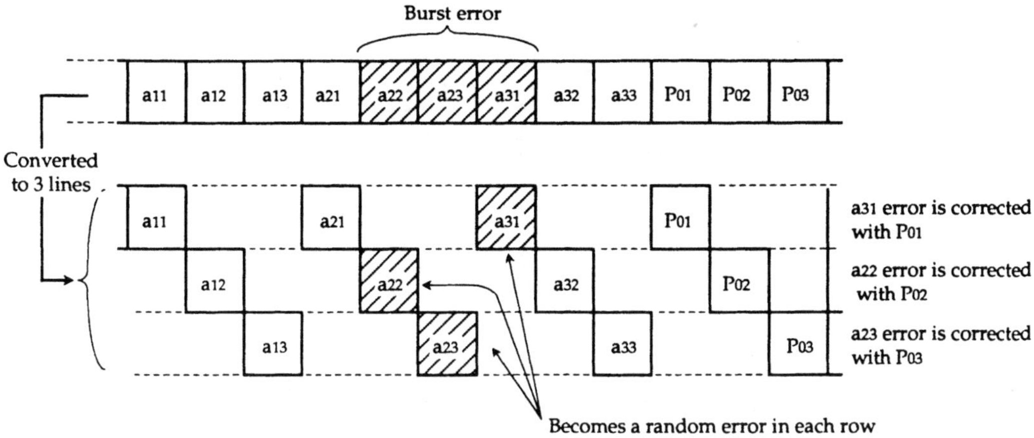

FIGURE 5.19. Interleaving; a-bits are data bits, and P-bits are parity bits.

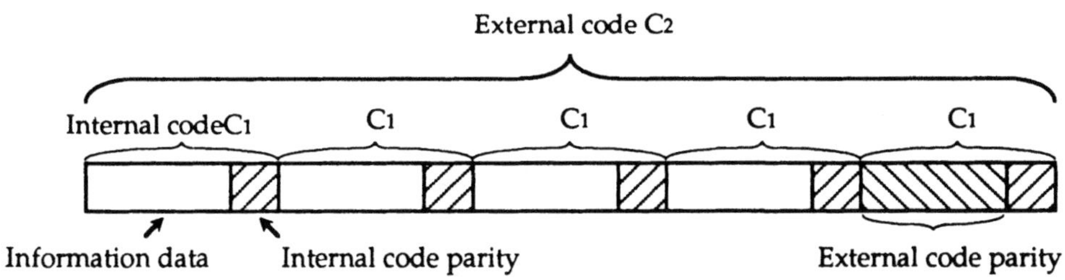

FIGURE 5.20. Composition of product code.

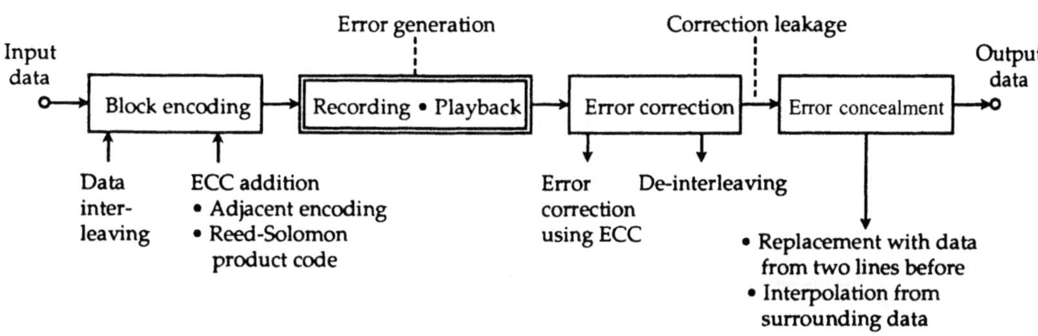

FIGURE 5.21. Error correction and concealment.

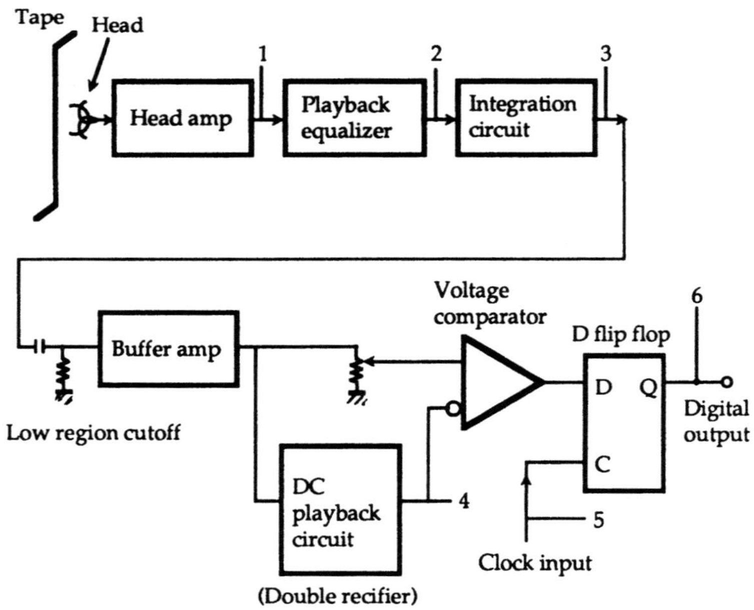

(a) Integration detection circuit

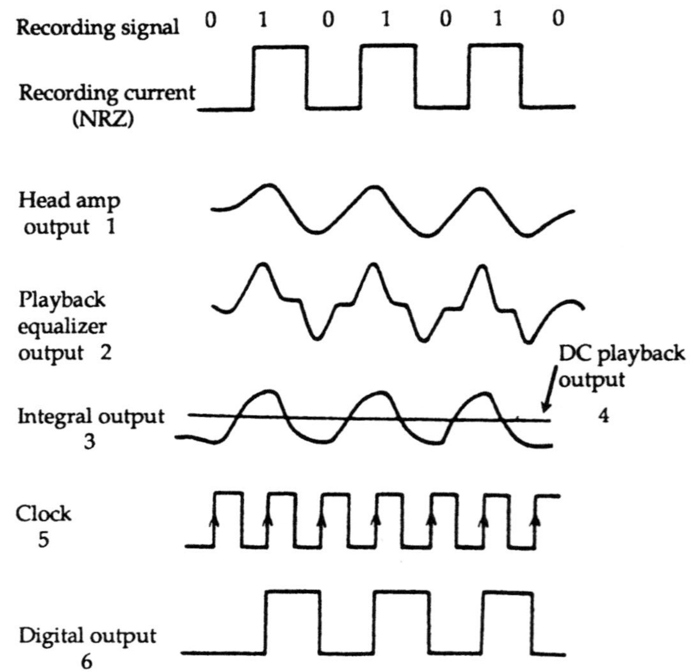

(b) Waveforms of integration detection circuit

FIGURE 5.22. Block diagram and waveforms of integration detection code.

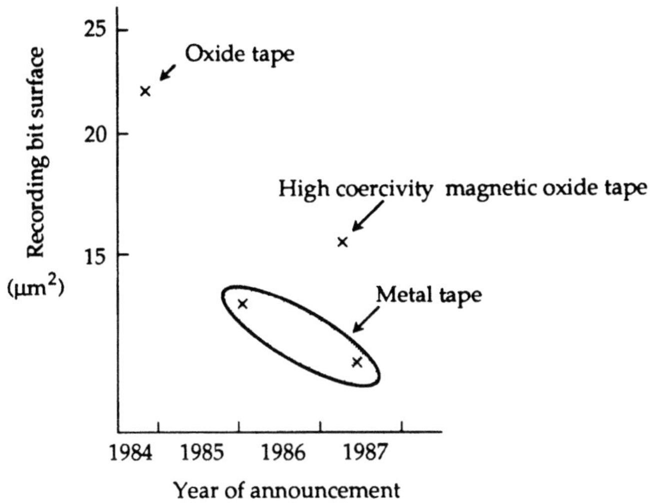

FIGURE 5.23. Trend in recording density of Hi-Vision digital VTR prototypes.

replaces data with data from two lines previous and interpolates from surrounding data.

The encoding process for the error correction and concealment system is shown from input to output in Figure 5.21.

(3) Modulation of Coded Signals

Code modulation refers to removing DC from the NRZ (non return to zero) encoding of output from an A/D converter, or matching the spectrum to the tape head characteristics so as to obtain a timing signal more easily.

The simplest modulation method, called scrambled NRZ modulation, involves multiplying with artificial random codes made from a fixed rule. Other methods used in digital VTRs include run length control, mapping (code replacement), partial response, and M^2 (mirror square).[8]

5.2.4 Playback Signal Waveform Equalization and Demodulation

After the tape output is amplified by the head amp, it is guided by the playback equalizer, and then combined with the demodulator to compose the decoder. The decoding method consists of oscillation detection, integrated detection, and

TABLE 5.10. Relationship between tape format, tape area, and recording time.

Tape format		Tape area	Recording time	Remarks
1-inch open reel (11.75-inch diameter)		78.5 m^2	64 minutes	Tape speed 805 mm/s
19mm cassette	D-1L	30.5 / 24.8 m^2	24.8 / 20.2 minutes	Cassette size: 366mm × 206mm × 33mm
	D-1M	13.6 / 11.1 m^2	11.1 / 9.0 minutes	Cassette size: 254mm × 150mm × 33mm
	D-1S	4.5 / 3.6 m^2	3.6 / 2.9 minutes	Cassette size: 172mm × 109mm × 33mm

Note: D-1 cassettes have two tape areas because they come in tape thicknesses of 13 μm and 16 μm.

TABLE 5.11. Hi-Vision digital VTR guidelines (framework proposed by NHK).

Video signal

Sampling frequency	Y	74.25 MHz
	P_B, P_R	37.125 MHz
Quantization		8 bits / sample
Effective scanning lines per frame		1035
Effective number of samples per line	Y	1920
	P_B, P_R	960
Number of users area lines per frame		At least 5 lines
Frequency characteristics	Y	0~27 MHz ± 0.5 dB ~30 MHz $^{+0\ dB}_{-1.5\ dB}$
	P_B, P_R	0~13.5 MHz ± 0.5 dB ~15 MHz $^{+0\ dB}_{-1.5\ dB}$
SN ratio	Y, P_B, P_R	At least 56 dB
Waveform characteristic	2T pulse Tilt Linearity	Less than 1 Less than 1% Less than 1%
Error correction code		Reed-Solomon product code

Audio signal

Sampling frequency		48 kHz
Quantization		At least 16 bits / sample Recording possible up to 20 bits
Number of channels	Digital	8 channels
	Analog	1 channel
	Time code	1 channel
Frequency characteristics	Digital	20 Hz ~ 20 kHz $^{+0.5\ dB}_{-1.0\ dB}$

Recording system

Mechanism	1-inch Type C format
Recording time	96 minutes with 14-inch reel 64 minutes with 11.75-inch reel
Tape	Metal particle coating H_c approx. 1,450 Oe B_r approx. 2,500 G

ternary detection. As an example, the circuit composition and waveforms of each component of the integrated detection method are shown in Figure 5.22.[10]

In this case, the output waveform of the integrator (3) is compared to the average DC level (4), and by latching to the rise of the separately reproduced clock signal (5), the NRZ digital waveform of the recording current can be restored.

With regard to Hi-Vision digital VTR prototypes, Figure 5.23 shows the year of introduction and recording density of various tapes. The figure shows that the recording density has been increasing annually. Further, it is clear that the recording density has increased with the introduction of oxide tape, high coercivity oxide tape, and metal particle tape. To record for 60 minutes with a recording area of 10 μm^2 per bit, it is necessary to use an open reel. A comparable recording time can be achieved with a cassette such as D-1M in Table 5.10 by using a combination of new high density recording tape (improved metal particle tape or deposition tape) and an appropriate bandwidth compression method, or if bandwidth compression is not used, a perpendicular magnetic recording method.

A Hi-Vision digital VTR that satisfies BTA standard S-0001 can thus clearly be achieved. Table 5.11 shows the framework of specifications and performance guidelines issued by NHK for this VTR. It is being developed as the second generation of VTRs for broadcast and production use, and will be used for Hi-Vision program production in the near future.

5.3 VTRs FOR INDUSTRIAL AND HOUSEHOLD USE

Hi-Vision VTRs for industrial and consumer markets will be instrumental in the diffusion of new video media, and are expected to have a wide range of applications outside of broadcasting. Industrial applications include areas such as video theaters, business, education, and medicine. VTRs will need to be operable by non-technicians and have a cassette format. Household VCRs will need to be even more user friendly, have a longer recording time, and lower cost. Further, they will need to be at least as versatile as currently available conventional VTRs.

5.3.1 Analog Recording

(1) Baseband and MUSE Recording

Since the recording of broadcast programs is expected to be a major use of Hi-Vision VCRs for households, VCRs should be capable of directly recording MUSE signals.[12] However, there are other important functions of VCRs besides viewing recorded television programs such as recording video camera output and viewing prerecorded videotapes, for which the direct recording and playback of baseband component signals is highly desirable. In industrial applications, baseband recording is essential.

(2) Image Quality

To obtain a high quality Hi-Vision image, the SN ratio must not fall below 40 dB. The desirable bandwidth for the baseband signal is around 20 MHz, and for MUSE signals it is essential that at minimum the 8.1 MHz bandwidth signal can be played back without degradation.

As evidenced by currently available VCRs in the consumer market such as S-VHS, ED Beta, and high band 8mm, one feature of analog baseband recording is that image quality can be enhanced by improving the head and tape characteristics, while at the same time maintaining compatibility.

(3) Chrominance Signal Processing

Chrominance signals require a minimum of one-third to one-fourth the bandwidth of the luminance signal, and in the case of Hi-Vision this ratio is one-fourth. A sufficient image quality can be obtained even if the two types of chrominance signals are incorporated line sequentially.

A widely used method for multiplexing the luminance and chrominance signals is TCI (time compressed integration). If the chrominance signals are one-third of the bandwidth, this method time-compresses the luminance signal

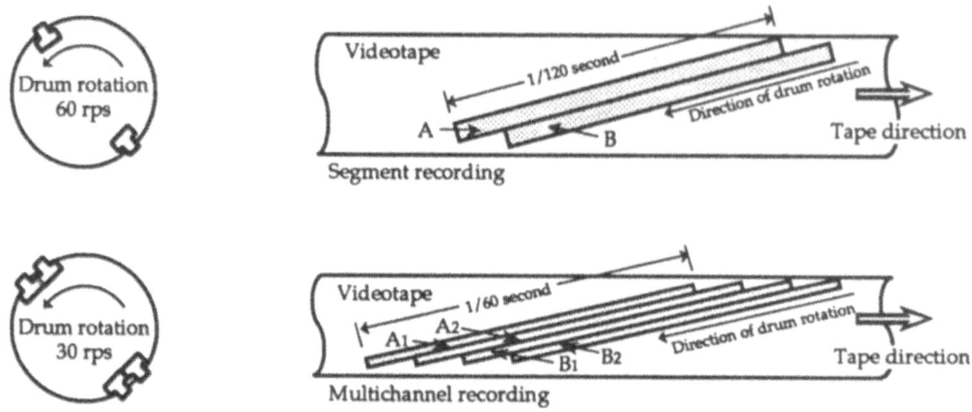

FIGURE 5.24. Comparison of segmented and multichannel recording.

to three-fourths and the chrominance signals to one-fourth in each line.

(4) Modulation Format for Recording

In magnetic tape recording, because band-pass characteristics become attenuated in the high and low frequency regions, analog VCRs in general use low carrier frequency FM recording.

The carrier frequency is set to avoid aliasing of the lower sideband and the intermixture of the demodulated moire signal, which is caused by high-order lower sidebands from the demodulator output. In concrete terms, the carrier frequency is set to be at least 1.5 times greater than the bandwidth of the demodulated image signal so that the second order sideband of the FM signal, whose frequency was doubled in the demodulator, is outside of the image signal band. Also, while magnetic recording has a nonlinear characteristic, a bias signal can be superimposed to support linearity over a relatively wide range. This suggests alternative methods such as AM recording, or using FM recording for low frequency regions and direct recording for high frequency regions.

5.3.2 Recording Wideband Signals

Attempting to record one field of the wideband Hi-Vision signal on a single track as is done with conventional VTRs would require a head drum several times larger in diameter and be quite impractical. Thus the signal for one field must be divided into several tracks. As illustrated in Figure 5.24, two-part division can be done either by segmented recording (dividing the screen), in which the drum rotational speed is increased, or using several heads and recording simultaneously on several channels.

(1) Segmented Recording

Segmented recording requires only one set of FM modulation and demodulation circuitry, heads, and rotary transformer for recording and playback. On the other hand, it requires signal processing to seamlessly record the signal for one field onto several sequentially recorded tracks, as well as the ability to shuffle signals in time for special playback features such as searching, slow motion, and still-image playback. We will discuss these types of signal processing in detail later.

While wideband circuitry becomes necessary, since the rotary transformer and playback head amp system resonate based on floating capacitance and inductance, the phase is greatly delayed in the high region. Thus to increase this resonating frequency, the amp is built into the rotary drum.

(2) Multichannel Recording

In multichannel recording, since each track scans for the duration of one field, special playback features are relatively easy to accommodate. However, this method requires one set of circuitry, head, and rotary transformer for each track. Not only is the scale of the circuitry increased, but differences in characteristics across channels becomes an important issue in analog recording. Thus a reference signal is recorded intermittently on each channel to compensate for the differences in characteristics across channels.

5.3.3 Example of an Analog VCR

(1) MUSE-VTR

The experimental MUSE-VCR developed at NHK Science and Technical Research Laboratories used a U-standard VCR mechanism and a cassette with 3/4-inch metal particle tape, and was able to achieve a recording time of one hour (Figure 5.25). Subsequently, the track pitch was narrowed and the recording time increased to three hours.

The MUSE signal bandwidth is almost three times wider than the 3 MHz signal of conventional household VCRs. Thus a Table 5.12 indicates, MUSE-VCRs introduced to date that use segment recording have a drum rotational speed that is two to four times as fast as conventional VCRs. Since the horizontal sync of the MUSE signal is a positive sync, separation is sometimes difficult in systems that have sudden time fluctuations such as VCRs. A common method of overcoming this in the recording signal is to time-compress the MUSE signal every horizontal cycle and insert a negative sync and burst Figure 5.26 shows an example of a block diagram for a MUSE-VCR.

(2) Baseband VCR

Since the bandwidth of the baseband signal is at least twice as wide as that of the MUSE signal, the recording method requires not only segmenting but multichannel recording as well. NHK Engineering Services has been developing a baseband VCR for industrial use jointly with nine electrical companies, and in December 1988 specifications for a cassette were established. Using a 205mm × 121mm × 25mm cassette with 1/2-inch metal particle tape, the recording time is more than one hour. The video bandwidth is 20 MHz for the luminance signal and 7 MHz line sequential for the color difference signals, and FM recording is used. As for the audio, a 20 KHz bandwidth signal is recorded with PCM. With a sampling frequency of 48 KHz and 16-bit quantization, a maximum of four channels can be recorded. The SN ratio is

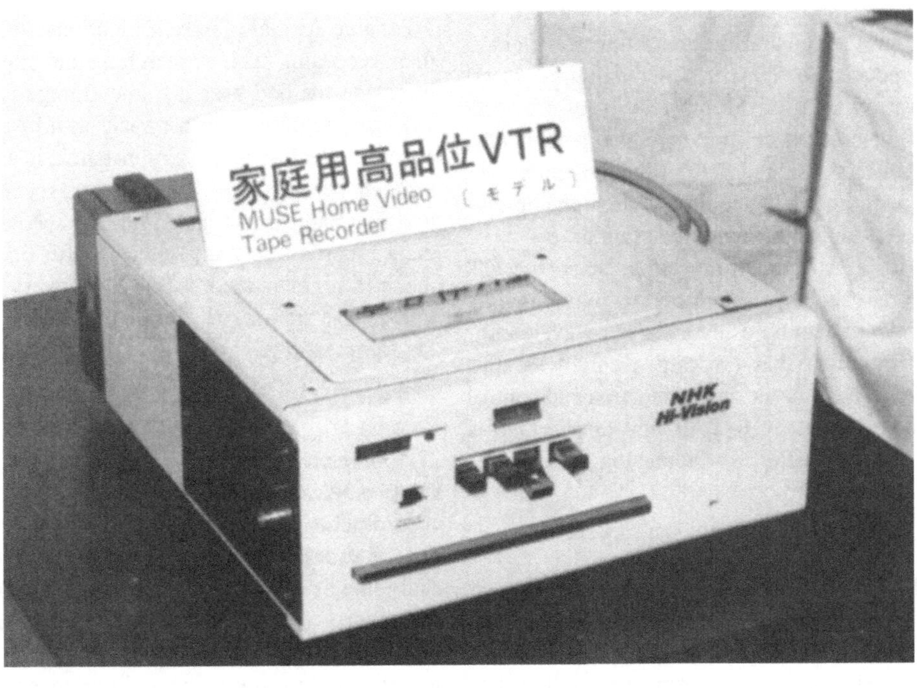

FIGURE 5.25. Experimental MUSE video tape recorder introduced by NHK.

TABLE 5.12. MUSE-VCRs that have been introduced.

	NHK	Hitachi	NHK, Matsushita	Mitsubishi	Toshiba	Sanyo
Tape	Metal	Metal	Metal	Metal	Metal, barium ferrite	Metal
Drum diameter (mm)	110	62	76	62	70	76
Drum rotational speed (rpm)	3,600	7,200	1,800	5,400	3,600	3,600
Head-to-tape speed (m/s)	20.5	23.3	7.1	17.4	13.1	14.2
Track pitch (μm)	34	60	42	58	24.5	17
FM allocation (MHz)	15~20	13.5~18	4.9~7.0	11.5~18	12.3~18.7	12.2~21.7
Recording time (minutes)	180	45	95	65	180	240

at least 41 dB for the luminance signal, at least 45 dB for the color difference signals, and at least 85 dB for the audio signal.

***(3) Increasing Recording Density of Tape*[13]**

To increase tape density and recording time, experimental high output metal particle tape and barium ferrite tape were used with the MUSE-VCR prototype, and the recording wavelength was reduced to 0.7 μm. In the future, metal deposition tape has the possibility of reducing the wavelength even further. In magnetic heads, a laminated sputtered Sendust head is under development that will accommodate high coercivity metal tape and a wide bandwidth.

With a fixed recording area, since the CN ratio can be increased more by reducing the track width than by using a shorter wavelength, it is important that tracking accuracy be improved. The track-position detection and control technology used in R-DAT and 8mm video is an effective method for increasing density.

5.3.4 Signal Processing Technology

(1) Blanking

Due to large time axis fluctuations with the track seams of a helical VTR, overlapping occurs in the playback signal of a segment recording. For this reason, a blanking interval is used as shown in Figure 5.27. The second segment is delayed by time T during recording, and when played back the image is restored by delaying the first segment.

(2) Shuffling

To enable special playback functions with segment recording such as searching and slow motion, one method uses a frame storage and divides one field into separate tracks as if by casting a screen over the image. For instance, in a three-segment recording, the first segment records the first line and every fourth line, the second segment records line 2 and every fourth line, and the third segment records line 3 and every fourth line. These signals are rearranged in time using the frame storage, and then rearranged again to their original sequence during playback.

(3) Compensating for Differences in Characteristics Across Channels

In multichannel recording, the signal for one field is usually distributed to the tracks sequentially line by line. A reference signal is recorded in the vertical blanking interval of each channel, and during playback the reference signal is used to detect output p-p value and nonlinear distor-

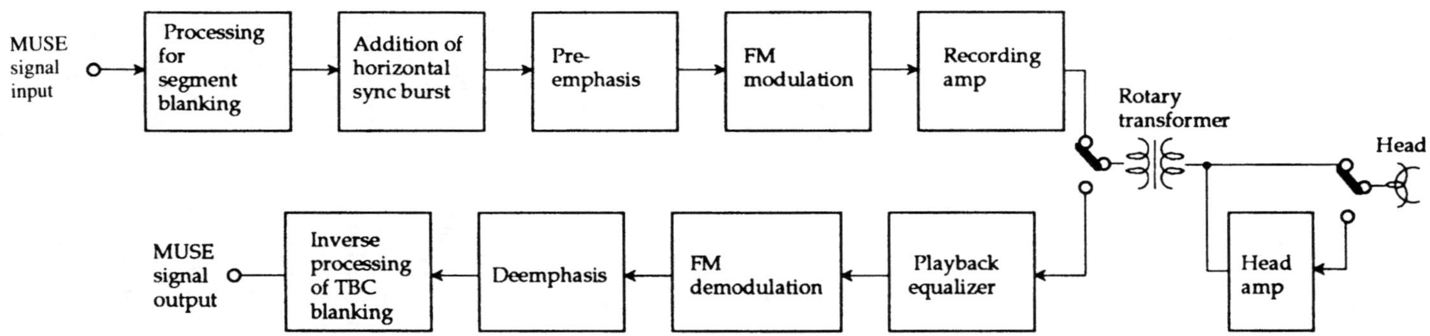

FIGURE 5.26. Block diagram of MUSE VTR.

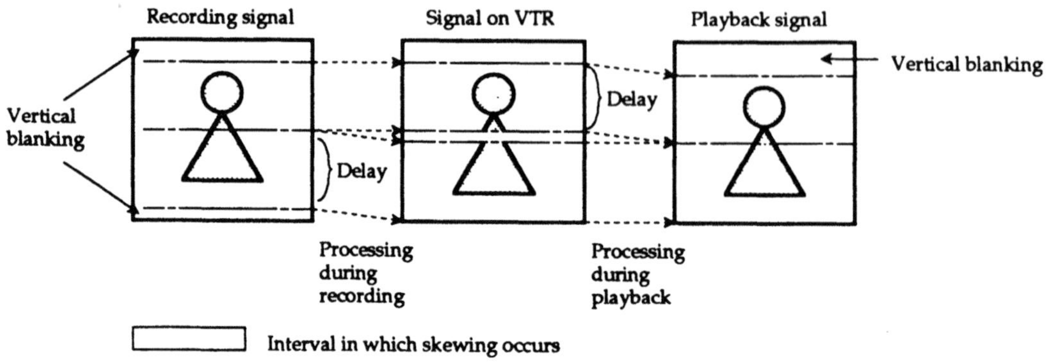

FIGURE 5.27. Signal processing for skewing.

tions. These distortions are then reversed with circuitry.

(4) Correcting Time Axis Error

The correction of time axis errors between tracks is indispensable in the Hi-Vision VTR, which records the signal of one field onto several tracks. This is especially true with MUSE, where the signal, which takes a subsampling of four fields is sent by analog transmission, and can tolerate time axis fluctuations of only a few ten nanoseconds when resampling to restore the image. To satisfy this condition, the MUSE-VTR requires a high speed, high precision method called TBC (Time Base Corrector). A feed-forward control is used in which horizontal sync and burst signals with negative polarity are added to the MUSE signal when recording, and a start pulse obtained from the playback sync and burst is synchronized with the writing clock to absorb sudden time axis fluctuations. A block diagram of the feed-forward TBC is presented in Figure 5.28.

In another method, the positive sync is used as is, and a pseudo-positive sync is recorded in the segment blanking interval to maintain the continuity of the horizontal sync.[14] With this method to detect the horizontal sync with certainty, the phase of the TBC writing clock is controlled by a separate superimposed recording of a pilot signal. An advantage of this method is that FM frequency deviations can be increased to the extent that there is no negative sync, thereby improving the SN ratio of the image.

5.3.5 Digital Recording

When the required luminance signal baseband bandwidth of VCRs for both industrial and household use is set at 20 MHz, the bit rate of the digital signal exceeds 600 Mb/s. If this volume of data were to be recorded on a conventional household VCR (with 1/2-inch tape) at standard speed, the recording area per bit would be less than 1 $\mu m.^2$ To achieve this value for digital VCRs, if we were to predict the future progress in recording density, is extremely difficult. Thus what is needed for a Hi-Vision digital VCR is a bandwidth compression method having a high compression ratio, with the condition that the data compression have a minimal degradation effect on image quality.

(1) Digital Recording of MUSE Signals

In an ordinary digital VCR, there is no need to record any information in the blanking interval of the video signal. However, with the MUSE signal, because the audio signal, control signal, and additional information are inserted, the MUSE signal must in principle record information not only in the video interval but in all intervals, The MUSE signal itself is compressed after these insertions, and even with the 10% addition of parity bits for error correction, the

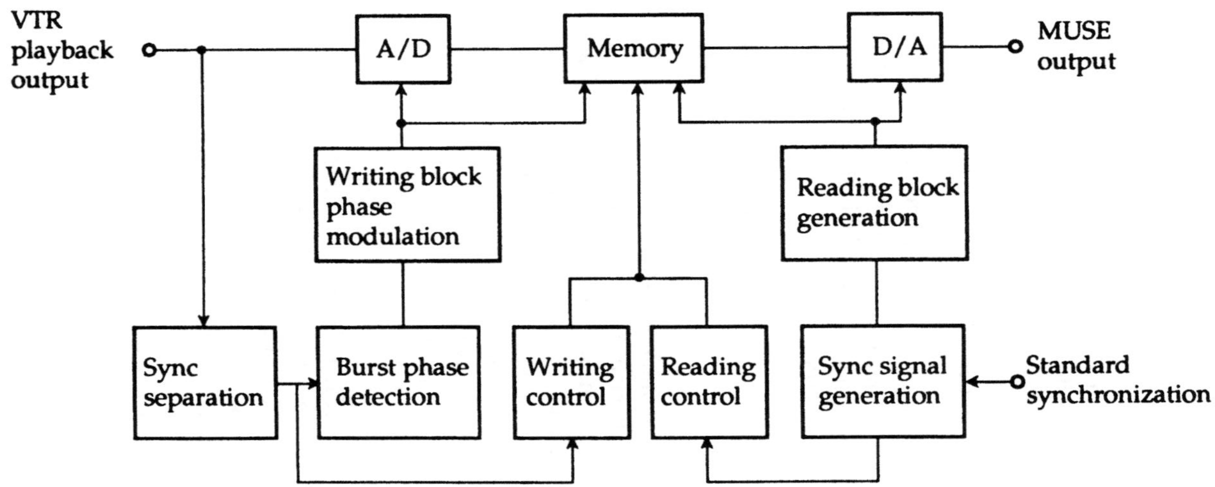

FIGURE 5.28. Feed forward TBC.

total bit rate is still 140 Mb/s. This is quite sufficient for a digital VCR.

(2) Band Compression of the Baseband Signal

Several methods have already been announced regarding the compression of digital recorded data for conventional television formats. Specifically, these include interfield sub-Nyquist sampling, DPCM (1-dimensional and 2-dimensional) Hadamard transform, and dynamic range encoding, all of which have possibilities for Hi-Vision VCRs.

When editing is done on a VCR, the interframe correlation can no longer be maintained at the edit seams. Thus there are difficulties in using a band compression method that uses interframe correlation without any modification.

5.3.6 Audio Recording

In the MUSE signal, audio and addition information is inserted in to the vertical blanking interval. Thus either two or four channels of near-instantaneous compressed and expanded DPCM are time compressed and multiplexed in ternary in lines 3 to 46 and 565 to 608. This digital audio signal has 16-bit interleaving as well as frame interleaving across fifteen frames to deal with burst errors generated by VTR dropout.

From the standpoint of the MUSE-VCR, it would be desirable to record this type of MUSE signal as is without any deficiency. However, this would take up the segment blanking interval and require the time compression of the MUSE signal.[15]

On the other hand, if the video signal were recording the digital audio in the case of baseband recording, a method using a rotating head could be adopted, as in the M-II format for conventional broadcasting or 8mm video for household use. This method is expected to be widely used in the future because of the possibility of high density recording and the elimination of a dedicated audio head. In the M-II and 8mm video formats, the track length is expanded by increasing the tape wrapping angle to over 200°, and digital audio signals are time-compressed either before or after the video signal.

5.3.7 Copy Protection

Prerecorded videotape sales are expected to be a major factor in the diffusion of Hi-Vision. Although the dubbing of prerecorded tapes is forbidden by copyright law, it is difficult to prevent individuals from copying tapes with the VCR's recording function. Thus from the viewpoints of copyright protection and the supply of prerecorded programs, the development of anti-dubbing technology is an important concern.

5.4 OTHER RECORDING TECHNOLOGIES

5.4.1 Disk Storage

Disk storage media feature many advantages, such as random access, fast access time, and the ability to play back recorded material in different ways such as fast forwarding and slow motion. These features are not available in tape storage formats, and as image information processing becomes more advanced and complex, applications are growing for disk media which has the ability to be searched and accessed.

There are three types of disks—read-only, writable, and rewritable. Since read-only disks can be produced in mass quantity, they can be distributed as a package media just as compact disks and laser disks are today. Writable disks are especially well suited to applications involving long term storage of image information and high speed search and retrieval. This application, which is being pursued in conventional broadcasting from the point of view of storing reference materials, is being developed in Hi-Vision for filing still images for household use. Rewritable disks, which can be used to edit and process images, are expected to be developed and commercialized for applications in areas such as broadcasting, printing, medicine, and computer graphics.

5.4.2 Read-only Disk

Read-only disks are of two types—the capacitance disk, in which the head comes in to contact with the medium, and the optical disk, which does not have contact with the head. They are cut with a short-wavelength (around 0.5 μm) gas laser, and the minimum pit length is about 0.4 μm. The track pitch is 1.35 μg for the former and 1.6 to 1.7 μm for the latter. For further information regarding the principles of recording and playback, the reader should consult the references at the end of the chapter.[4] The recording capacity of a CD (single sided, 12cm diameter) is 500 to 700 megabytes, and for the AHD disk described below (double sided, 26cm diameter) 2.54 gigabytes.

(1) Capacitance Disk

While a VHD is an analog recording of video signals, an AHD is a digital recording of audio and image data. Equipment has been developed to record and play back Hi-Vision still images on an AHD disk.[16] Since there is a large difference in the transfer speed between signals that can be recorded on the disk and Hi-Vision signals, frame storage was developed for time axis conversion between the two. Figure 5.29 is a block diagram of this frame storage. In the figure, the Hi-Vision input signal undergoes A/D conversion and only one frame is stored in the image memory. This signal passes through the low speed I/0 system and is transformed to the AHD recording system, then becomes a 15-second, 1-frame image signal and is recorded by a VTR. The disk is cut from this tape. The AHD has four channels with transfer speeds of about 0.7 Mb/s, two of which are used to record image signals. The audio signal is converted to a PCM signal and recorded on the other two channels. When played back, the signal enters the low speed interface in the reverse order from the above, passes through the image memory and high speed interface, and is reproduced as a Hi-Vision signal.

(2) Optical Disk

Since an optical disk is read with a semiconductor laser having a wavelength of 0.78 μm, the minimum bit length is about 0.55 μm. The disk rotational speed is limited by the servo response characteristic of the optical head. A plastic disk 30cm in diameter works best at a speed of about 1,800 rpm. Considering this minimum bit length and rotational speed, baseband recording and playback for Hi-Vision signals requires substantial improvements in the optical head and signal processing system. Thus optical disks and players for MUSE recordings must strive to be practical, including their playback duration, and a number of disks and players have been developed based on this.[17] An overview of these systems is presented below.

The signal recording system for producing the original disk is shown in Figure 5.30. After

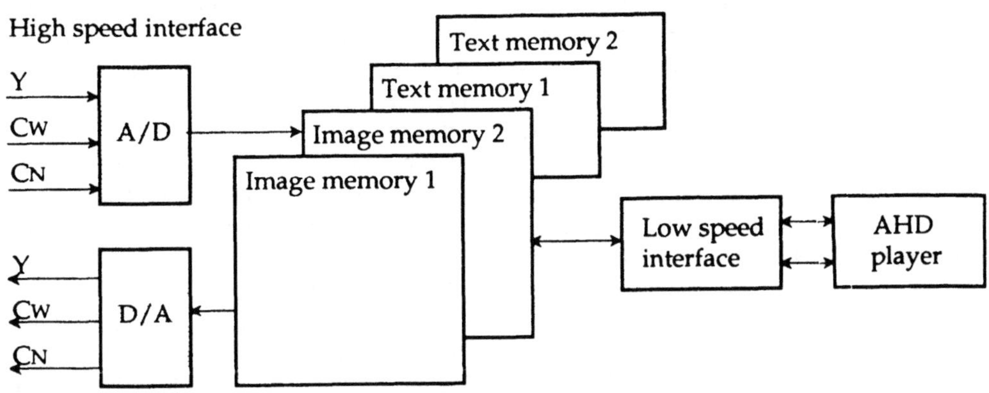

FIGURE 5.29. Block diagram of frame memory for time axis conversion.

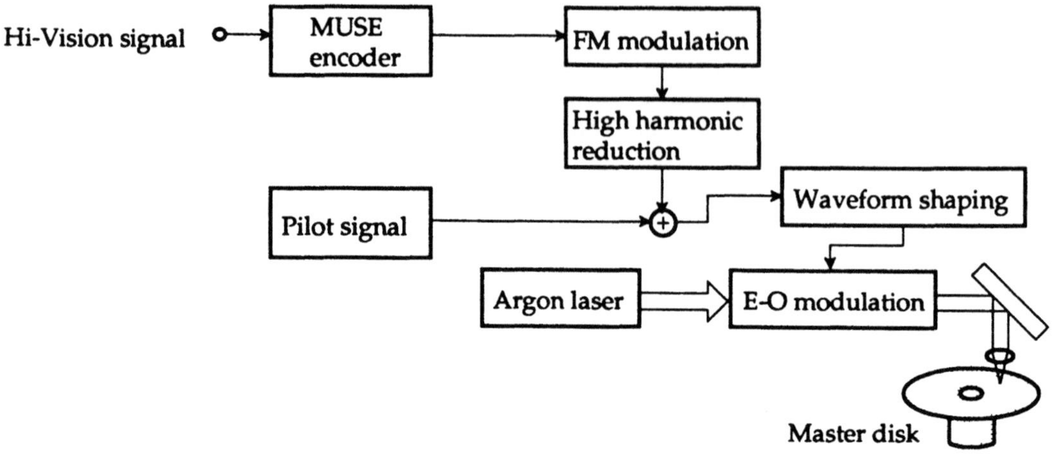

FIGURE 5.30. Recording on a read-only optical disk.

the MUSE input signal undergoes frequency modulation, a pilot signal to correct for jitter during playback is multiplexed in. The signal then enters the optical modulator and modulates the intensity of an argon laser. The process after this point is identical to the production of conventional video disk originals and copies.

Figure 5.31 is a block diagram of the playback system, which is essentially the same as for conventional video disk players. The signal is detected by a photodiode, passes through the RF equalizer, and enters the high-pass filter. Here the pilot signal is eliminated, and after demodulation in the FM demodulator, the signal goes through an 8.3 MHz low-pass filter and is outputted as a MUSE signal. The pilot signal, which was extracted by the band-pass filter, is used to control the disk rotation and correct jitter. In the demodulation of the MUSE signal, the presence of jitter causes the resampling phase to lag and increases interference between codes, making it impossible to restore the quality of the signal. The tolerable jitter level being aimed for is several nanoseconds. Also, since the recording and playback system, including the disk, is a nonlinear system, intermodulation occurs between the pilot signal and the video carrier. For this reason, the level and frequency of the pilot signal are selected so that the beat interruption is not visible in the reproduced image.

Reducing the light source wavelength in the playback system is an effective way to increase playback time. It makes possible the reduction of the minimum pit length and track pitch. Recently, an attempt was successfully made to increase the playback duration of MUSE signals by 1.5 times by using a semiconductor laser with a wavelength in the 0.65 μm range.

Another development is baseband recording and playback on a disk using parallel heads.[18] A 20 MHz bandwidth luminance signal and two color difference signals C_W and C_N (both having bandwidths of 5.5 MHz) are each time-expanded and compressed and divided into two sections. Two He-Cd lasers (wavelength: 0.44 μm) then simultaneously record the sections on two tracks using FM recording. The PCM audio is recorded by superimposing on the vertical retracing interval. The playback uses an optical head with a three-beam method. The central beam does the focusing and tracking, while the other two beams read the signal. The recording area on the disk is at least 18cm in diameter, and the playback duration when the disk's linear velocity is 17 m/s (fixed) is about 16 minutes.

(3) Still-image Storage

Since compact disks (CDs) have a large storage capacity, they are widely used as a package media for digital data and have been standardized in a format known as CD-ROM. Recently, a prototype disk called CD-HV was developed

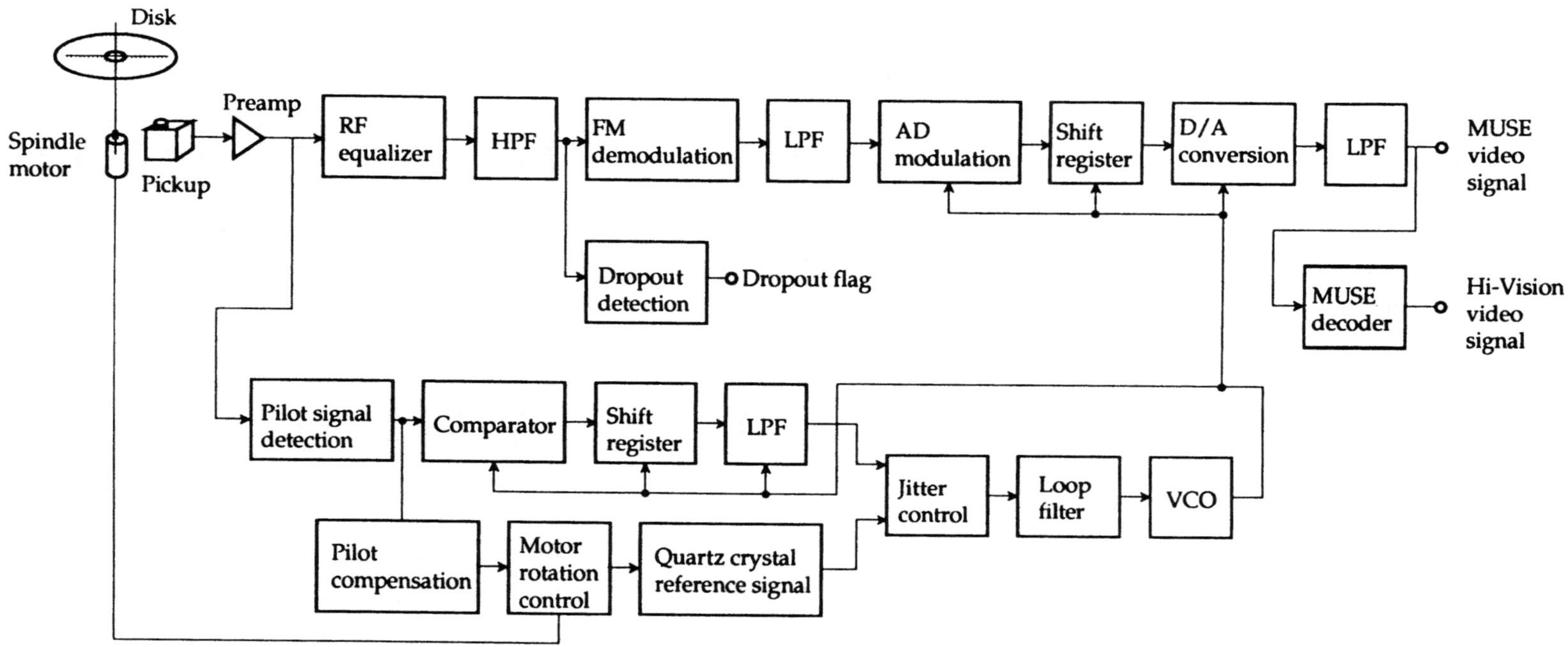

FIGURE 5.31. Block diagram of playback system for read-only optical disk.

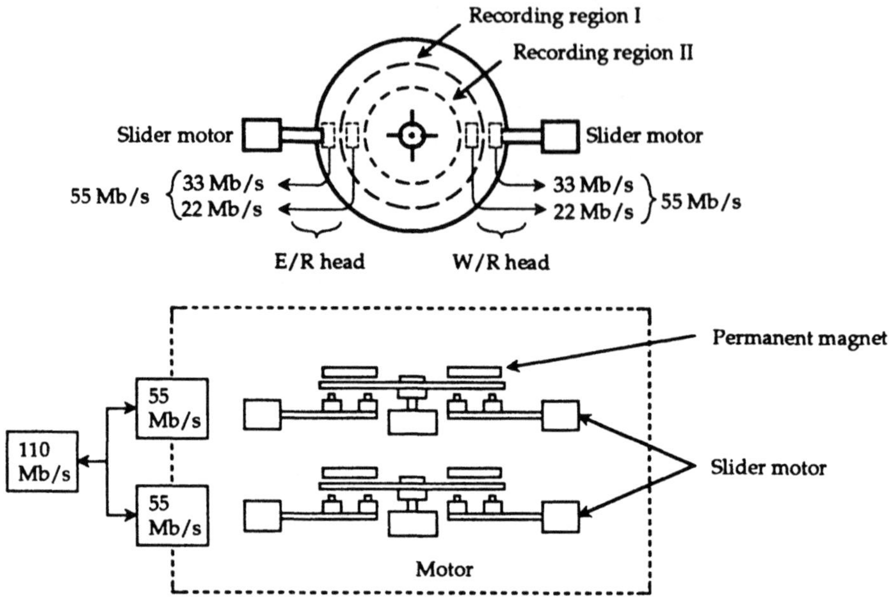

FIGURE 5.32. Diagram of high transfer speed optomagnetic memory disk.

in which Hi-Vision signals are recorded on a CD-ROM (refer to Section 4.4).

The CD-HV disk carries 640 Hi-Vision still images and 60 minutes of audio signals. It takes advantage of the random access feature of optical disks.

5.4.3 Rewritable Disk

Disk media that can record and play back Hi-Vision signals in real time are still under development. However, this can be realized for still images using magnetic disks. Furthermore, the storage capacity can be greatly increased with magneto-optic disks. Both these disks are used mainly in industrial applications, and can obtain high quality images by using baseband digital recording.

(1) Magnetic Disk

One example of the magnetic disk is a still image storage apparatus using four 5.25-inch Winchester disks. This apparatus has a capacity of 45 megabytes and can store seven color images. Because of its slow transfer speed, putting an image on the screen can take several dozen seconds.

(2) Magneto-optic Disk

In the magneto-optic disk an amorphous magnetic film is formed on either a glass or plastic substrate, and signals are recorded and played back using a laser. The reader should consult the references at the end of the chapter for detailed information on the principles involved in this technology.[4] Practical applications of this type of rewritable optical disk have already begun in some areas. At present, the lifetime for this medium is estimated to be at least ten years, with at least one million rewrites possible.

A digital image recording device with a magneto-optic disk was recently developed for use in conventional television.[19] Figure 5.32 is a diagram of this apparatus, which has two disk drives. The top view of the disk drives in the top part of the figure shows that the disk is divided into an inside and outside recording area. Since the disk rotational speed is fixed, the transfer speed is adjusted to the difference in linear velocity between the inside and outside recording areas. The combined transfer speed of the two drives is 110 Mb/s. Although there are four channels, this is quite high considering that most magneto-optic disks have a transfer speed of under 10 Mb/s.

TABLE 5.13. Still-image video disk parameters.

	Read only	Rewritable	
	Capacitance disk	Magnetic disk	Magneto-optic disk
Recording and playback signal	Baseband	Baseband	Baseband
Recording method	Digital	Digital	Digital
Sampling frequency (MHz)	Y: 51.79 C: 25.89	R, G, B: 64.8 for each	R, G, B: 64.8 for each
Quantize bit count	8	8	8
Number of channels	1	–	4
Transfer rate (Mb/s)	5.73	5	110
Disk diameter (cm)	26	13	30
RPM	900	3,564	2,250
Recording capacity (images / apparatus)	240	7	1,200
Access time (seconds)	15	~60	<1

A Hi-Vision image storage apparatus with a fast access time and large storage capacity has been developed based on this device. R, G, and B signals having a baseband bandwidth of 29 MHz are sampled at approximately 64.8 MHz and quantized at 8 bits. The storage required for one screen is about six megabytes. With a storage capacity of 6.4 gigabytes, the apparatus can store 1,200 images. The access time is less than one second. The device parameters for the disks discussed above are summarized in Table 5.13.

REFERENCES

1. Hayashi, "High Definition Television," *NHK Technical Report,* Vol. 21, No. 8, pp. 319–337 (1978) or Fujio, "High Definition Television and the Future of Broadcasting," *NHK Technical Report,* Vol. 24, No. 11, 12, pp. 434–440 (1981).
2. Shibaya, Yoshida, "Current State of Development of High Definition Equipment and VTR," *NHK Technical Report,* Vol. 24, No. 11, 12, pp. 483–486 (1981).
3. Yokoyama et al., "Development of a Hi-Vision VTR," *NHK Technical Report,* Vol. 38, No. 2, pp. 65–118 (1986).
4. NHK, editors, *Broadcast Technologies Series 5: VTR Technology,* pp. 60, 17, 284, Japan Broadcast Publishing Co. (1982).
5. Shibaya, Uehara, Abe, "Magnetic Head and Thin Film Technology," *NHK Technical Report,* Vol. 30, No. 11, pp. 383–391 (1987).
6. Umemoto et al., "Development of a Hi-Vision (High Definition Television) Digital VTR," *Hitachi Magazine,* Vol. 67, No. 5, pp. 55–58 (1985).
7. Tanimura et al., "HDTV Digital Tape Recording," *15th Montreux Television Symposium Records,* pp. 300–319 (1987).
8. Yokoyama, *Magnetic Recording Technology,* Sogo Denshi Shuppansha, pp. 542–548 (1981).
9. Doi, Iga, *New Version of Digital Audio,* pp. 220–230, Radio Technology Co. (1987).
10. Nakagawa, "High Density Magnetic Recording Technology Used in Digital VTRs," Digital VTR Special Edition, *Journal of the Institute of Television Engineers of Japan (ITEJ),* Vol. 37, No. 7, pp. 542–548 (1981).
11. Nakamura, "CCIR Standardization of Digital VTRs," *NHK Technical Report,* Vol. 29, No. 4, pp. 138–145 (1986).
12. Ninomiya et al., "Development of the MUSE System," *NHK Technical Report,* Vol. 39, No. 2, pp. 18–53 (1987).
13. Yokoyama, "Research on High Density Mag-

netic Recording Technology," *NHK Technical Report*, Vol. 28, No. 7, p. 36 (1985).
14. Shinmura et al., "Prototype of Hi-Vision VTR for Household Use," *ITEJ Proceedings* p. 181 (1987).
15. Yoshida et al., "Long Duration Recording with a MUSE VTR," *ITEJ Technical Report* VR 88-4, p. 24 (1988).
16. Shimada, "Hi-Vision Still-Image Disk," *NHK Technical Report*, Vol. 28, No. 12, p. 486 (1985).
17. Ninomiya et al., "MUSE System Home Video Disk," *ITEJ Technical Report* TEBS 99-4 (1984).
18. *Nikkei Electronics*, No. 393, p. 143 (1986).
19. Yokoyama et al., "High-Quality Magneto-Optic Disk Video Recording," *SMPTE Journal*, 96, p. 1062 (1987).

6

Applied Technology

Shin Ohno, Yukio Sugiura, Yutaka Tanaka, Yuji Nojiri

While current plans for Hi-Vision call for direct satellite broadcasting with the MUSE format, the technology can be applied in a number of other directions as well. With regard to current broadcasting, the ability to convert freely between Hi-Vision and NTSC or PAL formats will increase program availability as well as promote the diffusion of Hi-Vision. Moreover, the introduction of Hi-Vision is expected to have a new impact in fields such as motion pictures and printing, where existing television systems have failed to meet performance requirements. Under these circumstances, a major theme in the future will be to engineer the coexistence and co-prosperity of other media with Hi-Vision.

In this chapter, we will discuss some representative technologies which will be necessary for the application of Hi-Vision.

6.1 APPLICATIONS IN CURRENT BROADCASTING SYSTEMS

6.1.1 The Need for Format Conversion

Hi-Vision programs can be used not only in Hi-Vision broadcasting, but also in standard television broadcasting systems after undergoing a format conversion. In this case, the format conversion needs to produce a high image quality. In addition to the conventional filter technology, the image quality of the converted image can be improved with signal processing technologies such as motion adaptation processing and motion compensation technology.

Existing television formats can be divided into three major groups: NTSC, PAL, and SECAM. These three standard television formats are compared in Table 6.1. The NTSC format has 525 scanning lines and a field frequency of 59.94 Hz. The field frequency is slightly lower than the 60 Hz of Hi-Vision, the difference being only 1000/1001. The conversion from Hi-Vision to PAL is the same as for SECAM, as both these formats have 625 scanning lines and a 50 Hz field frequency, and differ only in chrominance signal modulation and multiplexing methods. Thus both the number of scanning lines and field frequency must be converted from Hi-Vision.

Conventional format converters for standard television basically consist of filter processing through linear interpolation, and use either repeated image conversion with CRT interposition, or direct conversion with delay lines. While the equipment for direct conversion has been become more complex and uneconomical, the image quality of the conversion is superior to the conversion method using a CRT. Recent advances in digital technology and lower prices for memory components have made this method

TABLE 6.1. Comparison of color television formats.

Item	NTSC	PAL	SECAM
Number of scanning lines	525	625	625
Line frequency (Hz)	15734.264	15625	15625
Field frequency (Hz)	59.94	50	50
Aspect ratio	4:3	4:3	4:3
Interlace ratio	2:1	2:1	2:1
Color subcarrier frequency	f_{SC} = 3.579545 MHz ±10 Hz	f_{SC} = 4.43361875 MHz ±1 Hz (I) f_{SC} = 4.43361875 MHz ± 5 Hz (B)	f_{SC} = 4.250000 MHz ± 2 kHz f_{SC} = 4.406250 MHz ± 2 kHz
Color signal modulation	Rectangular dual phase amplification modulation	Rectangular dual phase amplification modulation	Frequency modulation
Bandwidth of luminance signal	4.2 MHz	5.5 MHz (I) 5 MHz (B)	6 MHz (L) 5 MHz (B)
Color signal	Q = 0.41(B-Y)+0.48(R-Y) I= -0.27(B-Y) +0.74(R-Y)	U = 0.493 (B-Y) V = 0.877 (R-Y)	D_B = 1.5 (B-Y) D_R = − 1.9 (R-Y)
Transmission format	Simultaneous	Special simultaneous (180° inversion of SC phase every line)	Line sequential (R-Y and B-Y are transmitted every line)
Country	Japan, U.S.A., Canada	Great Britain, West Germany, China	France, Commonwealth of Independent States, Eastern Europe

the mainstream in format conversion. The image quality of conversions improved greatly with the announcement of the DICE (Digital Intercontinental Conversion Equipment)[1] converter in England in 1972, and of the ACE (Advanced Conversion Equipment)[2] converter, which performs field frequency conversion through higher-order interpolation. In Japan as well, format converters using direct conversion have been developed at NHK and KDD and are being used for international program exchange, but since both use linear interpolation, the image quality of the conversion needs further improvement.

The degradation in resolution has been a serious problem with scanning line conversions, while field frequency conversions have been afflicted by the deterioration in dynamic resolution and judder artifacts (where movements are jerky). These problems are being solved through motion adaptation processing technology and motion compensation technology using motion vectors. These techniques have permitted higher quality image conversion.

In this section, we will discuss the basic concepts involved in format conversion, as well as new technologies for converting between between NTSC and PAL (SECAM) formats.

6.1.2 Basic Concepts in Format Conversion

Television format conversion consists of the conversion of sampling frequencies such as the number of scanning lines and the field frequency. Sampling frequency conversion by filter processing is performed through a two-stage process involving interpolation and resampling. Figure 6.1 gives an example of digital–digital sampling frequency conversion in the temporal domain when F_{S1}:F_{S2} = 2:3 (F_{S1} and F_{S2} are the sampling frequencies before and after conversion). First, in interpolation, the F_S, which is the least common denominator of F_{S1} and F_{S2}, is designated as the sampling frequency for frequency conversion, and interpolation of the sampling point is performed, as shown in Figure 6.1 (b). Next the necessary sampling points are resampled as in Figure 6.1 (c). If the sampling frequency is smaller after conversion than before, the bandwidth is restricted in the interpolation process. When the hardware is developed, the interpolation and resampling will be performed simultaneously.

Figure 6.2 shows the sampling frequency conversion process in the frequency domain, and Figure 6.2 (a) and (d) are the sampling spectrums of sampled digital signals before and after conversion. In both cases, the analog signal spectrum bands are restricted to f_m or less. The digital input signal with sampling frequency F_{S1} is added to the low-pass filter (interpolation filter) shown by the dashed line in Figure 6.2 (a), and the spectrum of the output signal, which has been interpolated, will resemble Figure 6.2 (b). A new sampling point sequence is added to this signal, and the signal is inserted, as shown in Figure 6.1 (b). Finally, if the signal is resampled at T_{S2} (= $1/F_{S2}$) as in Figure 6.2 (c), a digitized signal with a sampling frequency F_{S2} as in Figure 6.2 (d) can be obtained.

The interpolation filter is designed with sampling frequency F_S (= $3F_{S1}$ = $2F_{S2}$) as shown in Figure 6.3 (a). The origin is inserted into the

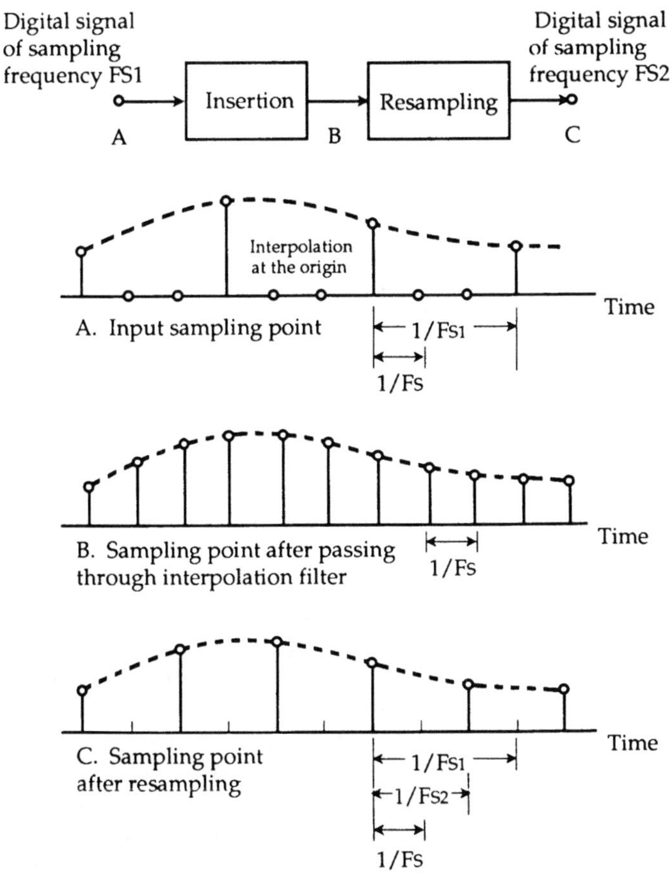

FIGURE 6.1. Basic configuration of sampling frequency conversion.

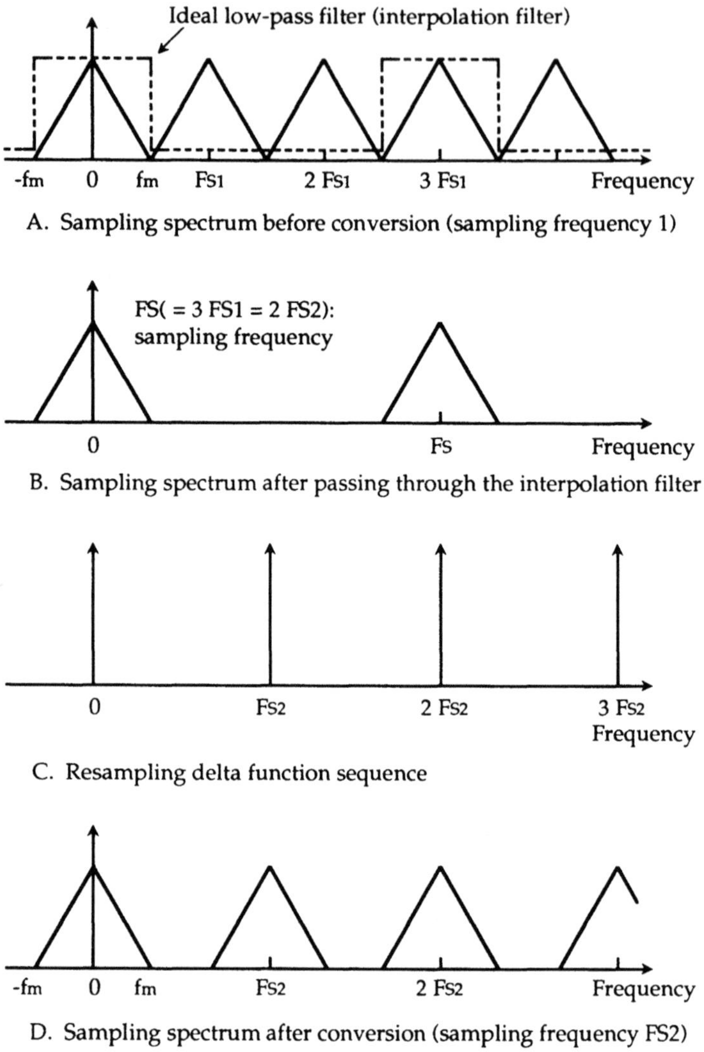

FIGURE 6.2. Sampling frequency conversion.

input point sequence of sampling frequency F_{S1}, resulting in $3F_S$. Therefore, as shown on the right of Figure 6.3 (b), it is equivalent to the three-circuit parallel processing of clock frequency F_{S1}. That is, when the impulse response of the interpolation filter transmission function $H_F(\omega)$ is $h_F(n)$:

$$h_F(n) = h_{F1}(n) + h_{F2}(n) + h_{F3}(n) \quad (6.1)$$

where n is an integer.

Here, $h_{F1}(n)$, $h_{F2}(n)$, and $h_{F3}(n)$ are the impulse responses of sampling frequency F_S, but because the origin exists as shown on the right side of Figure 6.3 (a), the actual sampling frequency is F_{S1}. There is a time difference of $1/F_S$ between $h_{F1}(n)$, $h_{F2}(n)$, and $h_{F3}(n)$, but if time axis compensation is performed at the output stage as shown in (b), the same operation as $H_F(\omega)$ at a clock frequency F_S can be achieved through the parallel processing of filters $H_{F1}(\omega)$, $H_{F2}(\omega)$, and $H_{F3}(\omega)$ (the Fourier conversions of $h_{F1}(n)$, $h_{F2}(n)$, and $h_{F3}(n)$) at clock frequency F_{S1}. In general, depending on how F_{S1} and F_{S2}

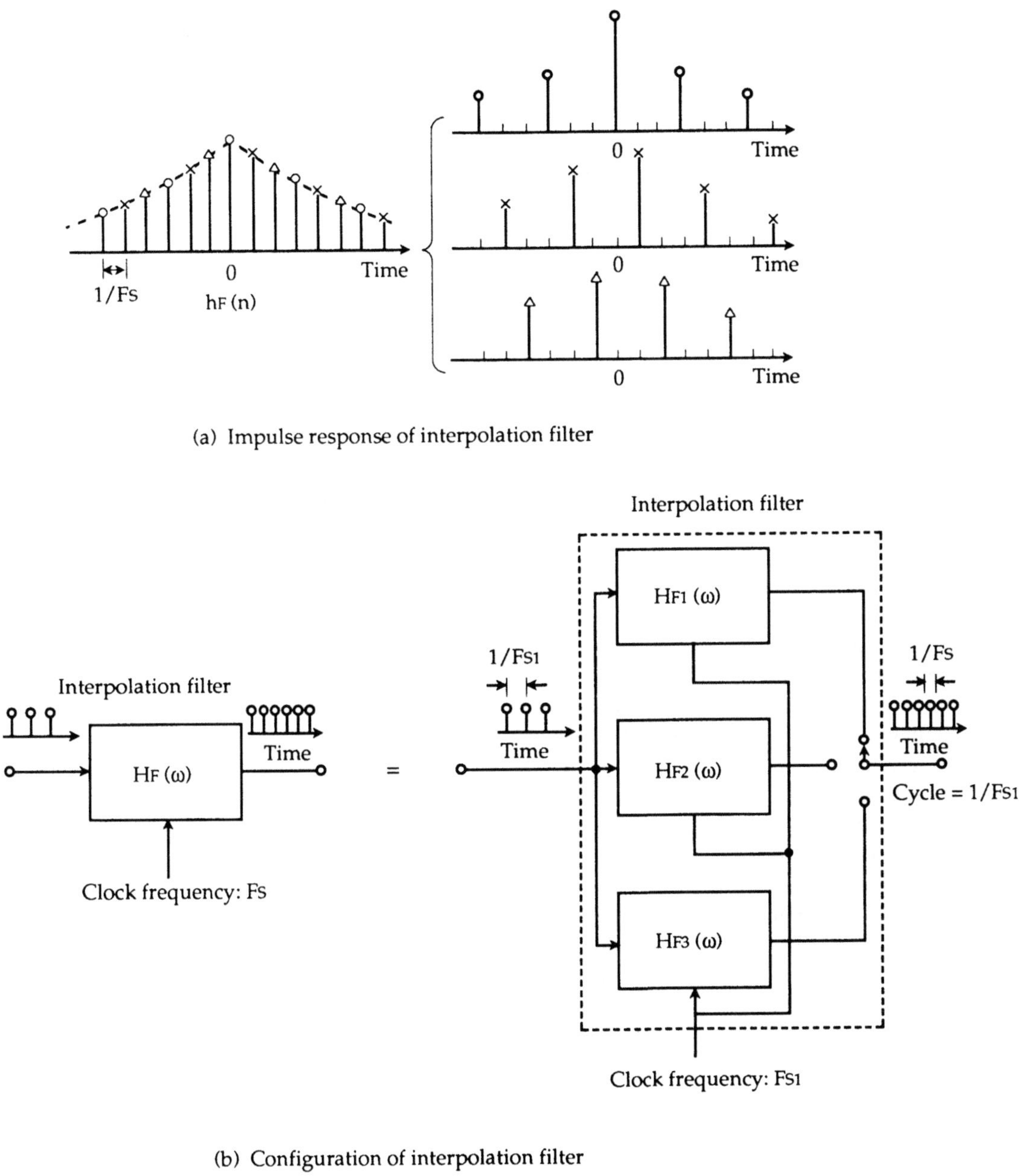

FIGURE 6.3. Parallel processing in the interpolation filter.

are taken, F_S is fairly high. This is a major obstacle in developing hardware, but if this method is used, high-speed computation using low-speed components is possible, and reference frequency sampling is simplified.

Television format conversion should be done through this kind of sampling frequency conversion of the number of scanning lines and the field frequency. The basic configuration of the conversion is shown in Figure 6.4. While the sampling frequency conversion takes place in the scanning line converter and the field fre-

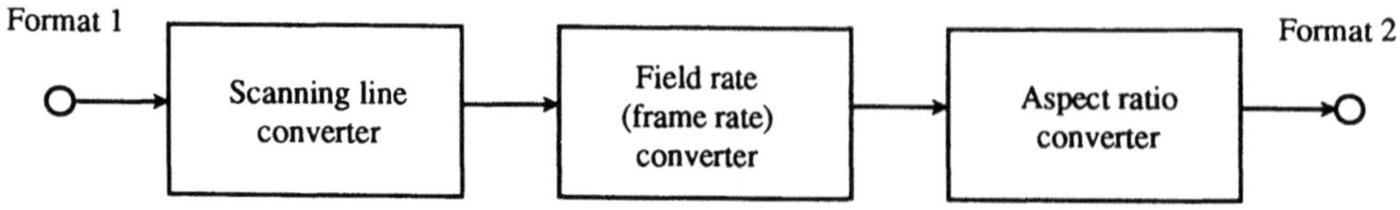

FIGURE 6.4. Basic configuration of format conversion.

quency converter, the sequence of the various signal processors is optional.

6.1.3 Hi-Vision / Pal (SECAM) Format Conversion

When converting a Hi-Vision program to PAL or SECAM formats, the linear interpolation method using fixed filters from existing standard television formats, cannot eliminate interference such as motion blurring of moving images and judder, especially in field frequency conversions.

Next, we will explain the operation of Hi-Vision-PAL format conversion equipment, which uses motion compensation technology to prevent this kind of degradation.[3] Figure 6.5 shows the configuration of the equipment. It consists of a converter for the number of scanning lines, a frame rate converter, and a scanning method converter. For Hi-Vision-SECAM conversion the last encoder converts the signal to the SECAM format. Conversion methods include the conversion of the number of scanning lines and scanning method using motion adaptation, and frame rate conversion using motion compensation.

Table 6.2 compares Hi-Vision and PAL standards. In format conversion processing, three types of parameter conversions are necessary: conversion of the number of scanning lines, the aspect ratio and the field frequency. In this equipment, the aspect ratio conversion uses a method that crops both sides of the Hi-Vision screen, so that the conversion involves the number of scanning lines and the field frequency. Seen in terms of sampling frequency conversion, in converting the number of scanning lines:

$$\left.\begin{aligned} F_{SL1} : F_{SL2} &= 1125 : 625 = 9:5 \\ F_{SL} &= 1125 \times 5 = 625 \times 9 \text{ (lines/screen)} \end{aligned}\right\} \quad (6.2)$$

and in field frequency conversion:

$$\left.\begin{aligned} F_{SF1} : F_{SF2} &= 60:50 = 6:5 \\ F_{SF} &= 60 \times 5 = 50 \times 6 \text{ (Hz)} \end{aligned}\right\} \quad (6.3)$$

(1) 1125 / 625 Scan Line and Progressive Scanning Converter

In this block the Hi-Vision signal input is converted into progressive scanning signals with 625 scanning lines and a frame frequency of 60

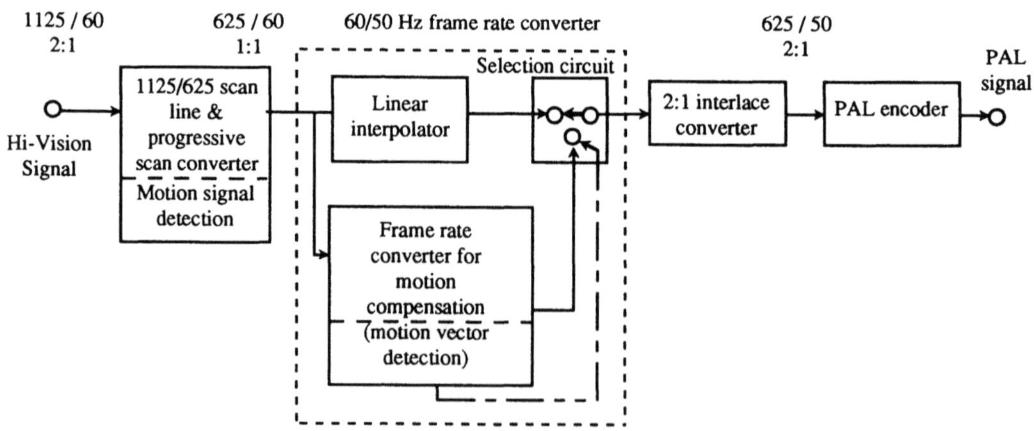

FIGURE 6.5. Configuration of Hi-Vision/PAL format converter.

TABLE 6.2. Comparison of Hi-Vision and PAL standards.

	Hi-Vision	PAL
Number of scanning lines	1125	625
Aspect ratio	16:9	4:3
Field frequency	60 Hz	50 Hz
Interlace ratio	2:1	2:1

Hz. The aspect ratio remains 16:9. In progressive scanning conversion, to improve the accuracy of motion vector detection and motion vector image compensation, an interpolation filter with a 2-dimensional time-space high-order filter is combined with the minimization of degradation in the conversion characteristic.

The configuration of the converter is shown in Figure 6.6. It consists of an interpolation filter which treats field memory and line memory as delay lines, and a motion detection circuit. The switching of the interpolation filter characteristic is controlled by motion detection signals, and the conversion to 625-line progressive scanning is performed based on interfield data for stationary areas and intrafield data for moving

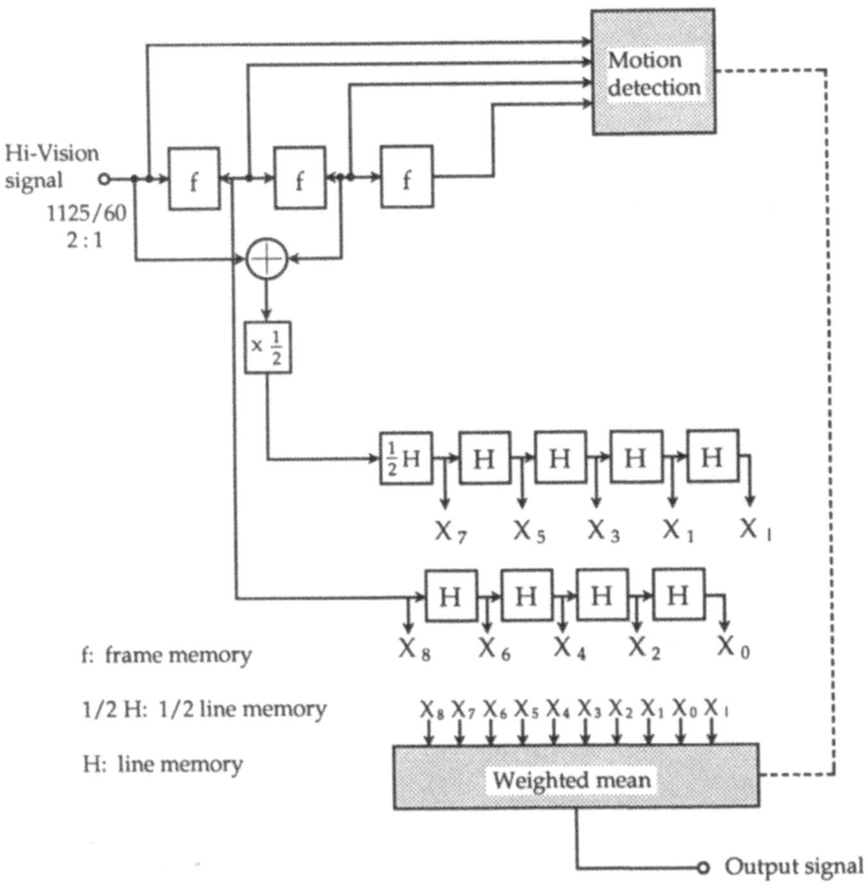

FIGURE 6.6. 1125-625 scanning lines/sequential scanning converter.

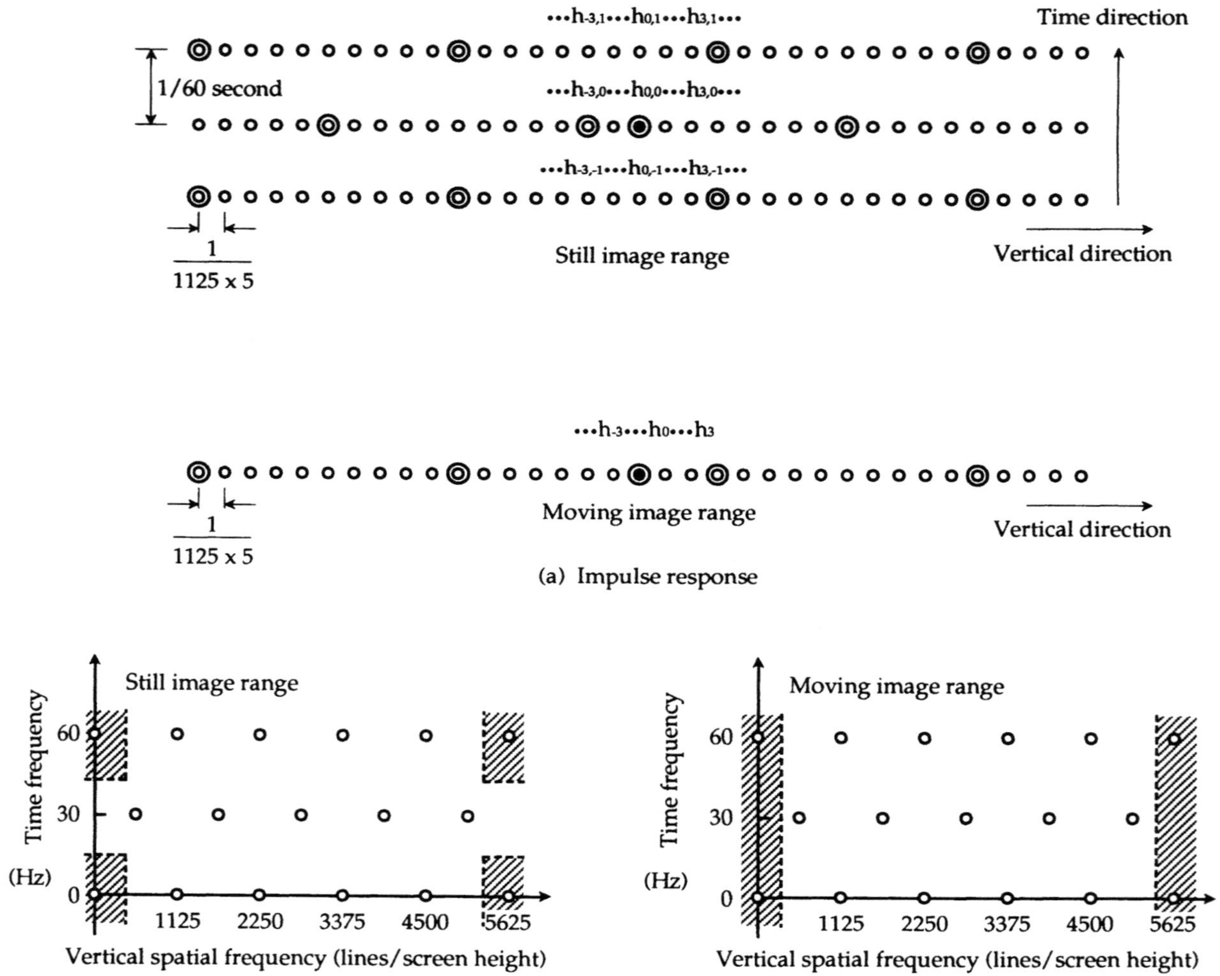

(a) Impulse response

(b) Frequency characteristic (two-dimensional display)

FIGURE 6.7. Interpolation filter characteristic (two-dimensional).

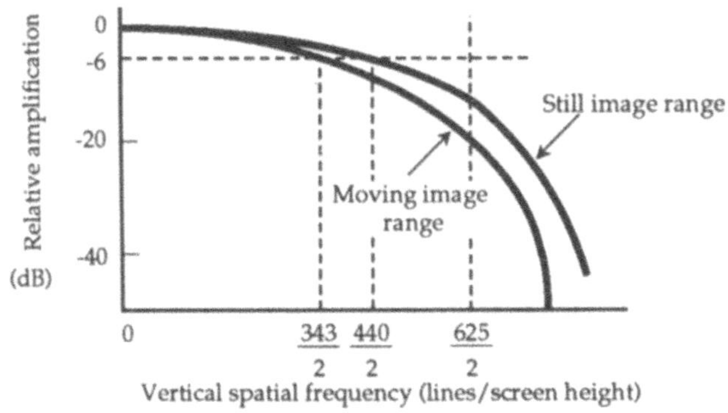

FIGURE 6.8. Interpolation filter characteristic (one-dimensional on vertical axis).

areas of the screen. In the figure, the even-numbered taps are the scanning conversion reference field, and the odd-numbered taps are data which are separated from that field by one field.

The impulse response of the interpolation filter is shown in Figure 6.7 (a). For the stationary area it has a 2 × 18th-order 2-dimensional time-space characteristic, and for the moving area it has an 18th-order vertical 1-dimensional characteristic. The sampling point interval in the figure is $1/F_{SL}$. Figure 6.7 (b) shows the sampling spectrum of input signals and the the characteristic of the interpolation filter. The circles indicate the carrier waves produced by sampling, and the shaded areas indicate the pass band.

The interpolation filter in the stationary area calculates the impulse response from the 1-dimensional frequency characteristic on the vertical axis through an expansion on the time axis with a weight coefficient of 1/4, 1/2, and 1/4. Therefore, the frequency characteristic in the time axis direction is a cosine characteristic in which 30 Hz becomes the origin. Figure 6.8 shows the frequency characteristic of the interpolation filter of Figure 6.7 seen from the vertical axis. The response in the still image range widens the band width, and prevents deterioration of the vertical resolution.

When the vertical bandwidth in the moving area is broadened, the reflection distortion from the carrier wave positioned at 30 Hz increases, and moire deterioration occurs in the converted image. Because conversion processing in the stationary area uses interfield data, a double image appears at the edges and significantly reduces the image quality. Thus in the moving area, the switching of the intrafield interpolation filter is performed with motion detection signals.

(2) 60/50 Frame Rate Converter

The conversion of the frame rate and aspect ratio is performed at this point. In addition to conventional frame rate conversion using linear interpolation, this equipment uses a positional interpolation method that detects motion vectors and generates the required edges of the moving image. Figure 6.9 shows the configuration of the frame rate converter.

(a) Linear Interpolator. The linear interpolator is a frame rate converter that generates interpolation frame signals using the weighted mean of contiguous frame signals. As Figure 6.10 shows, the weights of the weighted mean changes over a sequence of frames being converted from 0:1 to 2:8, 4:6, 6:4 and 8:2. The bottom part of the figure shows the process for generating frame No. 3 on the 50 Hz side. The weights are inversely proportional to the distance on the time axis. In this example, the interpolation ratio is 4:6, so the sum of 60% of the third frame and 40% of the fourth frame becomes the interpolation frame. The moving image shows significant degradation at the edges of the interpolation signal as indicated by the shaded areas. The peculiar discontinuity of motion which occurs in frame rate conversion (jud-

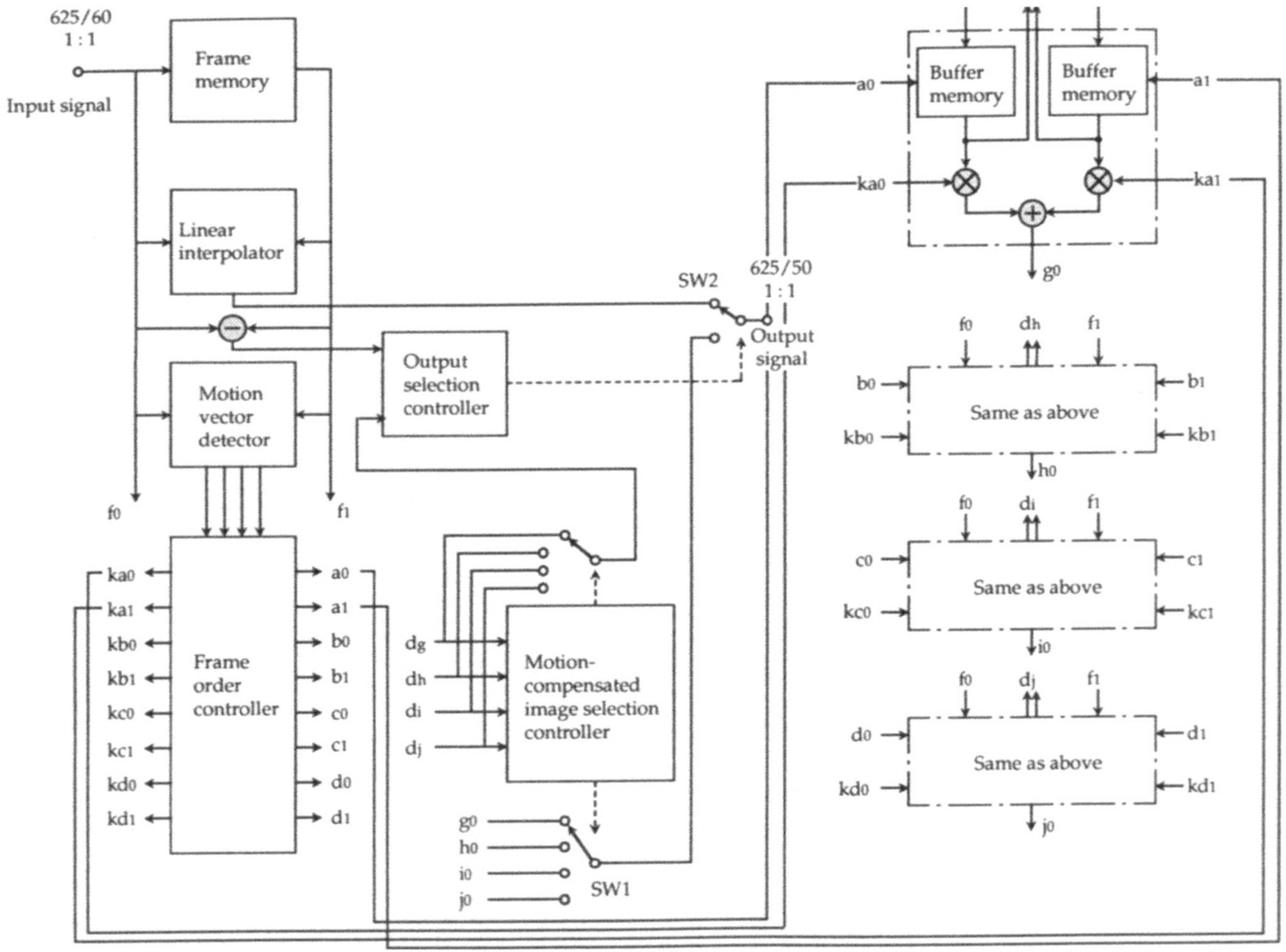

FIGURE 6.9. 60/50 frame rate converter.

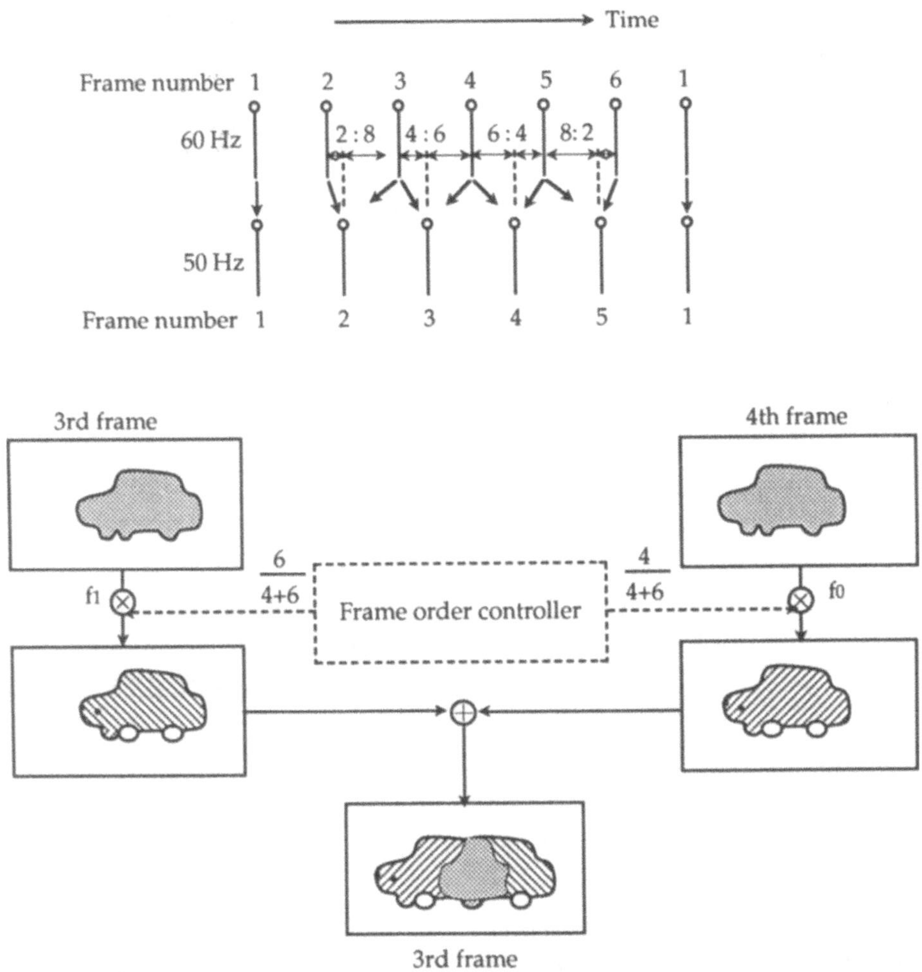

FIGURE 6.10. Linear interpolation.

der) results from the sequence-dependent degradation. The degradation also occurs in the form of motion fading.

Figure 6.11 shows the impulse response and frequency characteristic of linear interpolation. With a large attenuation in the normal band as well as residual reflection components, its conversion characteristics are insufficient. Figure 6.12 shows the relationship of frame order to the amount of degradation at the edges. As the figure shows, the amount of degradation has a period of 5 cycles, so a flicker with a 10 Hz fundamental frequency is produced at the edges of the moving image. This is seen as judder.

While linear interpolation is performed using a 1-frame delay line, higher-order linear interpolation methods that use more frames can be conceived. Higher-order interpolation has the effect of transferring judder into motion blur but does not increase overall image quality dramatically.

(b) Motion Compensation Frame Rate Converter. This frame rate converter is based on positional interpolation using motion vectors. The method has not been seriously researched in the past nor has equipment been developed for two reasons. First, unlike linear interpolation, it cannot be expressed in terms of a filter characteristic, and second, the capability does not exist to develop the hardware required

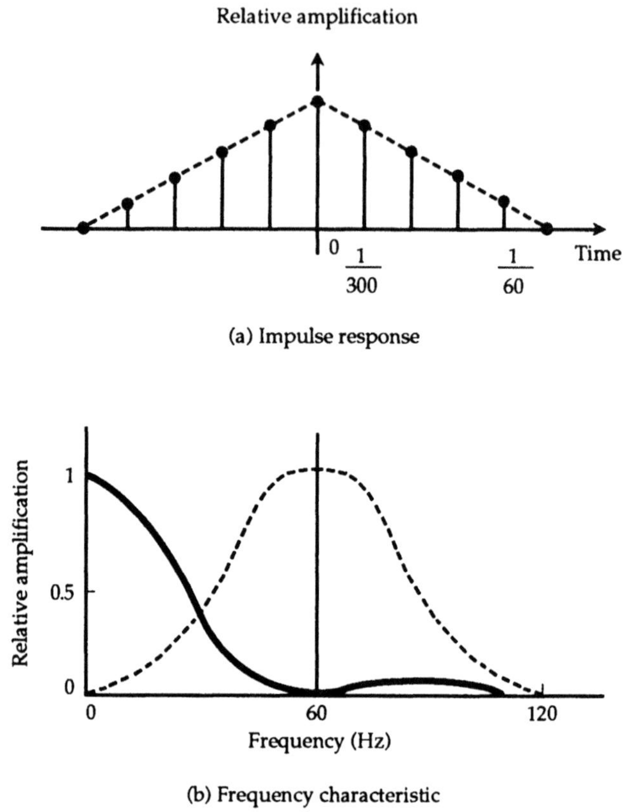

FIGURE 6.11. Linear interpolation characteristic.

for vector detection. However, positional interpolation is the ideal method of interpolation for moving images, and motion compensation interpolation using motion vectors is expected to be sufficiently effective.

The process of basic frame interpolation through motion compensation is shown in Figure 6.13. In this example, the interpolation ratio of the third frame is 4:6, the same as in the previous example. Whereas linear interpolation obtains the weighted mean of the image amplitude from the interpolation ratio, in motion compensation interpolation the position of the moving image moves according to the interpolation ratio. Image B compensates for the position of the third 60 Hz frame by 4/10 of its motion

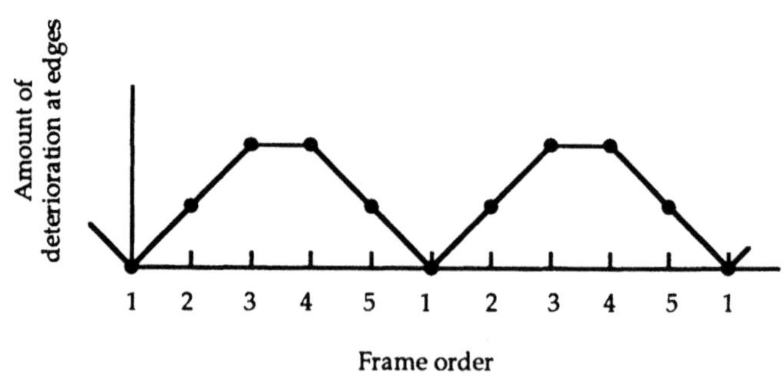

FIGURE 6.12. Amount of deteroriation at edges.

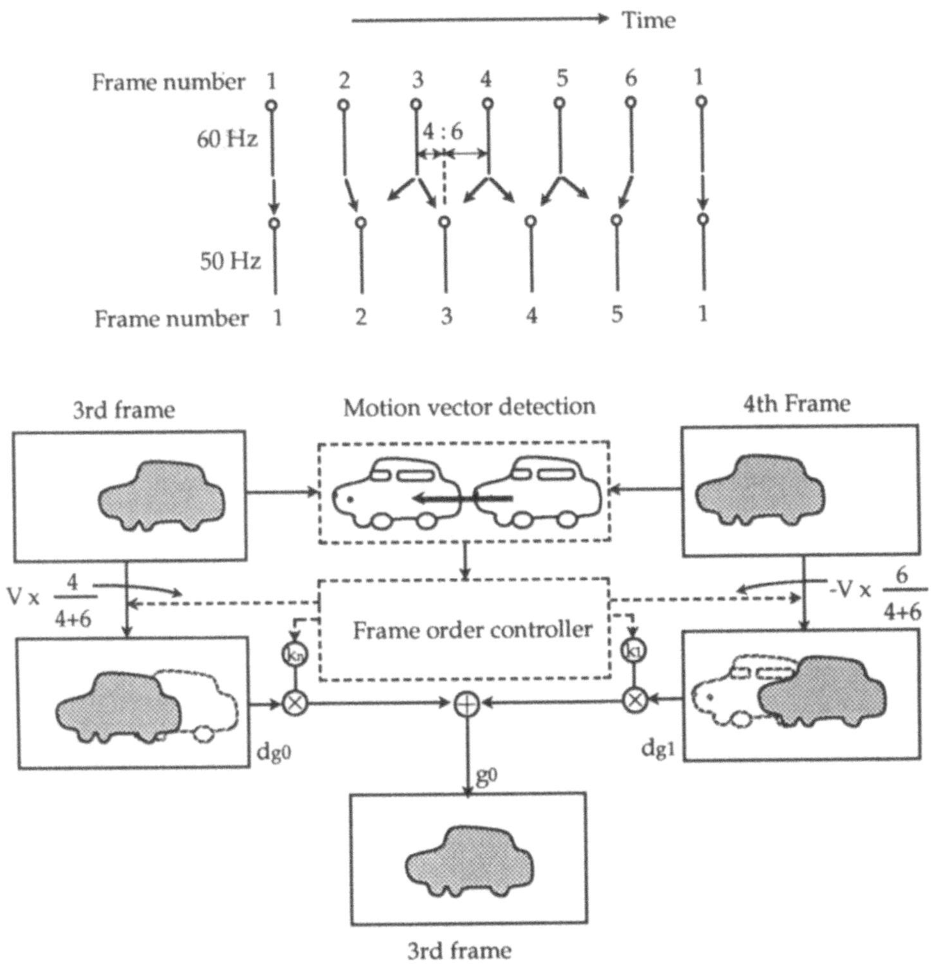

FIGURE 6.13. Motion compensation interpolation.

vector magnitude **V**, while image D compensates by taking −6/10 of the fourth frame. The weighted mean of images B and D then produces the interpolation image E. The weighted mean taken here is performed to smooth the motion within a single motion vector step.

As shown in Figure 6.14 (a), the screen is partitioned into four blocks, and motion vector detection is performed in each block. While a larger number of screen divisions would allow for the detection of smaller differences in movement, we decided to use four blocks to limit hardware requirements. In Figure 6.14 (a), cars A and B are moving to the left, and cars C and D are moving to the right. They are all moving at the same speed, and the first block is a still image. Therefore, the motion vectors detected for the first and third blocks are zero vectors, and those of the second and fourth blocks depend upon the speeds of the respective cars. In this way, four vectors can be obtained for the whole screen, and four kinds of motion-compensated images can be formed by the method indicated for image E in Figure 6.13. (However, in this case motion compensation is done for the entire screen. From these four images, which have undergone positional compensation, a pixel-unit comparison is performed with a method that uses frame difference signals (minimum frame difference addition) to obtain interpolation signals with optimal motion compensation. The frame difference signal used here is the differ-

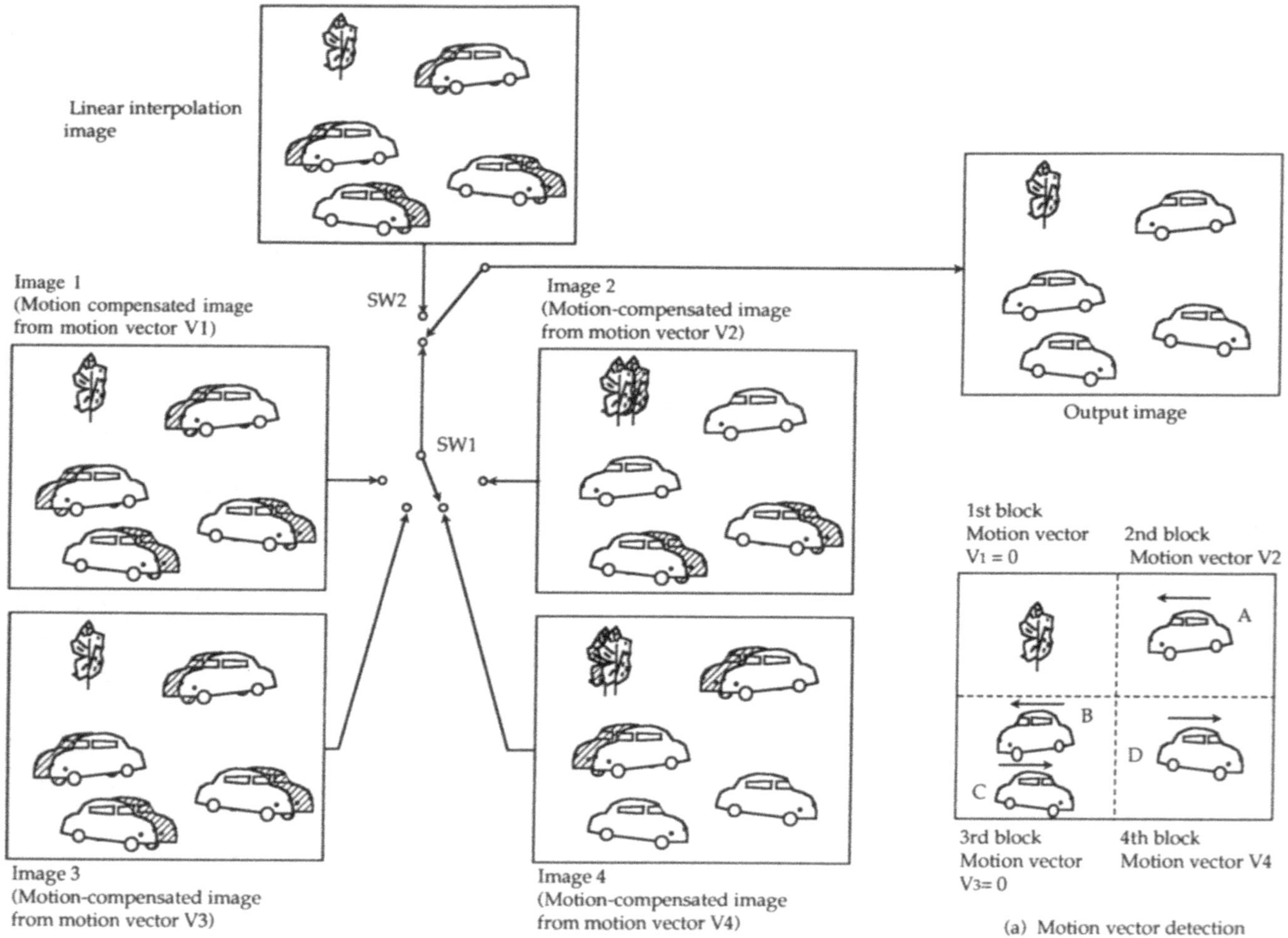

(b) Motion compensation with four vectors (Shaded areas indicate the positions where motion compensation must be performed.)

FIGURE 6.14. Motion compensation with four vectors.

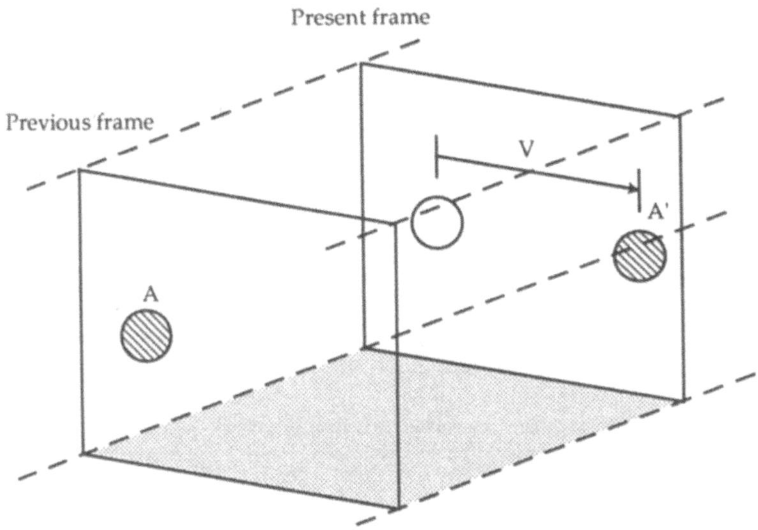

FIGURE 6.15. Pattern matching method.

ence between images B and D of Figure 6.13. The moving image frame difference that matches the detected motion vector is zero if there is an ideal parallel transfer, and this frame difference becomes the reference for selecting the optimal compensated image from the four compensated images.

Switch 1 (SW 1) in Figure 6.14 operates by using a frame difference minimum added value method as follows. For cars A and B, motion vector V_2 is detected in the second block and used for position compensation in image 2. Similarly, for cars C and D, image 4 is position compensated with motion vector V_4. The motion compensated interpolation frame signal is selected from these two images.

(c) Motion Vector Detector. In motion vector detection, the screen is divided into four blocks and a pattern matching method is used in each block. Pattern matching is a method of motion vector detection which uses the correlation between the frames of moving images. As shown in Figure 6.15, when the moving image A becomes A′ in the present frame, the displacement V that yields the greatest correlation with the previous frame becomes the motion vector.[4]

In the pattern matching method, the motion vector is the vector y out of a set of vectors which have been prepared in advance (called sample vectors) that minimizes $D_R(y)$ in the following equation.

$$D_R(y) = \Sigma\, w(y) \cdot f\,\{A^N(x) - A^{N-1}(x - y)\} \quad (6.4)$$

where

x: Pixel position vector of frame N

$A(x)$: Pixel level at point x

f: Correlation evaluation function

w: Weighted function of y

R: Aggregation of representative points

However, because of the need to reduce the scale of the hardware, the broadening of the image is taken into consideration and subsampled points (called reference points) are used in the detection.

(d) Motion-Compensated Image Selection Controller. Interpolation frame signals with optimal motion compensation can be obtained by applying the frame difference minimum added value method to the position compensation signal in Figure 6.9 ($d_g \sim d_f$: correspond to the opposing images B and D in Figure 6.13).

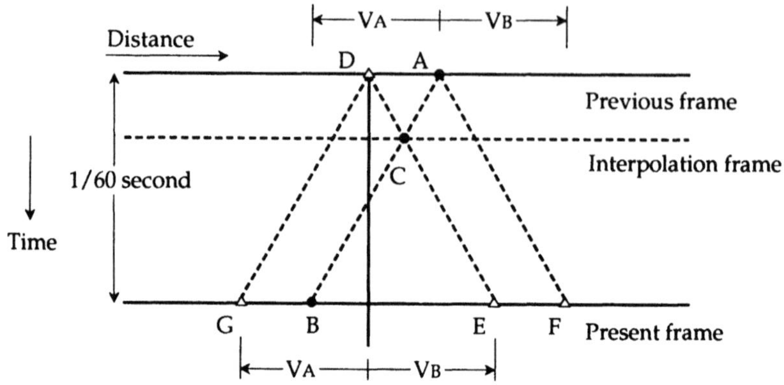

FIGURE 6.16. Frame difference minimum added value method.

Motion vector detection is performed independently in four blocks with the pattern matching method, with no relationship at the pixel level. The frame difference minimum added value method then reapplies the four-vector matching method at the pixel level (point-wise matching method) to the position-compensated images, which have been limited to four types.

We will now explain the frame difference minimum added value method using Figure 6.16.

A small black point with motion vector V_A is moving to the left from point A to point B across a screen with a uniform background. V_B is a motion vector detected in another block. The moving point's position in the interpolation frame is point C. For the moving point, V_B is a pseudo-vector. The compensated image derived from this motion vector is made from background-level points D and E, and must be eliminated from the selection.

With respect to interpolation frame point C, the frame difference added values L_1 and L_2, which are matching estimations, are made to correspond to motion vectors V_A and v_B respectively. Of these, the position-compensated image derived from the motion vector yielding the minimum added value is selected as the motion compensation interpolation frame signal. In this case, the frame difference added value for L_1 is V_A, and that for L_2 is V_B.

$$\left.\begin{aligned} L_1 &= |E_A - E_B| + |E_D - E_G| \\ L_2 &= |E_D - E_E| + |E_A - E_F| \end{aligned}\right\} \qquad (6.5)$$

Since in this example,

$$L_1 = 0 \qquad (6.6)$$

$$L_2 = |EA - EF| \neq 0 \qquad (6.7)$$

it follows that

$$L_1 < L_2 \qquad (6.8)$$

At point C, the interpolation frame signal position-compensated by V_A is selected, and the worm-hole phenomenon which results from selecting a compensated image based on a pseudo-vector does not occur.

We have explained the situation for the two vectors above. This method can be expanded for all four vectors by the same reasoning.

Noise can disturb the selection control signals obtained with the frame difference minimum added value method, causing a fine worm-hole type image quality degradation. The region for which a motion vector detected in the block can be selected as the optimal vector for each pixel in it widens according to the size and speed of the moving object. Based on this characteristic, the effect of noise is reduced using a 2-dimensional low-pass filter as the selection control signal, thereby forming an optimal region size.

(e) Output Selection Control. Lastly, selection control is needed for both the motion

FIGURE 6.17. Format conversion using output selection control.

compensation interpolation frame signals which have been selected from the position-compensated image of the four vectors, and of the interpolation frame signals from linear interpolation. In this case as well, while the frame difference minimum added value method basically should be applied, because there are only two alternatives, the equipment is simplified by comparing and choosing the frame difference signal with the smaller absolute value (in linear interpolation, it is the difference between f_0 and f_1 in Figure 6.10, and in motion compensation interpolation, the difference between images B and D in Figure 6.13.)

Figure 6.17 (a) is an example of an image for which the format has been converted, and (b) is a selection signal from the output selection control of the same scene. The television camera is following the running racehorses by panning, and the black and grey regions of (b) are linear interpolation regions. In this image, image quality degradation due to the synthesis of the linear interpolation image and the motion-compensated image was not detected.

6.1.4 Hi-Vision / NTSC Format Conversion

As the Japanese standard television format is the NTSC format, equipment for converting the Hi-Vision format to NTSC is highly useful. One advantage in converting from Hi-Vision into NTSC is that the image quality is superior to what can be obtained from an NTSC camera.[5]

The field frequencies of Hi-Vision and NTSC are 60 Hz and 59.94 Hz respectively, a difference of 1001/1000. If real time processing is not being performed, this difference is small enough that a 59.94 Hz VTR can compensate for it and play back a Hi-Vision tape. However, the sound pitch will be lower by 1/1000, and the program will run a little longer.

(1) Configuration of Hi-Vision / NTSC Format Conversion Equipment

Figure 6.18 shows the configuration of Hi-Vision-NTSC format conversion equipment. Conversion of the number of scanning lines is performed by a vertical insertion filter. As with Hi-Vision / PAL conversion equipment, while motion adaptive conversion can be performed using interfield data, the conversion ratio of 15:7 is

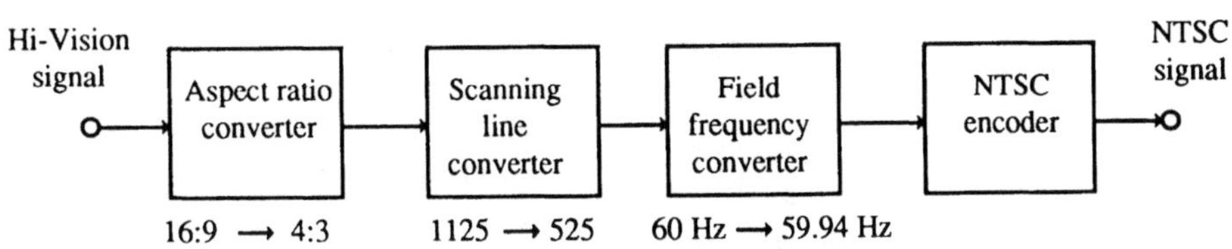

FIGURE 6.18. Block diagram of Hi-Vision/NTSC format converter.

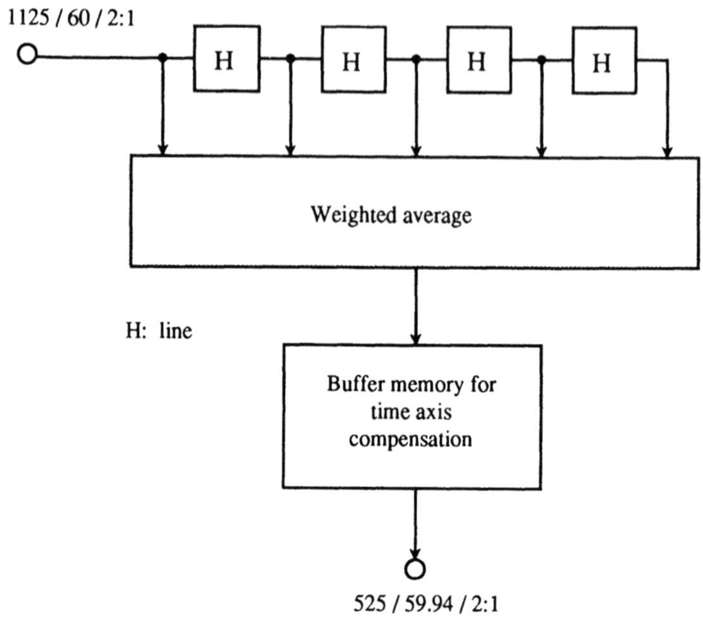

FIGURE 6.19. Vertical interpolation filter.

less than half. Considering the fact that the final image is a composite image, a simpler conversion method that uses intrafield processing is generally used. The configuration of this conversion is shown in Figure 6.19. The internal conditions of the weighted mean circuit are switched according to the line order, which has a period of 7 cycles. This is parallel processing in the sampling frequency conversion. Figure 6.20 shows the characteristic of the interpolation filter.

(2) Aspect Ratio Conversion and Conversion Modes

Figure 6.21 gives representative examples of conversion modes for aspect ratio conversions.

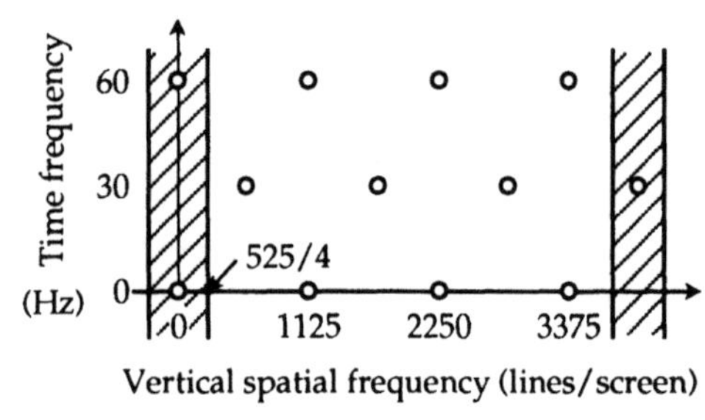

FIGURE 6.20. Vertical interpolation filter characteristics.

Mode A eliminates both sides of the 16:9 Hi-Vision image, and changes the aspect ratio to 4:3. Mode B changes the aspect ratio of the Hi-Vision image to 4:3 by compressing the image horizontally so that the converted image becomes vertically elongated. Mode C maintains the aspect ratio and displays a 16:9 screen inside a 4:3 screen. Thus the NTSC image has black borders on the top and bottom. Of these modes, the most commonly used is Mode C because it maintains the intended proportions of the Hi-Vision program.

(3) Field Frequency Conversion

After the number of scanning lines is converted, the buffer memory is used to convert the field frequency. The configuration of both of these conversions is shown in Figure 6.22. The area enclosed by the dashed line is the field frequency converter which has a motion adaptation frame synchronizer.

While the most commonly used frame synchronizers will skip one frame every 33 seconds, this causes a noticeable jump in a moving image. Thus in motion adaptation field frequency conversion, motion detection and scene-change detection are performed using frame difference

Parameter	Converter input	Converter output		
	Hi-Vision	Mode A	Mode B	Mode C
Aspect ratio	16:9	4;3	4:3	4:3
Image	---	Normal	Vertically elongated	Normal
Processing result	---	Left and right edges cropped	Full screen	Top and bottom blacked out

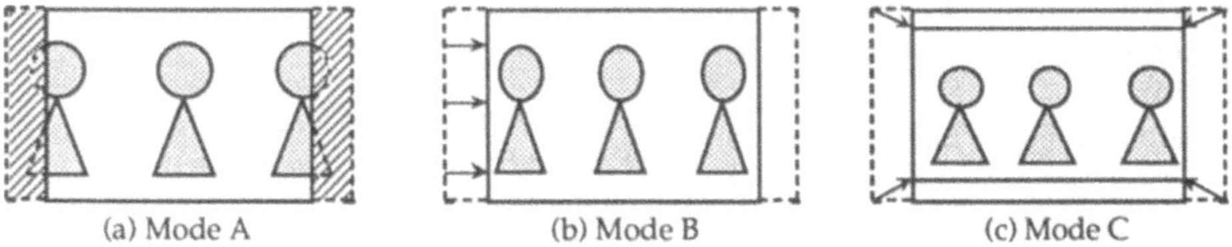

(Left and right edges are discarded. Image can be moved horizontally inside the 4 : 3 screen.)

(compressed in horizontal direction only)

(16 : 9 image is compressed horizontally and vertically into the 4:3 screen, leaving top and bottom of screen blank.)

FIGURE 6.21. Aspect ratio conversion modes.

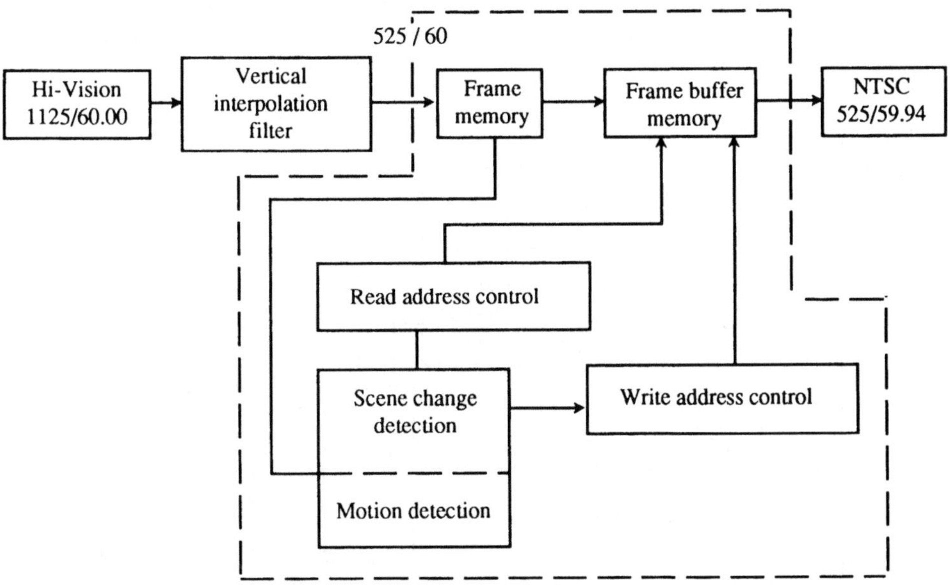

FIGURE 6.22. Scanning line/field frequency converter.

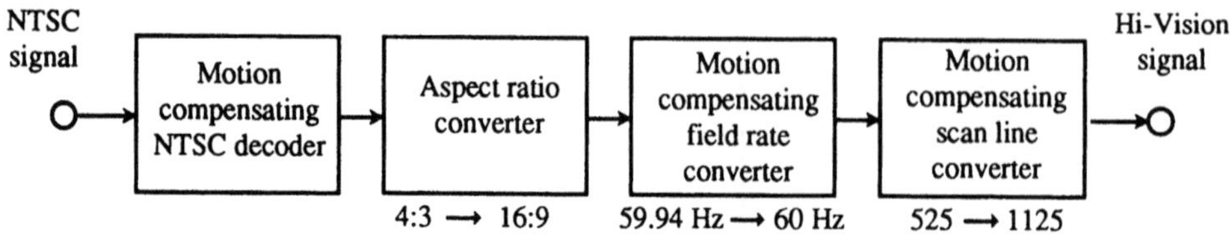

FIGURE 6.23. Configuration of NTSC/Hi-Vision converter.

signals, and a frame is skipped when one of the following four conditions is satisfied:

1. When it is a still image,
2. When a scene change has occurred,
3. When the area of the moving image is relatively small,
4. When no frame buffer memory remains.

Of these, condition 4 forces a frame skip regardless of the condition of the moving picture. However, when motion adaptation field frequency conversion is performed using these four modes, image degradation caused by frame skips is practically indiscernable. Increasing the capacity of the frame buffer memory will alleviate condition 4, but care must be taken to avoid problems with lip synching.

6.1.5 NTSC-Hi-Vision Format Conversion

NTSC programs, which have accumulated over many years and will continue to increase in number, are a valuable resource for Hi-Vision. Particularly in the initial stages of Hi-Vision broadcasting, as most video such as for news coverage will be limited to the NTSC format, conversion from NTSC to Hi-Vision will be crucial.

(1) Configuration of NTSC / Hi-Vision Format Conversion Equipment

Figure 6.23 shows the general configuration of NTSC/Hi-Vision format conversion equipment. Since the image is being converted into a high quality Hi-Vision signal, degradation of the NTSC signal characteristic during the conversion must be suppressed as much as possible. Thus in converting from 525 to 1125 scanning lines, the NTSC decoder, which performs Y/C separation, is used to detect the motion signals and in turn use these to control the motion adaptation processing.

(2) Motion Adaptation NTSC Decoder

Figure 6.24 shows the configuration of the motion adaptation NTSC decoder. It performs mo-

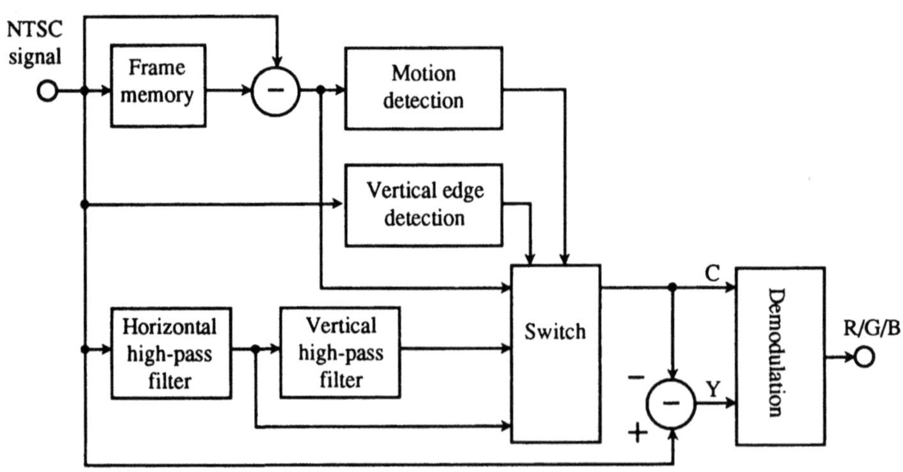

FIGURE 6.24. Motion compensation NTSC decoder.

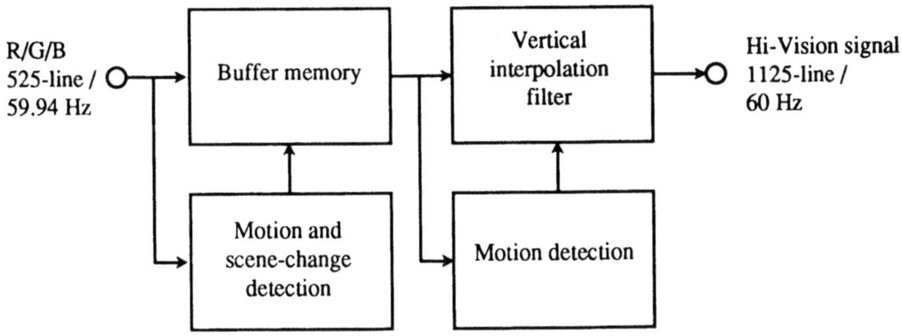

FIGURE 6.25. Motion compensating scanning line/field rate converter.

tion adaptation Y/C separation between frames in stationary image areas, and within the field in moving areas. It also detects vertical edge signals and eliminates dot artifacts in the moving areas.

(3) Motion Adaptation Scanning Line and Field Frequency Converter

The configuration of these units is shown in Figure 6.25. As with the Hi-Vision/PAL format converter, scanning line conversion is performed with interfield and intrafield motion adaptation processing. The vertical filter is an interpolation filter for scanning line conversion. The field frequency converter has a buffer memory which functions in the same way as in Hi-Vision/NTSC format conversion equipment, and eliminates image degradation caused by frame skips.

(4) Aspect Ratio Conversion Mode

Using horizontal sampling frequency conversion, the 4:3 aspect ratio is converted to 16:9 in two modes: the 4:3 image can be displayed in whole on the 16:9 screen, or it can be cropped at the top and bottom to fit the 16:9 screen. In terms of the resolution of the original image, the most effective mode is to display the NTSC image on part of the screen during a Hi-Vision program.

6.2 APPLICATIONS IN MOTION PICTURES

Television and motion pictures have enjoyed a long standing and close relationship. At its inception, television adopted and developed the more advanced technologies of the motion picture industry. Even now, a large number of movies are being broadcast as television programs. However, from the standpoint of movies, hardly any television technology is being used.

The development of Hi-Vision technology has generated a more active approach in applying television techniques for special effects such as chromakey synthesis. Below are some of the advantages of using high-definition in movie-making:

1. Special effects such as chromakey are very simple.
2. Costs can be reduced by shortening the production schedule.
3. The shooting can be viewed instantly.
4. Image degradation in the film-making process is minimal.

Experiments aimed to demonstrate these advantages have already been performed.

The Italian Broadcasting Institute (RAI) showed an interest in using Hi-Vision in motion picture production from its inception. In 1985, they produced the short film *Oneiricon*. In 1986, NHK produced the mini-drama *Autumn, Kyoto,* testing the expressiveness of the technology in movie-making. In 1987, RAI completed a 90-minute feature film called *Julia and Julia*.

In Japan as well, both the technological and economic aspects of applying Hi-Vision to movies are being studied.

6.2.1 Comparison of 35mm Movies and Film Recording

To successfully use Hi-Vision technology in movie production, the conversion from videotape to film must have a minimal image degradation. Laser film recording and electron beam recording, which we will discuss later, have been developed as film recording formats for use with Hi-Vision technology.[6] In contrast to conventional kinescope recording, in which images on a CRT are filmed, these methods record images directly onto 35 mm film.

Motion pictures which have been filmed and produced in these formats have the same basic scanning line structure as Hi-Vision, but are processed to make the line structure invisible by having adjacent scanning lines barely touching each other. Thus movies made in Hi-Vision can be shown in movie theaters as regular movies. However, movies made with the electronic image processing methods of Hi-Vision are technologically different from regular movies in a number of areas.

(1) MTF (Modulation Transfer Function)

In recording Hi-Vision onto film, the image is recorded onto a 35mm film frame with a 16:9 aspect ratio as shown in Figure 6.26. Since the height of the frame is 12.4 mm, a television image with 1000 TV lines converts to 40.3 lines per millimeter on film (a black and white pair is counted as one line). Based on this number,

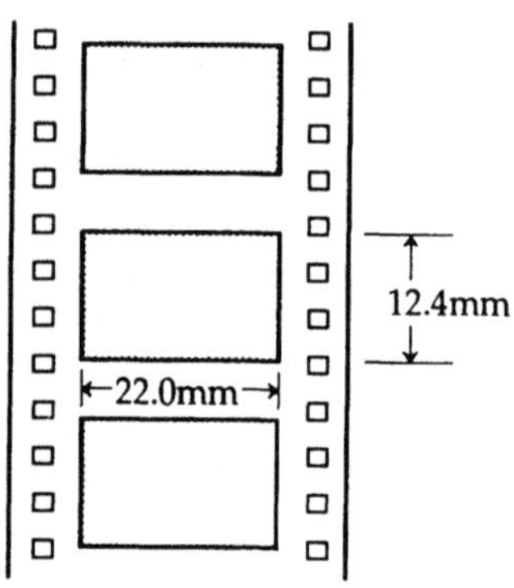

FIGURE 6.26. Screen dimensions for Hi-Vision 35 mm film reodering.

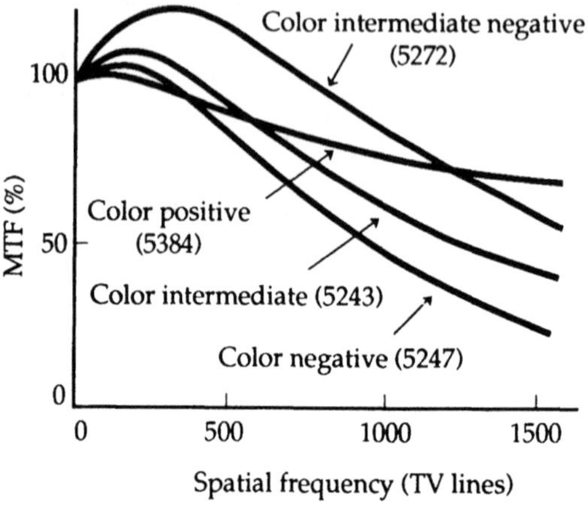

FIGURE 6.27. MTF of motion picture film.

Figure 6.27 shows the MTF[7] of several types of film which are used for motion pictures.

The fundamental process in movie production are recording images on color negative film, and printing a color positive film. The MTF in this case is a combination of the two, and that value is compared with high-definition in Figure 6.28. As shown in the figure, the MTF of Hi-Vision is restricted to the 20 to 30 MHz pass band of the equipment. However, so that the MTF with the band can be held constant during different types of Hi-Vision video processing, it can basically be controlled with the addition of compensation.

On the other hand, motion picture systems are not equipped to perform a systematic restriction and have a significant MTF even at 1000 TV lines or more. However, since the image goes through a series of systems including the camera lens, negative film, printer, positive film, and the projector lens, even if the MFT of each particular system is not extremely low, the MTF of the system as a whole is quite low. This characteristic cannot be controlled through compensation as it can be with video systems. Therefore, when complex special effects are done for dramatic effect, in some cases an extreme drop in MTF occurs when optical processing is repeated at many levels. This may also be accompanied by an increase in noise from granularity, which we will discuss later.

Taking the above points into consideration, the production of movies with Hi-Vision tech-

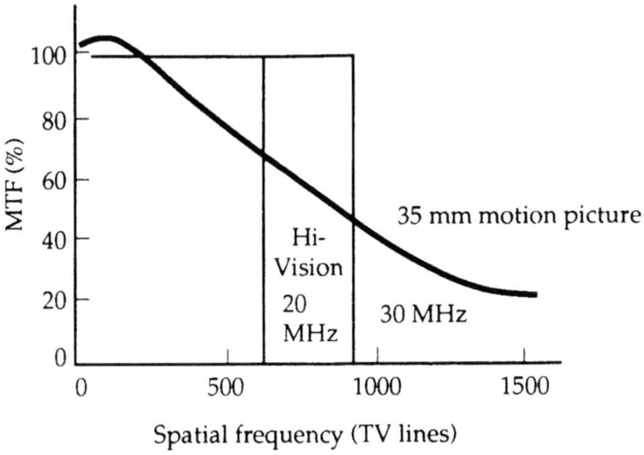

FIGURE 6.28. MTF of Hi-Vision and motion pictures.

nology has significant advantages in terms of complex special effects. That is, in this case, film is used as a medium only in the last film recording and positive printing processes, and the MTF can be compensated for electrically to some extent so that degradation is held to a minimum. Or, if high-resolution fine-grain film is used with the laser film recording which we will discuss later, these advantages can be put to even better use, resulting in improvements in the MTF.

(2) Noise from Granularity

Video noise and film noise differ in nature and also appear differently on the screen. That is, video noise (thermal noise) is generally visible in dark areas where signal levels are low, whereas film noise results from film emulsion granularity in phase tone areas, which are relatively bright.

Film noise is the accumulation of noise from input video signals and from film granularity, which comes from the positive film during film recording final printing. Video signal noise can be reduced somewhat by noise reduction, but since nothing can be done to rectify noise from film granularity during the final recording, fine-grain film must be used as much as possible.

Noise from granularity in movie film originates from both negative and positive film, and basically cannot be avoided. Moreover, when duplicated film is used to perform complex optical processing, the granularity noise can accumulate to rather high levels.

It is convenient to evaluate noise from film granularity by converting it into the signal-to-noise ratio used for electrical signals. That is, the SN ratio can be obtained by first converting RMS granularity G, which is defined as 10^3 times the standard deviation ΔD for a specified density D, into a transmission factor. Then, the ratio of this value to the transmission factor T_W, which corresponds to the white peak of the film, is found by the following formula:[8]

$$S/N = 20 \log \left[\frac{T_W}{(\ln 10) \cdot 10^{-1D} \cdot G \cdot 10^{-3}} \right] \tag{6.9}$$

The combined RMS granularity G_t for the negative and positive films can be expressed by the following formula where the RMS granularities are G_n and G_p for the negative and positive films respectively, and the gamma of the positive film is γ_p:

$$G_t = \sqrt{(\gamma_p \cdot G_n)^2 + G_p 2} \tag{6.10}$$

These formulas can be used to evaluate the granularity noise for motion picture and film recording methods.

Based on this method, Table 6.3 shows the SN ratio of motion picture and film recordings calculated from the values for granularity and gamma for different films shown in Table 6.4.[7] As the table indicates, the SN ratio of ordinary motion pictures printed from the original neg-

TABLE 6.3. SN ratios of motion picture and film recording.

Film format	SN ratio (dB)
Motion picture (5247 → 5384)	43.7
Film recording (5272 → 5384)	50.7

ative (5247→5384) is 43.7 dB. The SN ratio is, of course, the same for film recordings using the same film. However, the SN ratio of film recordings that use fine-grain color internegative film (5272→5384) is 50.7 dB, which is 7 dB higher than that of an ordinary motion picture. Therefore, Hi-Vision movies that are made by recording video signals with high SN ratios onto film are capable of achieving an image quality with SN ratios that are considerably higher than for conventional motion pictures. However, because the SN ratio must not be reduced by VTR dubbing, the commercialization of digital VTRs is much sought after.

(3) Aspect Ratio

Various aspect ratios are used for motion pictures, including standard, wide screen, and Cinemascope. Many movie screens are 400 inches or larger diagonally, which is larger than Hi-Vision screens. For screens of this size, a relatively wide aspect ratio is desirable. Figure 6.29 compares the aspect ratios of screens for movies and for Hi-Vision. The Hi-Vision aspect ratio approaches that of the wide screen. When a motion picture made with Hi-Vision is shown in an ordinary movie theater, the wide screen aspect ratio will become standard. When a Hi-Vision clip is inserted into a movie filmed with a wide aspect ratio, the aspect ratios do not match perfectly. Since the Hi-Vision clip will have a taller image, the top and bottom portions of the image will be cropped using a projection mask. This must be kept in mind when a Hi-Vision camera is used to shoot a scene.

(4) Gradation Reproduction

The gradation reproduction range of motion pictures and Hi-Vision differ by one order of magnitude. A typical motion picture has a film density range of about 3 and can reproduce 1000 to 1. In the Hi-Vision system, the SN ratio determines the reproduction range, which is normally 40 dB, and so the reproduction range is only 100 to 1.

Thus if the motion picture and Hi-Vision images were processed as is, the images would differ in contrast level. In motion pictures using Hi-Vision film recording, the image is processed to give it a film-like quality by performing black and white stretching with non-linear gamma compensation as well as implementing measures to improve the apparent contrast. When film recordings using Hi-Vision special effects are inserted into a movie, it is especially important to make sure that the film blends into the rest of the movie.

6.2.2 Laser Film Recording

Laser film recording is a recording method which forms a television image directly on film by exposing the film to laser light. Because television signals are one-dimensional time series

TABLE 6.4. Examples of RMS granularity and gamma for motion picture film.

Film	RMS granularity	Gamma
Color positive (5384)	6.0	3.3
Color negative (5247)	5.0	0.63
Color intermediate negative (5272)	1.6 *	0.51

* Observed value

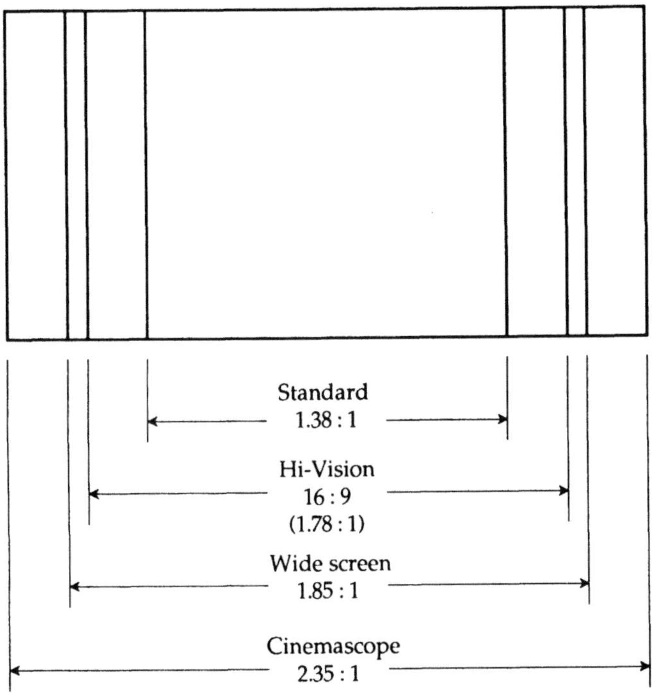

FIGURE 6.29. Comparison of motion picture and Hi-Vision screen formats (aspect ratio).

signals obtained by scanning, an image with the corresponding scanning line structure is formed on the film as well. Therefore the basic elements in laser film recording are a laser light source, a light modulator to modulate the laser light with the television signal, and a light deflecting system to deflect the laser beams. Also, a format converter is needed to record television signals onto film because of the difference in frame rates.

Figure 6.30 shows the basic configuration of laser film recording equipment. In addition to the components shown, a complete operating system needs a video process to regulate electrical signals, optical systems, and a film camera to advance the film. To record on color film, a system shown in the figure up to and including the light modulator is required for each of the RGB colors.

Since light deflection occurs after the RGB laser beam is synthesized, one light deflector suffices to record directly onto color film with the synthesized and deflected laser beam. In this section, the discussion of laser film recording will focus on the laser recording equipment developed by NHK.[9]

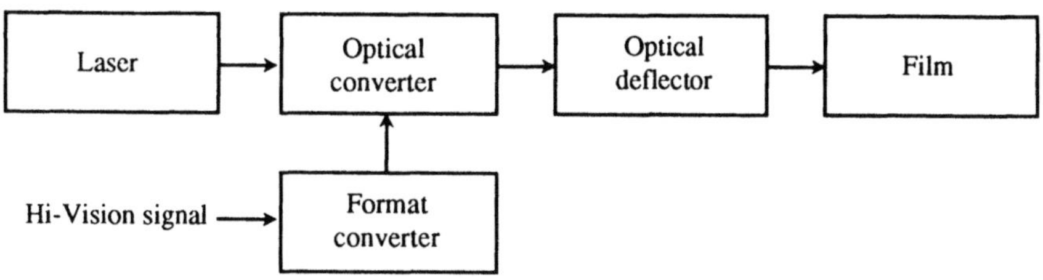

FIGURE 6.30. Block diagram of laser film recorder.

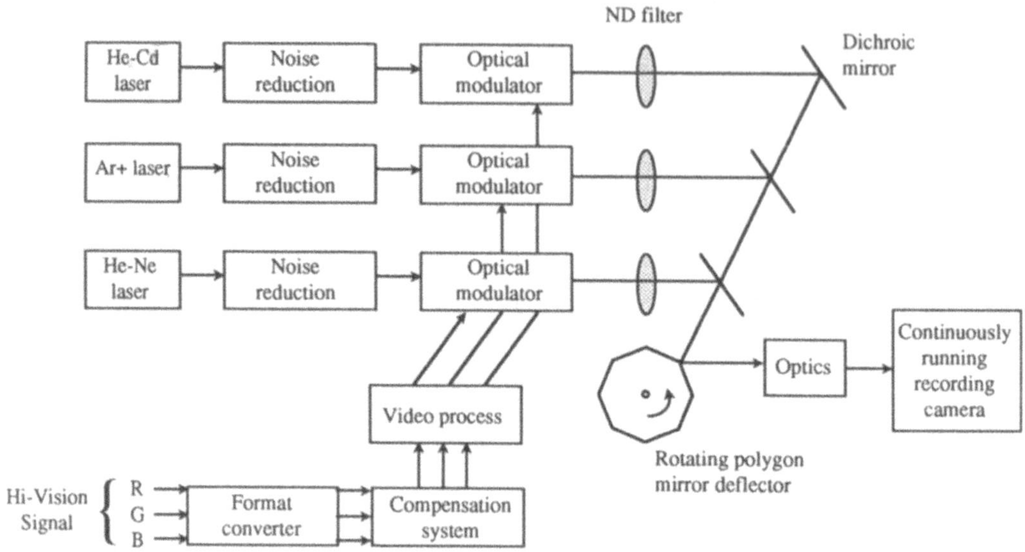

FIGURE 6.31. Laser film recorder.

(1) Laser Film Recording Equipment
Figure 6.31 shows the configuration of the laser film recording equipment developed by NHK.

(a) Laser Light Source. For best results, the laser used as the light source should function with stability over a long duration, and its wavelength should match the spectral sensitivity of the film being used.

Figure 6.32 shows the relationship between laser wavelength and spectral sensitivity. As shown in the figure, the light sources used are an He-Ne laser (wavelength 632.8 nm) for R, an Ar^+ laser (wavelength 514.5 nm) for G, and an $He\text{-}Cd^+$ laser (wavelength 441.6 nm) for B. In consideration of power loss inside the optical system, the lasers are set at 50 mW for R, 100 mW for B, and 15 mW for B to ensure sufficient exposure of the film.

(b) Noise Reduction Circuitry. Figure 6.33 shows the configuration of the noise reduction

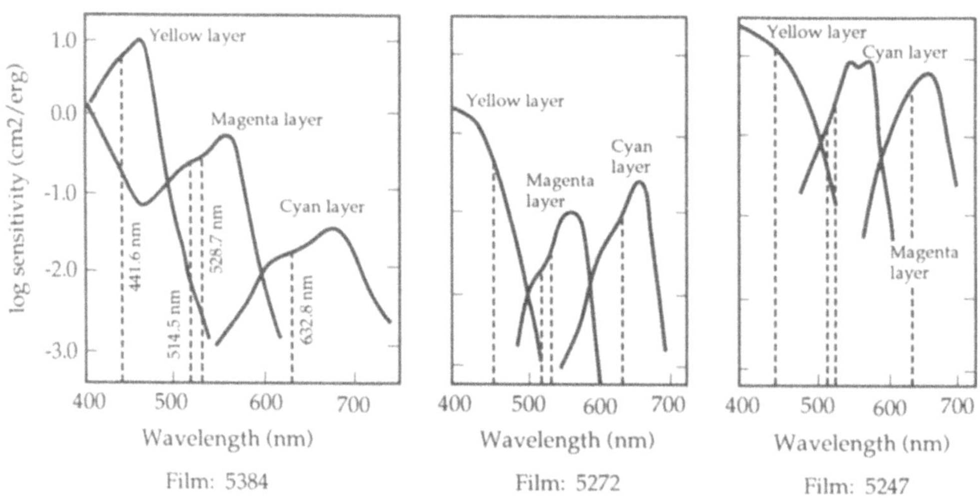

FIGURE 6.32. Laser wavelength and film spectral sensitivity.

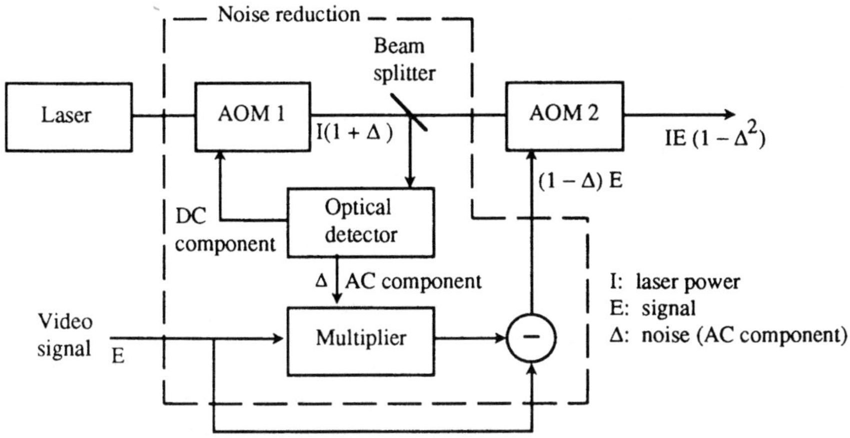

FIGURE 6.33. Noise reduction method.

circuitry, which reduces the laser's output fluctuation and noise.[10] Part of the noise-ridden laser light is split with a beam splitter and directed to a light detector. The DC component of the light detector output is fed back to AOM 1 (Acoustic Optical light Modulator—to be explained later) to reduce the fluctuation. After the multiplier takes the product of the AC component (noise) Δ and the input video signal, that product is subtracted from the signal. By using this signal to drive the AOM 2, which modulates the laser beam, the noise output becomes Δ^2 Since Δ is usually small, sufficient noise abatement (10 to 20 dB in practice) is possible when this method is used.

(c) Light Modulator. In laser film recording, it is necessary to modulate the strength of the laser beams with the input signal. Laser beam

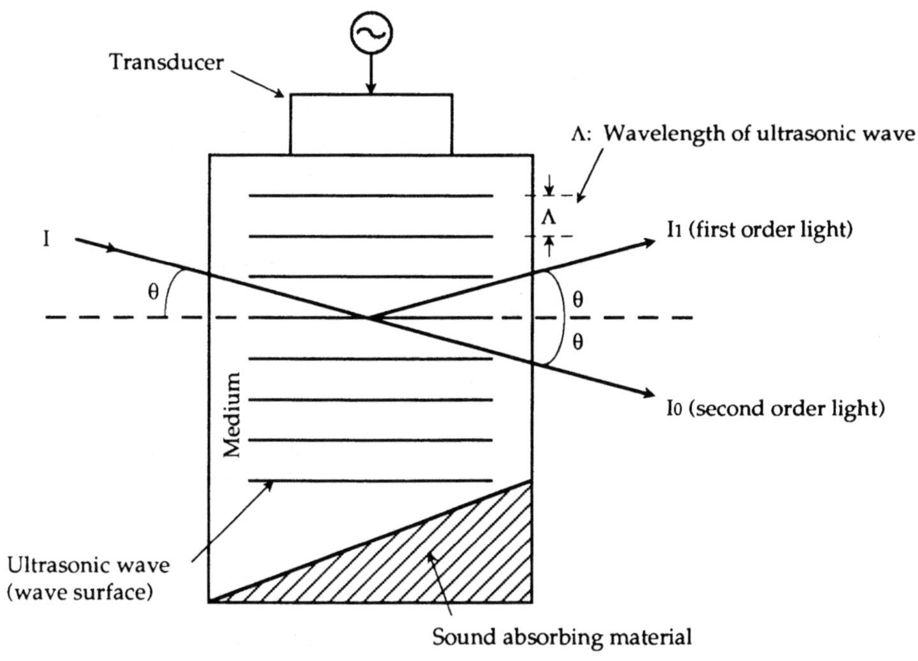

FIGURE 6.34. Light diffraction due to acoustic-optical effect.

modulators consist of two types. The electro-optical system uses the electric-field dependence of an electro-optical crystal's birefringence. The acoustic optical system generates ultrasonic waves across an ultrasonic medium in reference to a modulation signal. This alters the medium's refraction index and thereby modulates the intensity of the diffracted light from the incident laser beam (Figure 6.34).

The former system can have a high modulation frequency, but it needs a large driving voltage to produce an even lower contrast ratio. On the other hand, the latter system can obtain a high contrast ratio with a low driving voltage. Its modulation bandwidth is also sufficient for Hi-Vision. Therefore, Acoustical Optical Modulators (AOM) are the most suitable, at present.[11] In Figure 6.34, the diffraction efficiency for Bragg diffraction is high, and the output light I_1 is

$$I_1 = I \sin^2 (KV / \lambda) \qquad (6.11)$$

where V: Voltage applied to the transducer
λ: Wavelength of the laser beam

In laser film recording, it is necessary to take the $\sin^2$ characteristic of the AOM into account during tone reproduction. The luminous energy of the different laser beams which have been modulated by the R,G, and B image signals is regulated by the ND filter, and the light is deflected after being converged by a dichroic mirror.

(d) Light Detector. Because a laser film recorder forms images on film using a scanning line structure, it needs a way to deflect the laser beams both horizontally and vertically. Laser beam deflection can be performed using devices such as Acoustic Optical Deflectors (AOD), galvanometers, and rotating polygon mirror light deflectors. A galvanometer is sufficient for vertical light deflection, which is considerably slower than horizontal light deflection. The high speed required for horizontal deflection, especially in Hi-Vision, calls for either a rotating polygon mirror or an AOD. Rotating polygon mirrors are simple in principle but difficult to manufacture. However, they have superior characteristics, and do not suffer from color dispersion like AODs do, especially with a laser beam having three wavelengths for R, G and B. Unlike AODs, a rotating polygon mirror deflector deflects R, G, and B lasers simultaneously. Thus our laser film recording equipment uses a rotating polygon mirror for horizontal deflection.

The rotating polygon mirror deflector developed at NHK has a diameter of 40 mm, is prismatic with 25 surfaces, has static pressure pneumatic bearings, and rotates 81,000 times per second. Table 6.5 shows the manufacturing precision and the influence that that precision has on the image quality of recorded film. In order to record Hi-Vision signals onto film and to be able to enjoy the film as a motion picture on a

TABLE 6.5. Accuracy of the rotating polygon mirror and its influence on image quality.

Factor	Manufacturing accuracy	Influence on image quality
Angular distribution error	Within 10 seconds	Horizontal jitter in each scan line
Reflectivity error	Less than 1%	Banding noise (luminance non-uniformity in each scan line appears in a 25-line cycle)
Parallelism error	Within 20 seconds	Parallelism error between the reflective surface and rotational axis causes pitch irregularity of scanning lines→Luminance nonuniformity
Shading at the edges		Shading noise (depends on laser beam diameter)

big screen, error compensation must be performed again for the errors caused by the rotating polygon mirror deflector. We will discuss this later.

(e) Recording Camera. Instead of using a galvanometer for vertical deflection, this laser film recorder uses a recording camera that continuously advances the film. As the film advance feature functions as the recorder's vertical deflection mechanism, it needs to be extremely accurate to record the 1,000 or more scanning lines per frame. With regard to uneven film advance, the high frequency component causes non-uniformity in the scanning line structure on the film, which appears as expansions and contractions of the image. The low frequency component causes vertical jitter across the entire screen. In general, the latter is more pronounced. A camera that is accurate enough for Hi-Vision signals is needed.

(f) Format Converter. The format converter converts signals with Hi-Vision specifications into recording signals which meet the specifications of 35 mm movie film. Conversion is mainly from 60 fields per second to 24 frames per second, but when a continuously running recording camera is used, sequential scanning conversion of the Hi-Vision signal is also necessary. As frame memory must be used for these conversions, the signals must be digitized. Also, it is best to perform various kinds of digital signal processing which are thought to be effective along with the format conversion.

Figure 6.35 shows the configuration of the format converter.[12] This configuration includes a compensation system for the optical system and the deflector. The frame rate converter and the progressive scanning converter in this figure are configured using a frame memory. An example of a frame rate conversion method is given in Figure 6.36. In this example, five Hi-Vision frames (10 fields) are converted into four film frames. Hi-Vision fields 1 and 2 form the first film frame, while Hi-Vision field 3 is not used. Film frame 2 is formed from Hi-Vision fields 4 and 5, film frame 3 is formed from Hi-Vision fields 6 and 7, and field 8 is not used. By repeatedly converting the frames in this way, four frames are made from five frames as stated above, and the reference frequency of the conversion becomes 6 Hz. When this kind of conversion is done, the appearance of moving images becomes unnatural. This unnatural movement occurs as moving image judder, which is distinct from the unnatural movement which occurs during projection when each frame is projected twice.

Another format conversion method is linear interpolation. In this method, Hi-Vision fields are weighted by their distance along the time axis, then added to form one frame of film. The frame frequency produces an effect here as well, but in this particular case, the moving image area within a frame becomes a multiple image of the added fields, thereby causing the moving area to blur in such a way that the unnatural

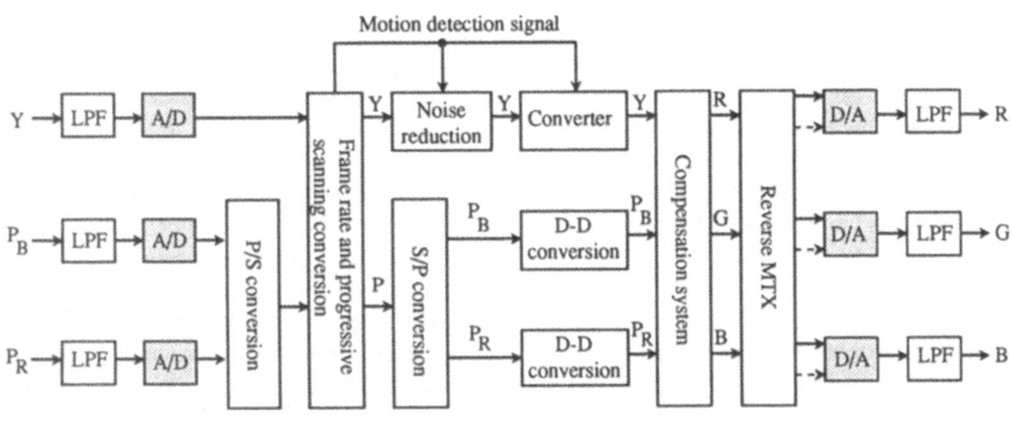

FIGURE 6.35. Block diagram of format converter.

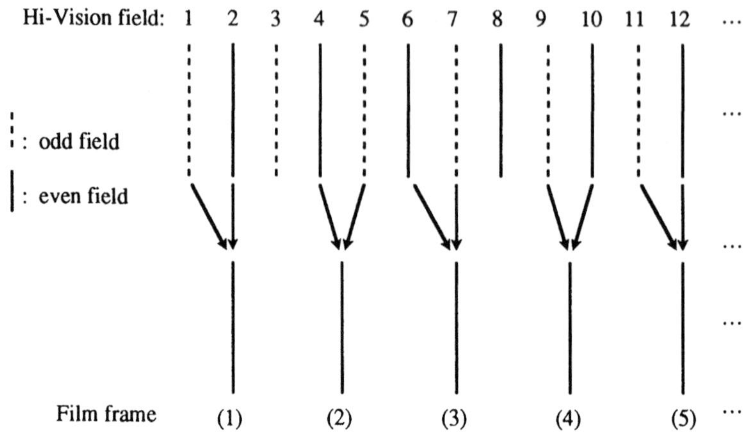

FIGURE 6.36. Frame rate conversion.

movement is not as noticeable as in the previously described conversion.

Next we will discuss progressive scanning conversion. In Figure 6.36, when film frame 1 is simply formed from Hi-Vision fields 1 and 2, an artifact (double image) is generated corresponding to the amount of movement that occurs in the 1/60 second between fields 1 and 2. Thus the moving image areas are detected, and for stationary areas an interfield conversion is performed between fields 1 and 2, while an interfield conversion is performed in field 2 for the moving image areas. In this case, the vertical resolution in the moving image areas is cut in half, but this is not very noticeable. While conversions like those mentioned above are presently being done, in the future frame rate and progressive scanning conversions will, like other format converters, use motion vector detection and position interpolation for moving areas.

Noise reducers are generally cyclical low-pass filters in the time direction. They improve the SN ratio using the correlation between the frames of the image through sequential addition of the previous frame image to the present frame image. Therefore, much improvement can be achieved since correlation between the frames in the stationary image area is large. On the other hand, in the moving image area, the image is degraded by an after-image artifact. In dealing with this, the after-image is controlled and noise is reduced to satisfactory levels by stopping the cycle of the previous frame image in the moving image area using the moving image area detection signals which are detected in the frame rate converter and in the progressive scanning converter.

The enhancer should adjust the amount of enhancement and the boost frequency according to the level and the image so as not to intensify the noise in the dark section significantly. The moving image area detection signal is used to ensure that only the completely stationary areas of the image are enhanced.

(g) Compensation System. The compensation system primarily compensates the optical system's chromatic aberration and errors of the rotating polygon mirror deflector mentioned above.[13] Also, because tone gradation in the dark section is important for the film system, a number of compensations for gradation are performed here. Figure 6.37 is a block diagram for chromatic aberration compensation and angular division error compensation. In chromatic aberration, the system shown in the figure is used on two of the R, G, B signals, and the delay is changed for each pixel. Compensation is divided into clock pulse level compensation and sub-clock level compensation. Clock pulse level compensation is done by delaying data, and compensation within one clock pulse is done by delaying the D/A converter clock (the register just before the D/A converter). By adding the delay caused by the angular division error per

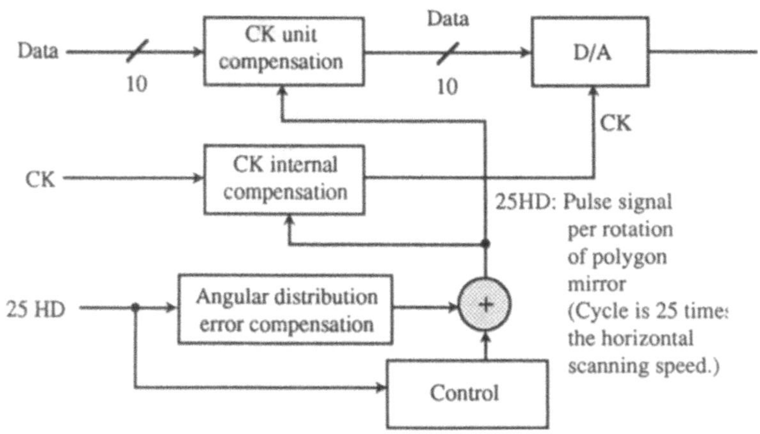

FIGURE 6.37. Color abberation and angular distribution error compensation system.

scanning line to the delay above, the angular division error is compensated for simultaneously as well.

Figure 6.38 shows the shading and reflection ratio error measurement system. Errors are detected on the film with a photo-multiplier, the scanning period is divided into 180 blocks, and 10-bit compensation data is stored in memory for each block. This compensation data is read from the memory in synchronization with the rotating polygon mirror deflector. Here, the noise reduction circuit becomes a cyclic noise reducer with a delay of 25 times the scanning cycle. This eliminates the effect of laser noise, and the correct compensation data can be obtained. This compensation is performed by multiplying the signal with the compensation signal in the video processor.

(h) Video Processing. Video processing involves aperture compensation, gain pedestal adjustment, and gamma compensation. These analog circuits have sufficiently wide bands to handle Hi-Vision signals. In gamma compensation, the gamma characteristic (γ_f), the AOM $\sin^2$ characteristic (γ_a) and the camera gamma characteristic for input signals (γ_c) are compensated for according to the different kinds of film.

(2) Laser Film Recording Characteristics

(a) Aspect Ratio. The Hi-Vision aspect ratio is 16:9 (1.78:1), which is close to that of wide screen motion pictures and Vista Vision (1.85:1). Even Hi-Vision laser film recording is done with an aspect ratio of 16:9. The screen dimensions of laser film recording are shown in Figure 6.39. The film frame is 22 mm wide,

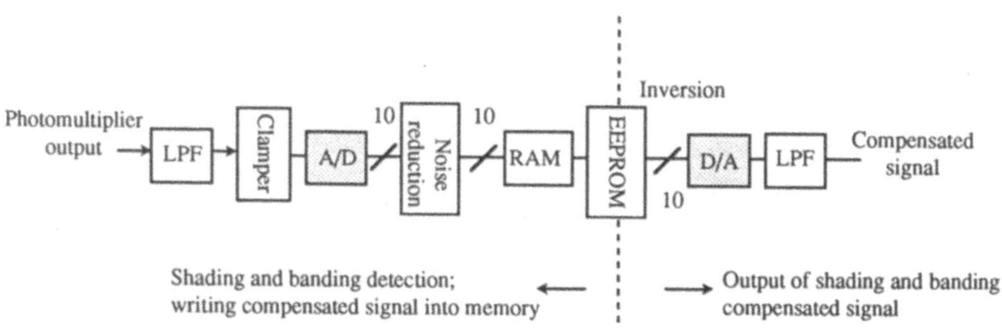

FIGURE 6.38. Shading and reflectivity error measurement system.

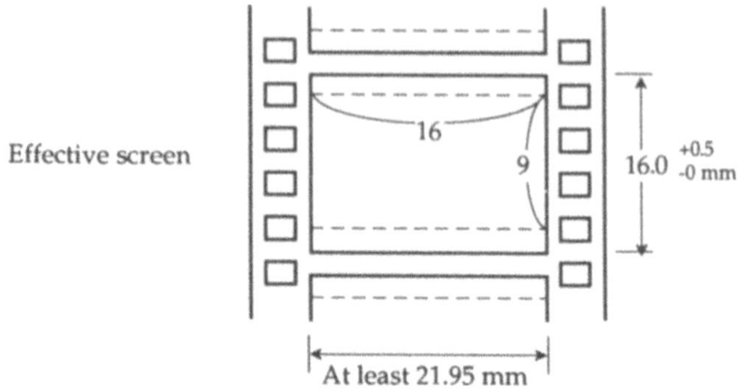

FIGURE 6.39. Aperture of 35 mm camera.

which means that with an aspect ratio of 16:9, the frame height is 12.4 mm.

(b) MTF (Modulation Transfer Function). The MTF of laser film recording is the product of the MTF of laser film recording equipment and the film MTF. The brightness distribution of the laser beam is Gaussian and is expressed as follows:

$$P(\gamma) = \exp\left(-2\gamma / (D/2)^2\right) \quad (6.12)$$

Here γ is the distance from the center, and D is the beam diameter which satisfies $P(D) = 1/e^2$. Therefore, the aperture response of the laser beam is a Fourier conversion of Equation 6.12:

$$\begin{aligned} Y(f) &= F\{P(r)\} \\ &= \frac{P}{2}\sqrt{\frac{\pi}{2}}\exp\left(-\pi^2 D^2 f^2 / 8\right) \quad (6.13) \end{aligned}$$

In this equation, f is the spatial frequency (cycles/mm). Figure 6.40 shows the aperture response of the laser beam and the film MTF when $D = 15\ \mu m$. The 5247 is 35 mm photographic color negative film, and 5384 is 35 mm color positive film for printing, but as shown in

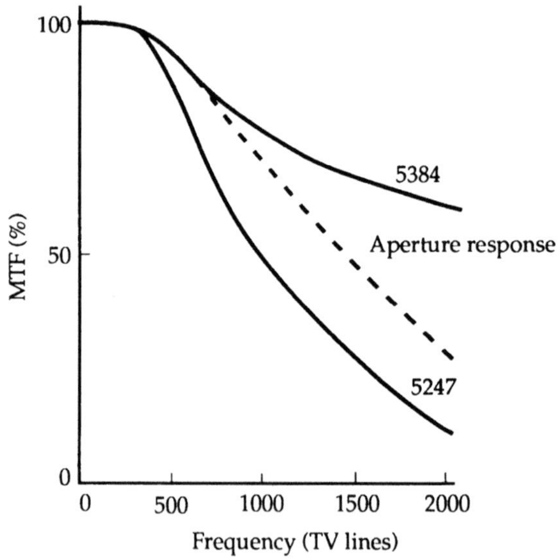

FIGURE 6.40. Relationship between aperture response and MTF of film.

Figure 6.27, there are several other kinds of film that can be used in laser film recording. For negative recording film, it is desirable to use high-resolution fine-grain film which is appropriate for laser film recording. A prototype for this film has actually been developed and is expected to be commercially available.[14] The MTF of laser film recording is the product of the MTF of the laser beam aperture response and the film MTF. When the MTF of printing and of the projector lens during the developing process is taken into account, the MTF will fall even lower. It is also necessary to take into account the MTF of the light modulator, but this has sufficient MTF to handle Hi-Vision signals.

(c) Gradation Characteristic. As stated above, gamma compensation takes into account the γ_f of the film, the AOM characteristic γ_a and the Hi-Vision camera γ_c. For the change in the transmissivity of the film to be linear during conversion to film, the gamma compensation of laser film recording must be in the $\gamma = 2.2$ range. However, the AOM is used in the $\gamma_a = 1$ range. In this case, gamma compensation is done by an analog process. Gamma compensation in the digital compensation system mentioned above is only an approximate compensation of the subtle gradation of the dark section during conversion to film, so the amount of compensation is quite small. At present, because the number of quantized bits of high-speed A/D converter data is limited, and because the number of quantized bits is limited to the 8-bit level by the scale of the digital hardware, it is difficult to process all of the gamma compensation with digital signals.

During the film printing process in motion picture making, each cut is adjusted for the exposure level of the positive film in what is called timing. But in laser film recording, regulation of the gamma compensation, gain, color compensation, and the amount of enhancement can be performed while repeating test shots electrically, through the use of time codes from the VTR signal source.

(d) Special Characteristics of Laser Film Recording. The brightness of the laser beam allows laser film recording equipment to use low sensitivity fine-grain film, as with direct exposure on positive film. Also, because the laser

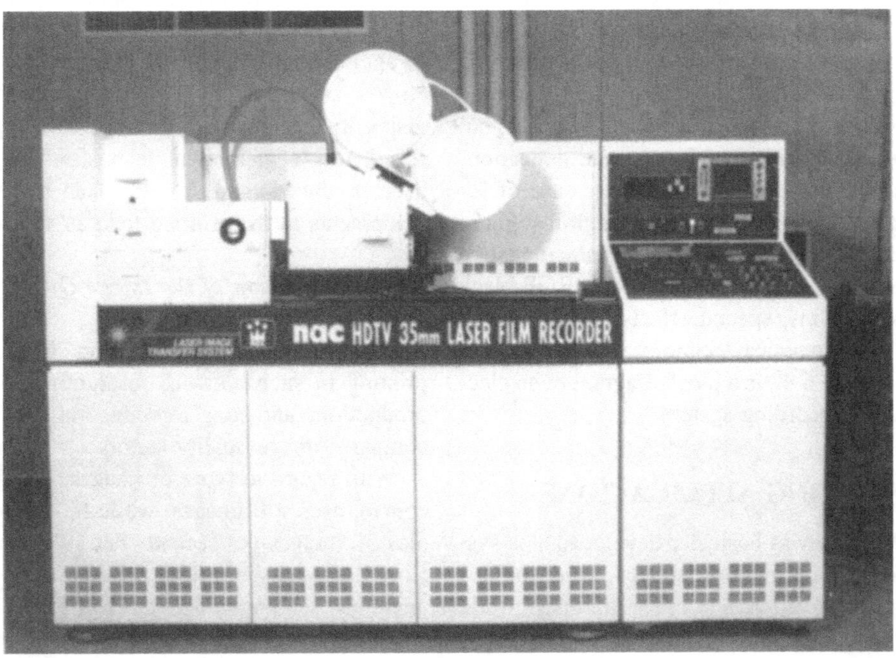

FIGURE 6.41. Laser film recorder HLR-350.

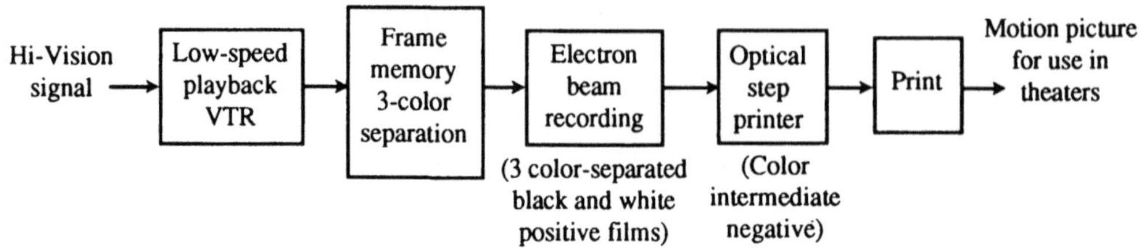

FIGURE 6.42. Electron beam recording system.

can focus on tiny spots with coherent light, it has the special quality of being able to achieve a resolution as high as that of an electron beam, making it sufficient for converting Hi-Vision signals to 35 mm film. Laser film recording equipment like that shown in Figure 6.41 has already been developed, and commercial applications have already begun.

(3) Electron Beam Recording

Although the discussion thus far has focused on the laser film recording equipment developed by NHK, there is another method that has been developed that uses electron beam recording. In this method film is exposed with an electron beam. When an electron beam is used, RGB images are recorded separately onto black and white film, and the color negative film is made from these. Therefore, processing in real time is a problem. Vacuum devices are absolutely necessary in order to use an electron beam, but advantages include highly accurate deflection, the elimination of an optical lens, ease of focusing on the film, and clear, high-resolution images. At the stage when the color negative film is being made from the separate RGB black and white films, special effects can be created using existing optical technology for motion pictures. Figure 6.42 is a block diagram of an electron beam recording system.

6.3 PRINTING APPLICATIONS

As television was basically developed as a system for sending moving images, a part of it is fundamentally different from the still-image world of printing and photography. However, it is common knowledge that the video images used in conventional television news and sports are also often used in print media such as newspapers and weekly magazines. This is because images have an instant and decisive impact even if the image quality is less than perfect. The high image quality of Hi-Vision makes it a valuable asset to the field of printing, and by all accounts Hi-Vision will become an increasingly important source of image information.

In this section, we will discuss printing and hard copy technology for Hi-Vision still images.

6.3.1 Printing Hi-Vision Images

The method of printing a television image as a still image is called video printing, and it is already in use and is achieving good results.[15] However, resolution is still insufficient even with the 525 scanning lines of present televisions, so its applications are limited. The development of Hi-Vision will bring the image quality of television images closer to that of photographs. A completely new kind of television image has been produced, and this will lead to new developments in the printing field as well.

(1) A Comparison of the Image Quality of High-Definition and Printing

The image quality of Hi-Vision differs from printing in such areas as resolution, color reproduction, and tone reproduction. Table 6.6 compares image quality factors.

With regard to types of images, printing, of course, uses still images, while Hi-Vision handles 30 images per second. For still images or slowly moving images, this is the same as sequentially sending images taken with a still camera at a shutter speed of 1/30 second. However, for moving images, it corresponds to sending each image twice with a camera shutter speed

TABLE 6.6. Image quality factors in Hi-Vision and printing.

Item	Hi-Vision	Printing
Type of image	Moving image	Still image
Number of pixels	1035 × 1920	150, 170 lines / inch
Color reproduction	RGB additive color mixing	YCMBk subtractive color mixing
Gray scale reproduction	Linear, contrast ratio 30:1	Including non-linear, contrast ratio 30:1
Noise	Random noise	Granularity, dots
Viewing distance	3 × screen height	Variable

of 1/60 second because of the interlaced scanning. In printing these images, there are no major problems with still images. But when one frame of a moving image is printed, the two superposed field images which have shifted because of movement, cause the image to blur. If the movement is fast enough, a double image is printed. Printing only one field of the frame to avoid this problem causes the vertical resolution to drop when the image is printed.

Resolution is dependent on the number of pixels. The resolution of printed materials is expressed as the number of lines per inch, and the number of pixels can be calculated from this figure. The total number of pixels on printed matter changes according to the size of the paper. For a resolution of 150 lines per inch with a 16:9 aspect ratio, the number of pixels would correspond to 638 × 359 for an image the size of a business card, 1754 × 987 for an A4 paper size (296 × 210 mm^2), and 2480 × 1395 for an A3 paper size (420 × 296 mm^2). The resolution for existing digital televisions is 720 × 483, comparable to the business card sized image. Thus the images from conventional televisions cannot be enlarged when printed without a significant drop in sharpness. On the other hand, Hi-Vision has 1920 × 1035 pixels, which is about the same resolution as the A4-sized printed image. Since this is the size of paper used in magazines, Hi-Vision images can be printed and handled just as other general printed materials without sacrificing sharpness. Incidentally, when A4-size paper is represented in terms of a Hi-Vision display, it corresponds to 13 inches diagonally. At this size, there is no noticeable image blurring even when viewed as close as can be distinctly seen.

The reproduction range differs because of the difference between the light emission characteristics of the CRT phosphor and the ink characteristics of printing. Figure 6.43 compares Hi-Vision and printing ink in a chromaticity diagram. Hi-Vision covers almost all of the color reproduction range of printing. Therefore, Hi-Vision images can be printed with virtually no major problems. However, there are slight differences in the actual color reproduction because Hi-Vision has an additive color mixing system with R, G, and B, while printing uses a subtractive color mixing with Y, M, C, and Bk. In general, in additive color mixing, bright colors tend to be rendered vividly, while dark colors tend to be rendered vividly in subtractive color mixing.

Hi-Vision and printing systems also differ with regard to noise. In the Hi-Vision system, in addition to noise such as random noise from the electrical system, the scanning line structure is also considered to be a kind of noise. In the printing system, the dot pattern itself is thought to be noise, but since this is considerably smaller than the size of the image, it is generally not necessary to treat it as noise.

As for tone, Hi-Vision is on the whole designed as a linear system as shown in Figure 6.44. But the printing system generally performs nonlinear compensation to achieve the

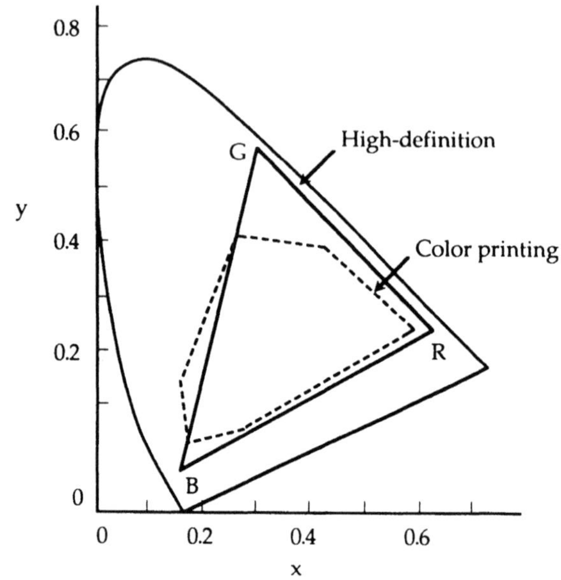

FIGURE 6.43. Color reproduction range of Hi-Vision and printing.

desired reproduction in the print. When a Hi-Vision image is printed, the output signal from the camera or VTR is printed either after recording it on film as described later, or directly as electrical signals. For this reason, gamma-compensated images are used, and so it is necessary to perform nonlinear compensation as post-processing for compatibility with printing systems.

One of the conditions in looking at an image is the viewing distance. The Hi-Vision system is designed to look best at a distance three times the height of the screen, but in printing, the standard condition is the least distance for which the image is distinctly visible. Therefore, it is possible that the same image will appear differently in Hi-Vision and in print.

(2) Printing Methods

There are now two methods of printing a Hi-Vision image, as shown in Figure 6.45. The first method is shown in the upper portion of the figure. The Hi-Vision image is recorded onto color film (film recording), converted into electrical signals by a scanner, processed by a computer, and finally made into color separation plates using a scanner.

As for film recording methods, in place of using surface photography on a conventional television CRT, methods have been developed

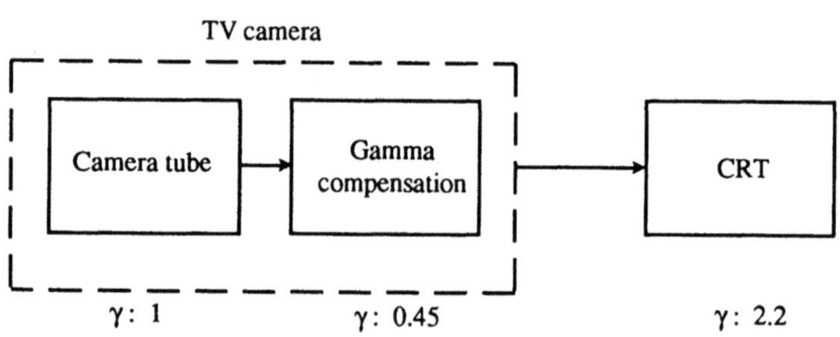

FIGURE 6.44. Gamma values in a Hi-Vision system.

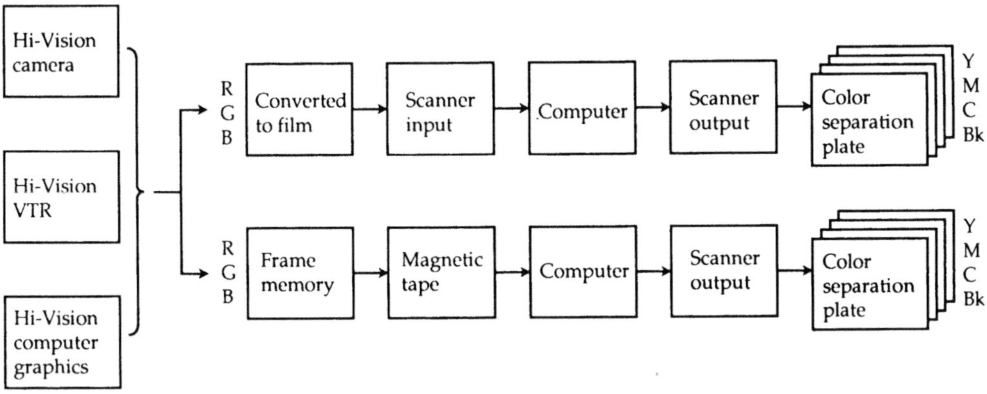

FIGURE 6.45. Techniques for printing Hi-Vision images.

that use either a laser beam or an electron beam to record images directly onto film.[16] With the laser beam method, 35 mm synthesizer recording is now being done, but for printing, a 35 mm Leica plate or a 4 × 5 plate frame recording is also possible. Hi-Vision images which have been converted into film using this method can be handled in exactly the same way as ordinary print materials. Figure 6.46 shows an example of an image taken with a Hi-Vision camera and converted to film by laser beam recording. Figure 6.47 is an image of about 200 characters converted into film in the same way. From these examples, we can see that Hi-Vision can be used to print fairly detailed pictures and text.

The second technique for printing Hi-Vision images is an electronic system which does not use film. As shown in the lower portion of Figure 6.45, Hi-Vision signals are recorded on magnetic tape using a frame memory, processed by a computer, and output by a scanner as color separation plates ready for printing. Since all stages except the final printing are done with electronic signals, there is little image degradation, and high quality images can be printed.

(3) Image Processing for Printing

Because of the differences in image quality between Hi-Vision and printing, the following image processing is necessary when printing Hi-Vision images.

The image processing is done through the

FIGURE 6.46. Image recorded on film with laser beam for printing.

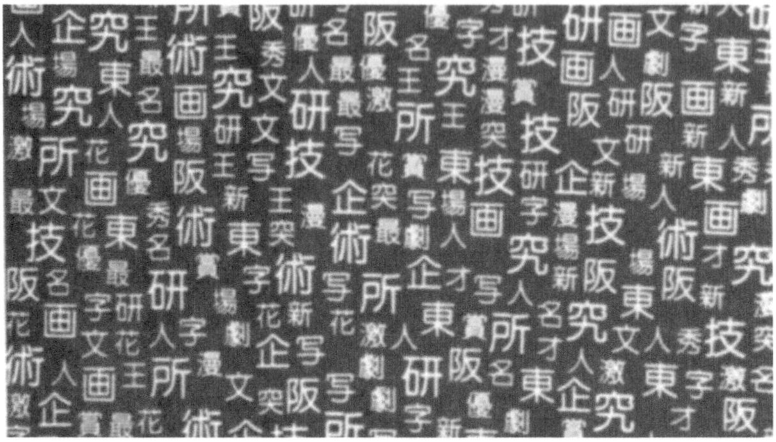

FIGURE 6.47. Chinese characters recorded on film with laser beam for printing (approximately 200 characters).

same RGB system as is used in H-Vision. One of the ways to print attractive Hi-Vision images is to use tone processing to expand the tone in the dark parts of the image and contract it in the bright parts. At the same time, the saturation is raised to prevent color turbidity, and edge processing is done to guard against unnatural emphasis of the edges of the image. Video noise, which is relatively unnoticeable in television images, is particularly noticeable in still-image printing. This noise is dealt with through filter processing and noise reduction.

After the above types of processing have been performed, RGB signals are finally converted into YMC signals by a matrix. Bk (black) signals which supplement the tone are added, and are used as signals to make a color separation plate for printing.

The next area of concern is the problem of processing moving images. In printing Hi-Vision images, frames which are appropriate for printing must be chosen from among the frames of a moving image. However, blurring invariably occurs in fast moving images. For this reason, motion adaptation processing technology, a common form of digital processing of television signals, is used As far as the moving image area is concerned, the image is made from only one field recorded in 1/60 second. It will also be necessary to employ methods which reduce motion blurring.

(4) Printing Examples

There are already actual examples of the two printing techniques mentioned above. The use of Hi-Vision in publishing is expected to grow as Hi-Vision proliferates.

The first example of a publication of images from a film recording is *Okhotsk no Kodomotachi.*[17]

The second example, *Mitsuko,* used an electron beam system.[18] *Mitsuko* has 235 mm × 210 mm pictures printed across two pages and effectively uses a Hi-Vision 16:9 wide screen format (Figure 6.48). Pamphlets and picture postcards are also printed with these methods. Such methods will continue to be used in the future.

6.3.2 Hi-Vision Hard Copy Technology

While permanent Hi-Vision hard copy images had their first use as supplemental printed matter for broadcast programs, in the future it will be possible to make hard copies from Hi-Vision images received in the home. Applications are also anticipated in new electronic image systems such as electronic still photography systems.

A hard copy is defined as an image which has been printed on a recording medium based on data transmitted through electronic signals. Hard copies have progressed from simple text images for business use to beautiful color pho-

FIGURE 6.48. Example of a two-page spread.

tographs. If made from television signals, the hard copy is called a video hard copy. Hi-Vision hard copies have improved on the quality of video hard copies made with conventional television technology.

We will now discuss the combination of Hi-Vision and hard copy technology. First we present a basic discussion of the present state of video hard copies and the capabilities of Hi-Vision hard copies, followed by a discussion of the prototype printer system developed by NHK Engineering.

(1) Present State of Video Hard Copy Technology

Video hard copies are original photographs which are used to print images output from new electronic still photography and television systems. Because the output signals from existing electronic still cameras are compatible with NTSC signals, the output images of these new photographic systems are equivalent to video hard copies made from conventional television broadcast signals.

Video hard copy technology is based on compact, high quality printer technology, for which the development of a recording medium is vital. Today, optical printers which use photographic materials and thermal transfer printers which use thermal sublimation pigment ink have been developed, allowing video hard copies which are rich in tone.

However, with video hard copies from conventional NTSC television and from new photographic systems patterned after these specifications, the number of pixels per screen is, at most, the digital television standard of 720×483. For this reason, the screen size must be reduced to less than half the size of a business card to produce an image which does not have conspicuous blurring. Thus the high data density and pixel count of Hi-Vision is being greatly anticipated.

(2) Capabilities Required of Hi-Vision Hard Copies

Since hard copies of still images are subject to close scrutiny, there are rigorous requirements for image quality. As stated previously, the specifications of input signals have a great influence on image quality.

Table 6.7 shows the specifications of Hi-Vision signals and their effect on the image quality of video hard copies. The primary concern is resolution, which is significantly better in Hi-Vision signals than for NTSC signals and makes Hi-Vision the preferred signal source for video hard copies. Signals from the MUSE system, which will be used to transmit future Hi-Vision broadcasts, also have characteristics conforming to studio standard base band signals in the still image mode.

Among the characteristics of television signals, gradation has continuity but cannot be unlimited because the signals include noise, and also because the dynamic range is restricted. However, this is the same kind of restriction that applies to granularity noise in photographic images, and so it is not a problem as long as the signal's SN ratio does not drop sharply. Photographic film noise occurs when the SN ratio exceeds the 50 dB level.

The color on video hard copies depends on the bandwidth of the R, G, and B signal components supplied to the printer, and on the capabilities of the stylus and the color reproduction of the recording medium. In CRT printers, the phosphor stylus supplies a colored light beam that exceeds the color reproduction range of the film recording medium. For this reason, the color reproduction of the image depends on the color reproduction range of the recording medium.

Next, let us consider the scanning line structure and scanning line interpolation, which are related to resolution in a number of ways. In television images, the scanning operation samples the image along the vertical axis, and if the band restriction is sufficient during image-pickup, the images which are sent can be totally restored using the sampling theorem. In order to perform restoration, the area between the scanning lines can be interpolated by the sampling coefficient $\sin(\theta)/\theta (\theta = \pi/p$, where p is the scanning line pitch, and y is the vertical distance from the scanning line position). That is, if the distribution of the beam intensity of the scanning lines is the sampling coefficient, the original image is recreated.

In practice, the sampling coefficient has a negative value that continues without limit, making it impossible to use with hard copies. However, the space between scanning lines can be smoothly filled even with a Gaussian distribution coefficient, which is often seen in the expansion of pixels. In this case, when the ratio (h/p) of the peak width at half height, h, and

TABLE 6.7. The Hi-Vision studio standard and its effect on hard copy image quality.

Item	Standard	Effect on hard copy image quality
Images per second	30.0 frames, 60.0 fields	Motion is blurred
Scanning method	2:1 interlace	Double image artifact due to motion
Number of scan lines	1125 (1035 effective scan lines)	Vertical resolution, scanning line structure
Signal bandwidth	Luminance Y : 30 MHz Color P_B : 30 MHz P_R : 30 MHz	Horizontal resolution Color resolution, color blurring
Aspect ratio	Horizontal : vertical = 16:9	Aspect ratio of image

the scanning line pitch p for the distribution is 1:3, the scanning line structure can no longer be discerned.[19] However, the resolution is lower than when interpolation is done through the sampling frequency coefficient.

(3) Hi-Vision Printers

Hi-Vision hard copies must display rich tones and true color reproduction. For this reason, printers that can be used are limited to those which employ a recording medium that can directly control changes in the color density of the pixels through the amount of coloring material. Therefore, either optical printers that use silver salt photographic film, or thermal transfer printers that use thermal sublimation pigments can be used for Hi-Vision hard copy printing.[19]

Next, we will discuss the Hi-Vision hard copy production systems developed by NHK Engineering, which consist of CRT printers that use instant color film and thermal transfer printers. In the prototype CRT printer shown in Figure 6.49, after television input signals are stored in the frame memory, the signals are processed and sent to the printer. The CRT is a flat color tube that reportedly emits only one scanning line each of R, G, and B. Each color passes through the image-forming lens and exposes the film to one scanning line. The image-forming lens moves parallel to the film in steps according to the scanning lines of the screen to make a two-dimensional image. Because the light emission positions of R, G, and B differ on the CRT, misalignment occurs on the film, but that mis-

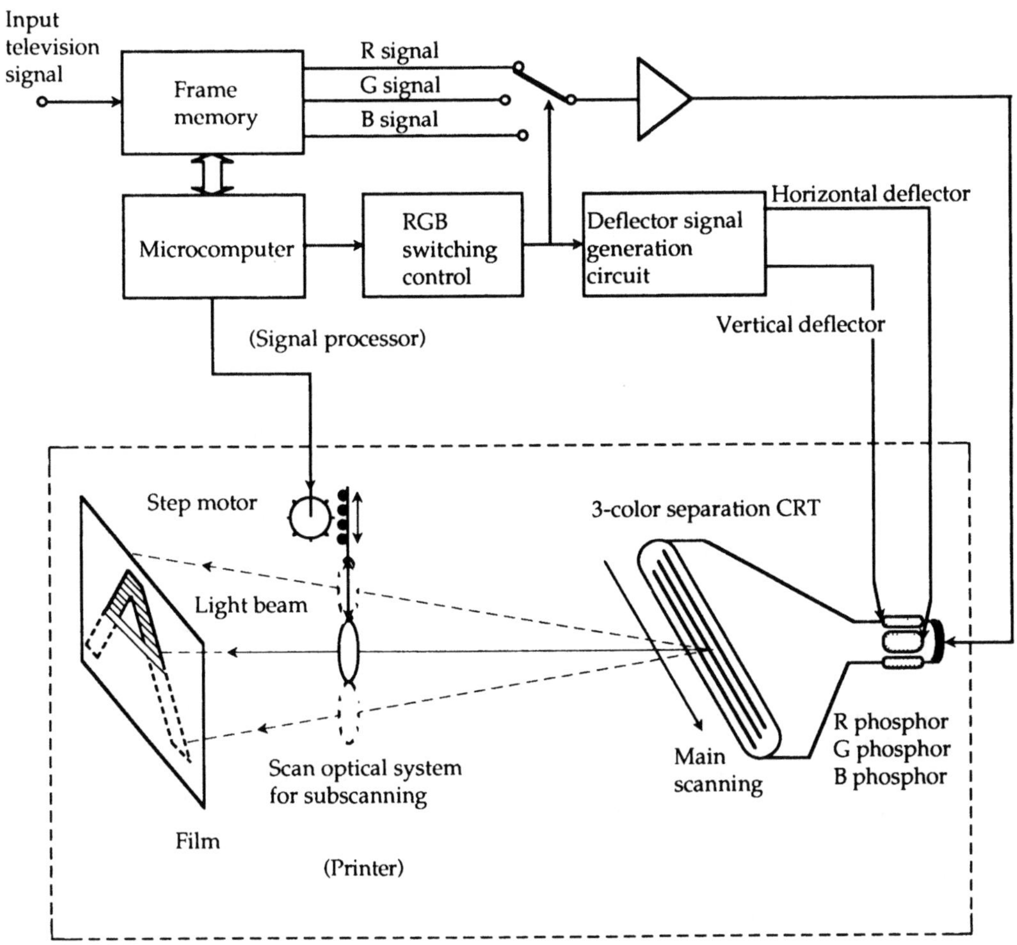

FIGURE 6.49. Video hard copy printer prototype developed by NHK.

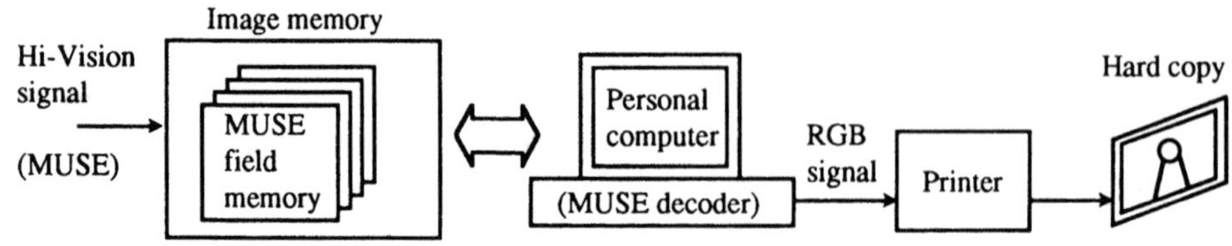

FIGURE 6.50. Hi-Vision hard copy printer system for MUSE signals.

alignment is compensated for by shifting, in advance, the scanning line numbers of the signals input to the CRT. The amount of light emitted by the CRT is controlled by a digital modulation system according to the number of low-amplification pulses. The use of this method improves resolution.[21]

The thermal transfer printer shown in Figure 6.50 can accept the direct input of MUSE signals. The printer has 8 dots/mm and 1,024 thermal heads. It transfers three pigments from the inksheet, Y (yellow), M (magenta), and C (cyan) onto imaging paper, and a color image with a screen size equal to or smaller than the size of the cabinet (16 × 12 cm^2) is formed. Decoding of the MUSE signal is done off-line on a microcomputer. The frame memory, which stores input signals, has space for four fields of MUSE signals. Its capacity is only one-fourth that of a full-band RGB memory. While the printer's characteristics will be listed later, the MTF of the output image is at least 0.3 at a resolution of 1,000 TV lines, which corresponds to the upper limit of the Hi-Vision bandwidth (11 dots/mm: equivalent to 95 × 7-inch size). Gradation and color reproduction are comparable to that of ordinary photographs.

Figure 6.51 (a) shows a Hi-Vision hard copy photograph from a thermal transfer printer. Parts (b) and (c) of the figure show photographs of a blown-up portion of the same image which were made using Hi-Vision signals and today's television signals respectively. The Hi-Vision signals are at studio specifications and the television signals are at NTSC specifications. The resolution of the image is higher with the Hi-Vision signals and there are no noticeable dot artifacts even in the red flower area where the color level is high. The image from the Hi-Vision signals produces smoother texture and a more natural-looking image than that which was made from NTSC signals.

(4) Hi-Vision Hard Copy Characteristics

When cabinet Hi-Vision hard copies are made from Hi-Vision images, the pixel resolution is

FIGURE 6.51. Comparison of Hi-Vision hard copy to conventional video hard copy. (a) Hi-Vision hard copy. (b) Hi-Vision hard copy. (c) NTSC video hard copy.

5.4 lines/mm (10.8 dots/mm). Among present video hard copy printers, the only printers which can produce images with a gradation that gives 0.09×0.09 mm^2 pixels a continuous and wide range of densities are the two printers mentioned above: the optical printer which uses color photographic materials, and the thermal transfer printer which uses sublimation pigments. Figure 6.51 is a Hi-Vision hard copy from a thermal transfer printer with an 8 dot/mm thermal head.

Figure 6.52 shows the MTF (sharpness) characteristic for an image from a sublimation pigment thermal transfer printer with 8 dot/mm thermal heads, and for an instant color film print from a CRT printer which has a light beam stylus with a half-bandwidth 80 μm in diameter. Presently, Hi-Vision hard copies printed on a thermal transfer with an 8 dot/mm thermal head have been image quality than those copies printed onto instant color film by a CRT printer. This is because the MTF of instant color film is inferior.

Thermal heads, which are the stylus of today's thermal transfer printers, are being developed to be sharper, with either 12 or 15 dots/mm, and the MTF value in the spatial frequency range of 4.4 lines/mm is expected to improve. Therefore, Hi-Vision hard copies from thermal transfer printers will have better resolution and sharpness, and neater, clearer images. At present, thermal transfer printers using pigmented inks, which can produce images without development processing, produce the Hi-Vision hard copies with the finest image quality.

(5) Comparison of Hi-Vision Hard Copies and Color Photographs

Hi-Vision hard copies can be considered electronic still photographs which are produced through a Hi-Vision system. What follows is a comparison of the image quality of Hi-Vision hard copies and ordinary color photographs.

In general, color prints are made by photographing the subject with color negative film, and the negative image is enlarged and printed onto positive color photographic paper. Because the grains of a photograph are minute, and because no restrictions are imposed by the equipment, it is difficult to think in terms of pixels. Also, since the photographic paper used for enlargements has grains that are smaller than those of the negative film, in theory there is no theoretical image degradation caused by printing.

In order to compare photographs with Hi-Vision hard copies, it is necessary to introduce the concept of pixels into photographic images. The size of one pixel is calculated as 16×16 μm^2 based on the fact that a resolution of 30 lines/mm corresponds to MTF-0.5 for negative film. When an image with an aspect ratio of 16:9 is recorded with a pixel of this size, a full-sized frame of 35 mm negative film would consist of $2{,}250 \times 1{,}266$ pixels, and half-sized negative film would consist of $1{,}500 \times 844$ pixels. Therefore, the number of pixels in Hi-

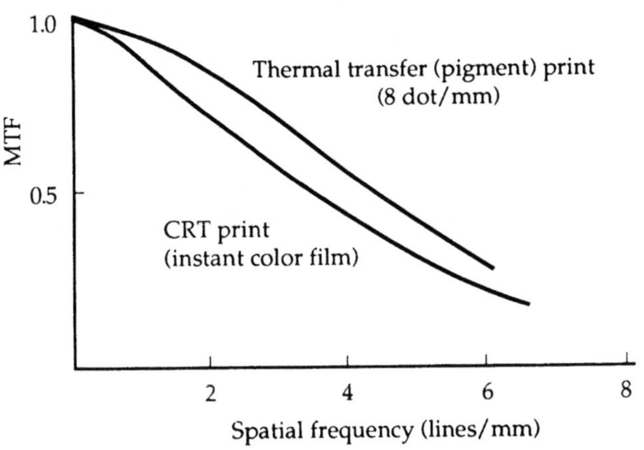

FIGURE 6.52. Spatial frequency characteristics of two kinds of Hi-Vision hard copy prints.

Vision is almost halfway between the half-sized and full-sized negative film.

(6) The Future of Video Hard Copies

Subject to certain conditions, Hi-Vision hard copies have characteristics which are comparable to those of color photographs. And, the areas in which they will have applications are expected to increase in number.

First Hi-Vision hard copies will be applied as still image program material in the broadcasting world. Also, the sharp images of Hi-Vision hard copies will work remarkably well in data transmission, and it is even possible that images from television programs will be received as color prints right in the home.

The use of program images for photographs in magazine photogravures and book illustrations is expected to become more and more popular. The diffusion of Hi-Vision and the development of Hi-Vision hard copies will probably cause electronic still image systems to be changed to Hi-Vision specifications in the near future. In this way, Hi-Vision hard copies will function with the new image services of the future.

REFERENCES

1. "New IBA Picture Converter for Transatlantic Colour Television Uses Computer Technology," *IBA Engineering Information,* (1972).
2. Clarke et al., "Digital Standards Conversion: Interpolation Theory and Aperture Synthesis," *BBC RD* (1984).
3. Tanaka, "The Configuration and Image Quality Evaluation of Hi-Vision–PAL Format Conversion Equipment Using a Motion Compensation Frame Rate Conversion Method," *IEICEJ Proceedings* (D), J 70-D, 8, pp. 1535–1545 (1987).
4. Ninomiya, Otsuka, "Motion Compensation Predictive Methods in Interframe Encoding," *IEICEJ Proceedings* Vol. J 64-B, No. 1, pp. 24–31 (1981).
5. Okada, Mihashi, "Hi-Vision–NTSC Format Converter," *NHK Technical Report,* Vol. 28, No. 9, pp. 359–364 (1985).
6. Sugiura, "Film Image Transmission and Recording," *Journal of the Institute of Television Engineering (ITEJ)* Vol. 39, No. 8, p. 690 (1985).
7. Kodak Nagase Catalogue
8. Sugiura, "35 mm Film Laser Recording, *ITEJ Technical Report,* OPT 169, p. 13 (1982).
9. Nojiri et al. "35 mm Laser Film Recording For Hi-Vision," *NHK Engineering Monthly Report.* Vol. 28, No. 11 (1985).
10. Sugiura, "Laser Applications in High Quality Television," *Laser Research,* Vol. 12, No. 5, p. 240 (1984).
11. Sugiura, "Acoustic Optical Modulators and Deflectors," *O plus E,* No. 63, February, 1985, p. 86.
12. Nojiri, "Laser Recording Equipment for Hi-Vision," *ITEJ Technical Report,* OPT-227, p. 1 (1986).
13. Nojiri, "Image Quality Improvement in 35 mm Film Laser Recording for Hi-Vision," *ITEJ Proceedings,* p. 317 (1986).
14. Urata, Honjo, "Color Film For Laser Recording," *Scientific Publication of the Fuji Photo Film Co., Ltd,* No. 33, p. 9 (1988).
15. Hirao, "Video Printing," *ITEJ Journal,* Vol. 39, No. 8, p. 690 (1985).
16. Sugiura, "Film Image Transmission and Recording," *ITEJ Journal,* Vol. 39, No. 8, p. 690 (1985).
17. Okazaki, *Ohoushiku no Kodomotachi,* Japan Broadcast Publishing Co. (1985).
18. Yoshida, *Mitsuko,* Japan Broadcast Publishing Co. (1987).
19. Okuda, Ohno, "Reproduction Characteristics of Television Signal Hard Copy Images From CRT Printers," *ITEJ Journal* Vol. 40, No. 11, pp. 1140–1146 (1986).
20. Ohno, "The Color Hardcopying Technologies—Video-Hardcopy Printers and Digital Color Copiers," *Optoelectronic Devices and Technologies,* Vol. 3, No. 1, pp. 47–69 (1988).
21. Okuda, Tadenuma, Ohno, "CRT Digital Drive Systems For Hard Copy Printers," *ITEJ Journal,* Vol. 42, No. 10, pp. 1057–1063 (1988).

Appendix

Taiji Nishizawa

Four years have passed since the original Japanese version of this book was published. In order to bring the book up to date with developments in Hi-Vision technology during that period, we have obtained permission from the NHK Science and Technical Laboratories to reproduce excerpts of newly written articles based on their journals, in this Appendix.

A.1 DIGITAL TRANSMISSION OF HI-VISION SIGNALS[1]

The practical application of Hi-Vision broadcasting will entail the transmission of program materials between broadcasting stations as well as between countries. In the future, digital transmission to households is conceivable. However, when component signals used in Hi-Vision studios are digitally transmitted in their normal form, the bit rate can exceed 1 Gb/s. Because transmission at this rate would make the bandwidth and transmission costs prohibitive, band compression becomes necessary.

Alternatives to transmission paths for digital Hi-Vision signals include the H4 rate (140 Mb/s) of broadband ISDN, INTELSAT satellite (120 Mb/s), and group 4 (100Mb/s) of Japan's digital hierarchy. Several high-efficiency coding systems (see Table A.1) have been proposed that would be compatible with these transmission paths. While high-efficiency coding for Hi-Vision is basically an extension of standard television coding, what is needed is of higher quality and higher efficiency. We will discuss five representative systems.

(1) Interfield/Intrafield Adaptive Prediction Coding With MUSE Signals[2,3]

Having already undergone band compression, MUSE signals have less redundancy than Hi-Vision source signals, making it difficult to reduce their bit rate by any significant amount. However, unmodified MUSE signals have a bit rate of 129.6 Mb/s when digitally transmitted. This is compatible with the wideband ISDN H4 rate. Transmission is also possible with the INTELSAT satellite and the group 4 digital hierarchy through various bit rate reductions. Presently, methods are being studied to compress the bit rate to 60 Mb/s using interfield/intrafield adaptive prediction coding systems. One such system diagram is shown in Figure A.1.

Interfield and intrafield prediction systems use the prediction function shown in Figure A.2. Interfield prediction is used for still images, and intrafield prediction for moving images. The decoder switches between interfield and intrafield prediction according to the size of the difference between the decoded value of the previous line pixel and the interfield predicted value (inter-

TABLE A.1. High-efficiency coding systems for Hi-Vision.

System	1 Interfield/ intrafield predictive coding	2 Intrafield DPCM coding	3 Interframe/ intrafield predictive coding	4 Interframe/ intrafield predictive coding	5 Hybrid DCT
Signal type	Y / R-Y / B-Y (20, 7, 7 MHz)	Y / C_W / C_N (20, 7, 5.5 MHz)	Y / P_B / P_R (20, 7, 7 MHz)	Y / C_W / C_N (20, 7, 5.5 MHz)	Y / P_B / P_R (20, 7, 7 MHz)
Sampling frequency	Y 48.6 Mhz C 16.2 MHz	Y 48 Mhz C 16 MHz	Y 44.55 Mhz C 14.85 MHz	Y 48 Mhz C 16 MHz	Y 74.25 Mhz C 37.125 MHz
Pre-processing	MUSE band compression	Thinning out of pixel spacing by 1/2 using line offset subsampling	• Noise elimination by time-space adaptive filter • TDM	• Noise elimination by time filter • TDM	None
Coding algorithm	• Interfield/ intrafield adaptive prediction • Predicted value adaptive quantization • 3/6-bit semi-variable-length coding	• Previous value prediction • 4-bit fixed coding • Reflected quantization • Noise correcting filter	• Interframe extrapolation/ intrafield interpolation/ intrafield interpolation adaptive prediction (pixel units) • Variable-length coding • Scalar quantization	• Interframe extrapolation/ intrafield interpolation/ intrafield interpolation adaptive prediction (block units) • Variable-length coding • Vector/scalar quantization	• Motion compensated interframe DPCM and DCT • 2~18-bit variable-length coding
Bit rate	60 Mb/s	120/140 Mb/s	100 Mb/s	100/140 Mb/s	70 Mb/s

field difference). In addition, an average bit compression of 3.7 bits/pixel is performed by a predicted value adaptive quantizer and a 3/6-bit semi-variable-length encoder.

(2) Intrafield Predictive Coding[4]

The intrafield predictive coding system was developed to obtain a satisfactory image quality with less hardware. Figure A.3 shows the system diagram of this coding equipment. In preprocessing prior to encoding, color-difference signal progressive scanning and line offset subsampling are performed. The sampling pattern of this subsampling is shown in Figure A.4, and the spatial frequency range for which transmission is possible is shown in Figure A.5. Although the diagonal resolution decreases, this is not likely to result in visible image degradation. Also, a field memory such as that used in interframe prediction systems is not required because signal processing is performed within one field. This also makes the equipment more compact.

The DPCM encoding system is shown in Figure A.6. One-dimensional previous value prediction is performed, and the prediction error with respect to the moving image is smaller than in interframe prediction. However, since the prediction error with respect to a still image increases, a noise correcting filter is used to convert the encoding noise spectrum to triangular noise, and to reduce perceived noise power to a minimum. However, in some cases, the noise correcting filter causes a larger quantiza-

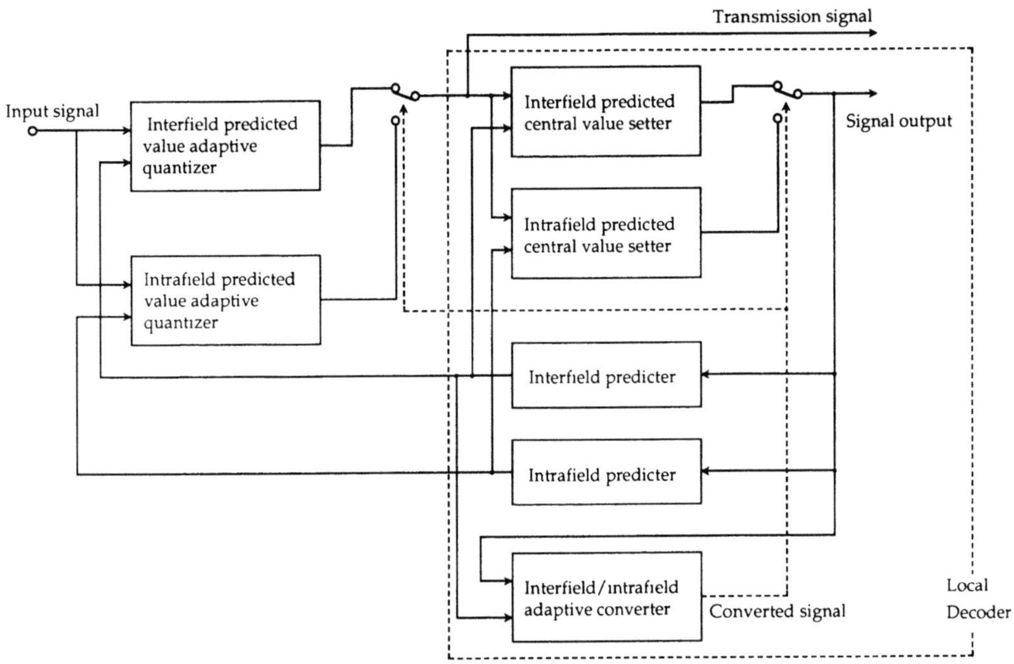

FIGURE A.1. MUSE DPCM block diagram.

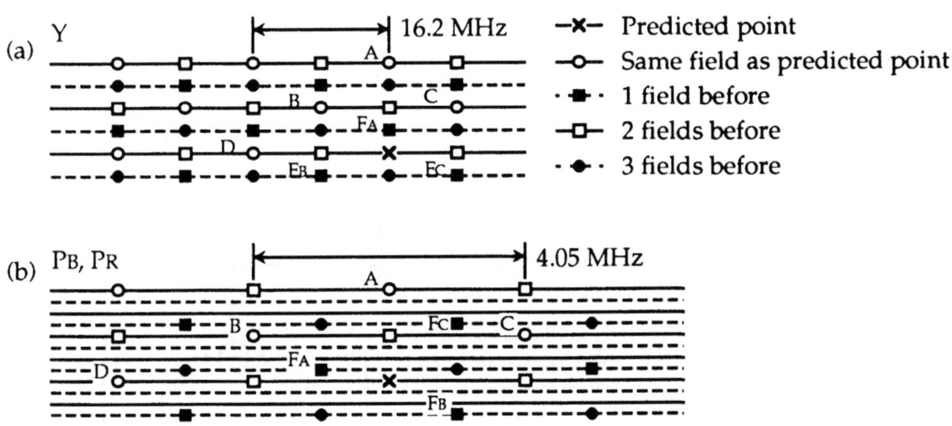

(c) Prediction Function

	Intrafield prediction	Interfield prediction
Luminance signal	0.48 C + 0.51 D	0.5 FA + 0.25 FB + 0.25 FC
Color-difference signal	0.47 A + 0.28 C + 0.22 D	0.52 FA + 0.35 FB + 0.12 FC

FIGURE A.2. MUSE-DCPM prediction method.

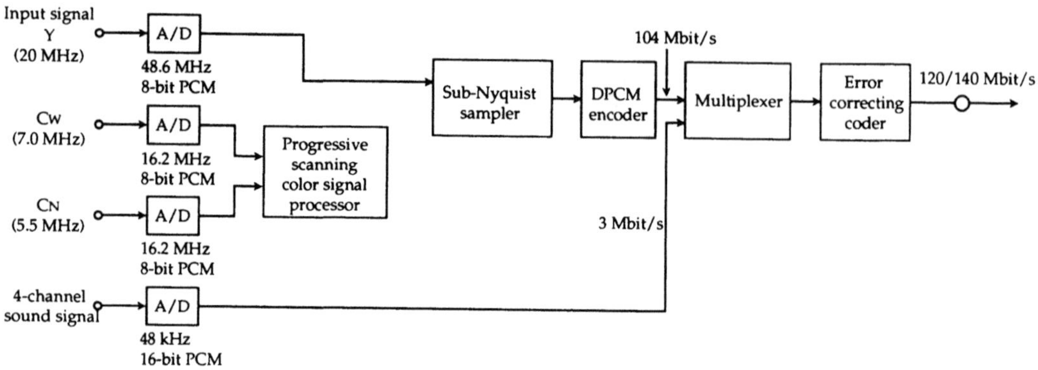

FIGURE A.3. Configuration of 120/140 Mbit/s encoding system.

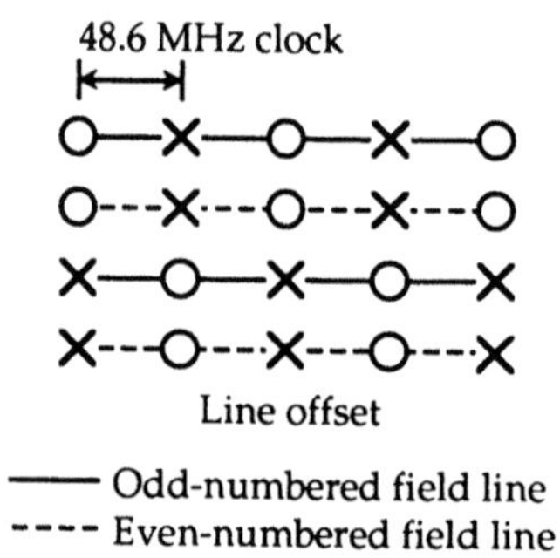

FIGURE A.4. Line offset subsampling pattern.

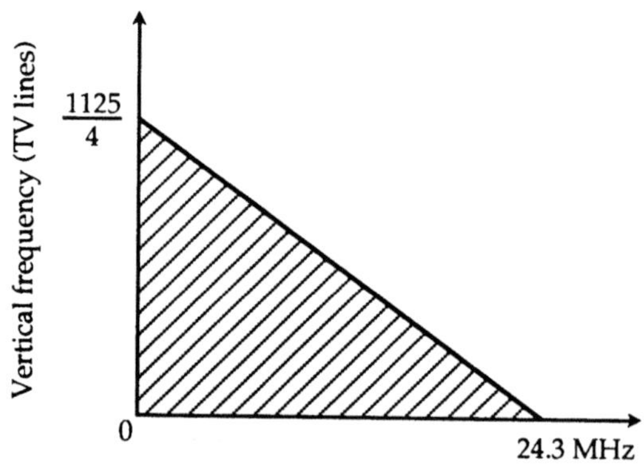

FIGURE A.5. Spatio-frequency range at which transmission is possible.

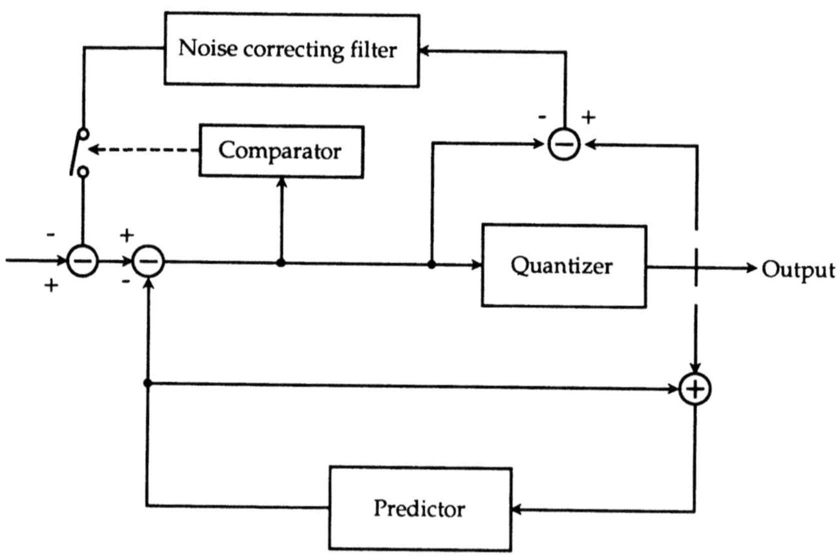

FIGURE A.6. DPCM coder with adaptive noise correcting filter.

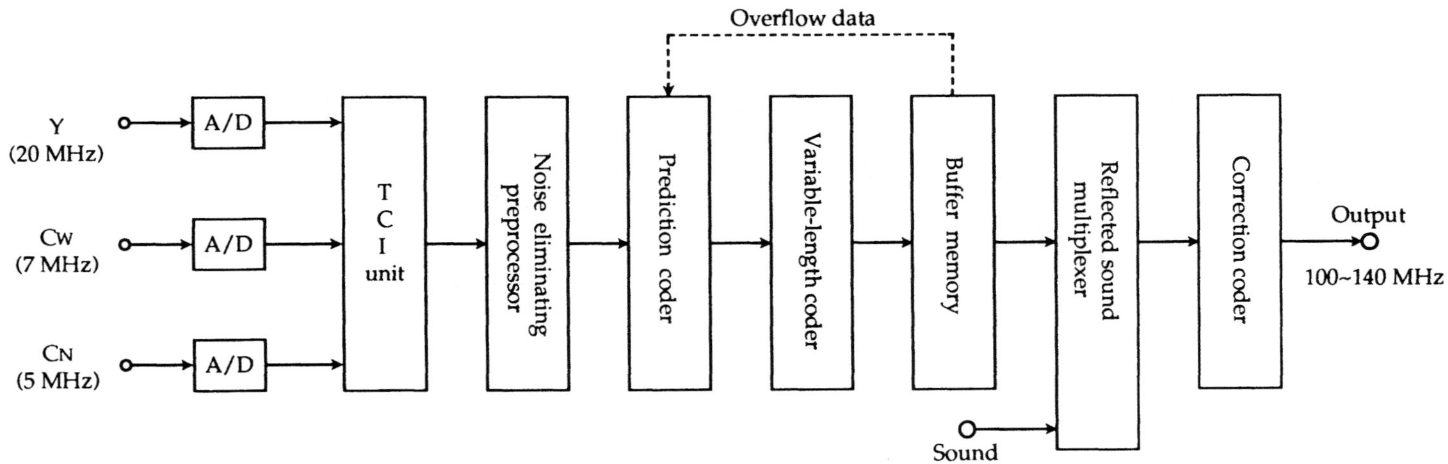

FIGURE A.7. Basic configuration of coder.

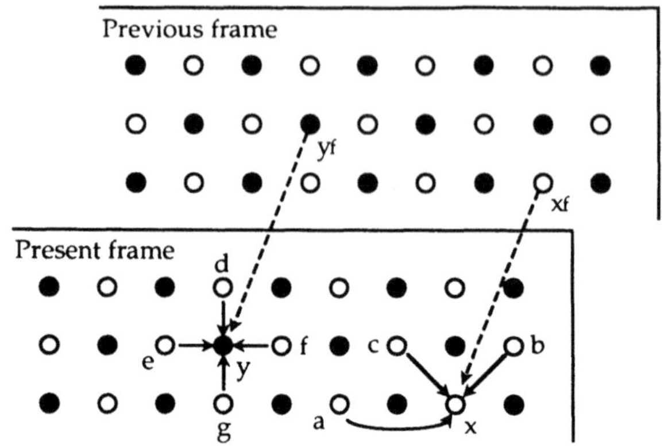

FIGURE A.8. Principle of interframe/intrafield interpolation and extrapolation prediction.

tion error to be propagated in the horizontal direction, which causes image degradation. In this case, adaptive processing is performed to stop quantization error addition, the quantizer switches the quantization characteristic according to the size of the predicted value, thus keeping the prediction error in the low range.

(3) Intrafield/Interframe Prediction Systems
Intrafield/interframe prediction systems set the bit rate at 140 Mb/s or less to achieve as high an image quality as is possible. Here we will discuss two representative systems.

Figure A.7 shows the basic configuration of system 3 from Table A.1.[5] In the preprocessing for encoding, the color difference signals are changed to a progressive scan format and then multiplexed with the luminance signal. The signal is then handled as a TCI signal. To reduce the drop in predictive efficiency due to noise in the Hi-Vision signal, and to reduce detection error in the moving area, the noise eliminating preprocessor has a spatio-temporal adaptive smoothing filter to eliminate noise.

In the encoder, as Figure A.8 shows, the pixels are divided into two quincunx-shaped groups composed of solid and empty dots. For each group, interframe prediction is performed in the still area, and intrafield prediction in the moving area. For the solid-dot pixels, interframe prediction is done by extrapolating from the previous frame's solid-dot pixels, which have already undergone encoding and decoding. In intraframe prediction, efficiency is increased by performing an interpolation prediction using the surrounding four empty-dot pixels.

A subsampling mode has been prepared to handle buffer memory overflow, which can occur when variable-length encoding is performed. In this case, the time-space adaptive

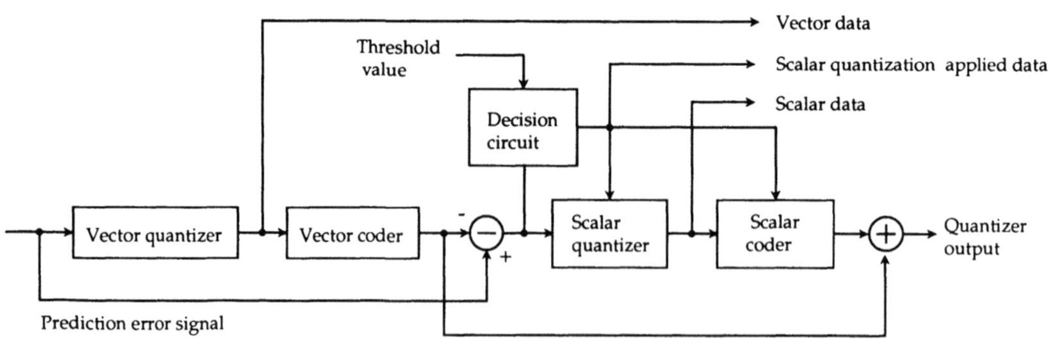

FIGURE A.9. Configuration of hybrid quantizer.

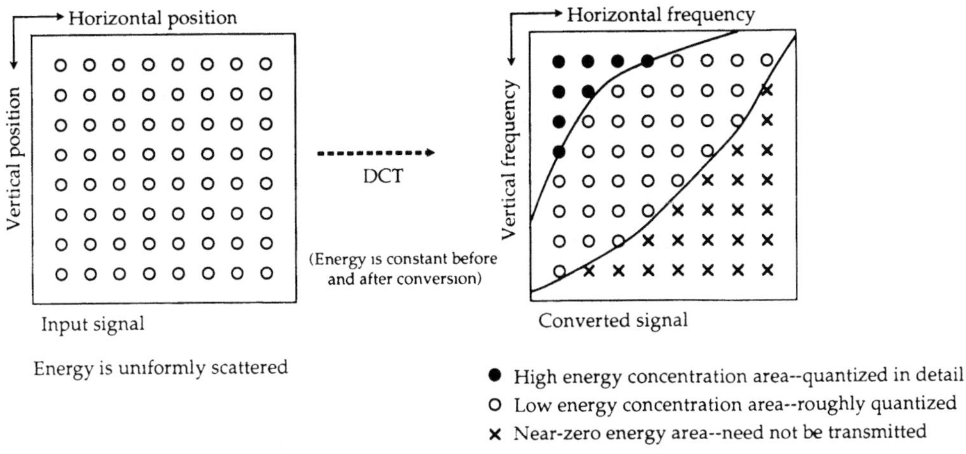

FIGURE A.10. DCT principles.

smoothing filter acts as a prefilter for the subsampling, and only the empty-dot pixels are encoded.

System 4 from Table A.1 is basically the same as system 3, but uses a hybrid quantizer consisting of a quantizer, vector quantizer, and scalar quantizer.[6] In this system as well, the pixels are divided into two groups, and coding is performed in units of 4-line × 4-pixel blocks.

The configuration of the hybrid quantizer is shown in Figure A.9. The predicted errors which have been grouped into blocks first undergo vector quantization and are then transmitted as vector data. If the errors from vector quantization are deemed large, the quantization errors then undergo scalar quantization and are transmitted. This means that only the vector quantizer is used for blocks which have relatively low resolution, while a supplemental scalar quantization is added if the resolution is high.

(4) DCT[7,8,9]

In recent years, orthogonal conversion coding with DCT (Discrete Cosine Transform) has been a focus of attention. As shown in Figure A.10, orthogonal conversion is performed using the DCT for each 8 × 8 or 16 × 16 pixel block, and image data is converted into a spatial fre-

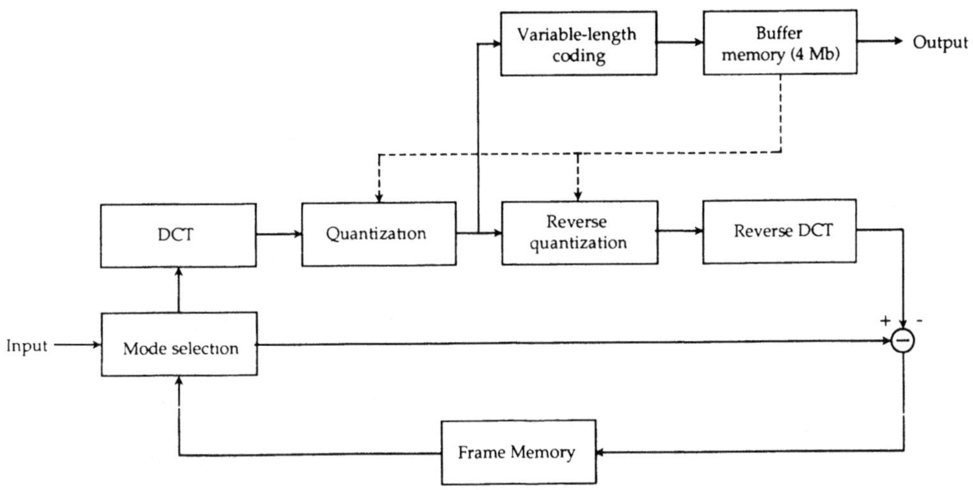

FIGURE A.11. Example of hybrid DCT.

quency spectrum. In general, image energy is concentrated in the low spatial frequency range, with little energy in the higher regions. Therefore bit compression is performed by thoroughly quantizing the high-energy areas, while the low-energy areas are either roughly quantized or deemed to be zero. However, because this measure alone will not produce a sufficient compression ratio, research is being done on hybrid DCT systems which combine DPCM with DCT.

Figure A.11 shows a block diagram of a hybrid DCT system which combines motion compensation DPCM with DCT. In this system, DCT is performed either for the actual pixels or for the motion compensating interframe difference. The system switches between these two modes to attain the maximum bit compression ratio. The DCT output is further quantized and undergoes variable-length encoding. It is then output as a 70 Mb/s signal.

A.2 THE 1/2-INCH UNIHI VCR FOR INDUSTRIAL USE

Of the wide range of audio and video applications being considered for Hi-Vision, the demand is especially strong for industrial applications. This puts a high priority on the development of a VCR that excels in cost effectiveness and operability. In response to this interest, NHK Science and Technical Research Laboratories created specifications for industrial VCRs based on our latest research. Manufacturers then developed and manufactured VCRs and compact VCRs satisfying these specifications, and first put them on the market in 1989.

A.2.1 Required Specifications

The required specifications for VCRs for industrial use were set to reflect both the opinions of users and the results of research.

(1) Recording Time

The program recording time had to be 60 minutes, or 63 minutes if additions such as test signals are included.

(2) Image Quality

In consideration of the nature of industrial applications, an analog (FM) recording system would be used under the assumptions that special effects processing will not have to be performed and that large numbers of copies will not have to be made. The frequency characteristics are 20 MHz for luminance, 7 MHz for color difference (P_B, P_R progressive scanning), and SN ratios of 41 dB and 46 dB.

(3) Sound Quality and the Number of Channels

Considering the presence of DATs (Digital Audio Tape recorders) in the consumer market, the sound quality required a PCM system with a sampling frequency of 48 kHz quantized at 16 bits/sample.

The number of channels was set assuming nonstudio uses, and to accommodate the four channels of the 3-1 system recommended for Hi-Vision audio.

(4) Operability

A cassette tape format gives the VCR operability at the same level as is now customary in home video. We took the recording time and the equipment into consideration in determining the size of the tape. The tape is 1/2-inch wide, which is 20% larger than video tapes used in the home.

(5) Compactness

The cassette, electrical circuitry, and tape transport system, including the head drum, all influence the size of the equipment. We assumed the use of the tape transport mechanism of a broadcast-quality VCR using a 1/2-inch cassette. The electrical circuitry could be miniaturized with large scale integrated circuits, so that the machine could be made as compact as existing 1/2-inch VCRs for broadcasting.

An outline of the specifications based on these requirements is shown in Table A.2.

A.2.2 Cassette

The required specifications for the cassette were as follows:

(1) Tape Length

With a 63-minute recording time and a tape advance speed of approximately 120 mm/s, 453 m would be needed.

(2) Hub and Flange

A nominal value of 13.5 μm is assumed for the tape thickness. This value is used to determine the size of the hub and flange.

(3) Guide Post

In high-density recording, the accuracy of the tape advance mechanism is extremely critical. For this reason the guide post, which affects the accuracy of the tape transport, is positioned not on the cassette but in the VCR. The structure of the cassette mouth was designed accordingly. This arrangement eliminates restrictions imposed by the tape loading mechanism, and also allows companies to differentiate their methods and reduce costs.

(4) Dust protection mechanism

In high-density recording, special measures are necessary to protect the tape from dust and incidental contact. For this reason, the cassette has an airtight construction with features such as a two-layered lid through which the tape passes.

(5) Automatic mode identification hole, accidental erasure protection

The cassette is equipped with mechanisms expected to be essential for its use, such as an identification hole for tape type, thickness, and anticipated future recording mode. It is also equipped with mechanisms to prevent accidental erasure and tape slackening. The specifications of a cassette made according to the required specifications indicated above are shown in Table A.3. Figure A.12 is a photograph of the cassette.

A.2.3. VCR Design

VCR design consists of a mechanical design for the tape transport system including the head rotation mechanism, and an electrical design that determines the machine's electrical performance.

At the beginning of the project, the mechan-

TABLE A.2. Required UNIHI VCR specifications.

Recording format	
Video	Baseband FM recording
Audio	PCM digital recording
Performance	
Frequency characteristic	
Luminance	20 MHz
Color difference	7 MHz (progressive scanning)
Sound	20 kHz
SN ratio	
Luminance	41 dB
Color difference	45 dB
Sound	85 dB
Mechanism	
Based on variance of 1/2-inch VCR for broadcasting	
Tape	
Coated metal tape (fits in newly developed cassette)	

TABLE A.3. Cassette specifications.

Dimensions	205 mm (w) x 121 mm (d) x 25 mm (h)
Functions	Uses identification bit to identify four formats (variable). Identifies 16 types of tape lengths and thicknesses (variable).

ical system was not completely redesigned because of considerations such as development time, cost, and dependability. Instead, the transport mechanism was adapted from a 1/2-inch VCR for broadcasting, with modifications in only the drum rotation speed and tape transport speed. However, the recording format depends on the recording pattern recorded on the tape and is not directly related to the diameter of the drum or the rotational speed. Thus the recording pattern was selected to give as much freedom as possible to the mechanical design. The latter system is explained in detail below.

(1) Recording Frequency Band and Number of Channels

The problem of channel separation of the recording frequency is a basic consideration in wideband recording. Table A.4 compares minority channel wideband recording and majority channel narrowband recording in the recording signal processing of wideband signals.

A minority channel system was adopted for UNIHI based on research results in wideband, high-density recording technology. This system made the recording and playback systems more compact. Problems raised by this selection, such as divisions within the screen, would be solved by technologies such as TCI (Time Compression Integration) and shuffling. These will be discussed next.

FIGURE A.12. UNIHI cassette.

(2) TCI

The TCI system was designed at NHK Laboratories to separate out color signals and perform baseband time division multiplexing so that the signal can be transmitted over a narrow-band transmission path.[10] Later it was used to prevent deterioration of the SN ratio of the color signal and cross-color interference, both of which result from the triangular noise generated during composite video signal FM transmission. The European MAC system is a variation on this system. In the UNIHI design, TCI is used to record the total 27 MHz for both the luminance and color signals on two channels with the following two goals in mind: (1) to treat the two channels as the same type of signal and unify the circuit configuration; and (2) to make the signal amenable to shuffling to reduce intertrack interference and visible screen division.

A signal that has undergone TCI is shown in Figure A.13.

The frequency band of the TCI signal is 24 MHz. This was set for the following reasons:

1. Using a coated metal tape, and the shortest recording wavelength of existing 1/2-inch broadcast VCRs, a recording band of 12 MHz, three times that of present systems, will be realized by tripling the relative speed of the tape head and widening the band of the rotary transformer.
2. By doubling the time axis, the 24 MHz bandwidth is reduced to 12 MHz, which can be accommodated by a 2-channel recording and playback system.

TABLE A.4. **Comparison of wideband recording signal processing formats.**

Item	Minority channel	Majority channel
Band/channel	Wide	Narrow
Record/playback system	Few	Multiple
Screen division	Yes	No
Signal processing	Simple	Complex
Tape/head relative speed	Fast	Slow
Head characteristic	Wideband	Narrow band

3. Whatever cannot be accommodated in the 27 MHz total band can be recorded using part of the horizontal and vertical retrace line period.

The signal processing including TCI is shown in Figure A.14. For comparison, it is shown together with the Tsukuba Expo VTR specifications.

(3) Shuffling

Since UNIHI accommodates six tracks in one field, head switching points appear on the screen when recording and playing back in real time, and channel differences in recording and playback characteristics become visible as "banding." To avoid this, the beginning and end of each track are positioned at the top and the bottom of the screen, and the banding is made less noticeable by shuffling during every horizontal scanning period. Consideration is also given to an H-alignment that makes the image viewable even when transporting at a nonconstant speed.

The block diagram of a VCR video processing system that uses the methods mentioned in items (1), (2), and (3) is shown in Figure A.15.

(4) Frequency Quadrupling FM Demodulation

In the FM demodulation of broadcasting VTRs, pulse counter demodulation systems, which have superior linearity, have traditionally been used. In the low-carrier frequency FM used in VCRs, artifacts from FM sideband waves during demodulation cannot be avoided.[11] One method of reducing this is to raise the FM carrier frequency during demodulation to reduce the amount of the lower sideband component that leaks into

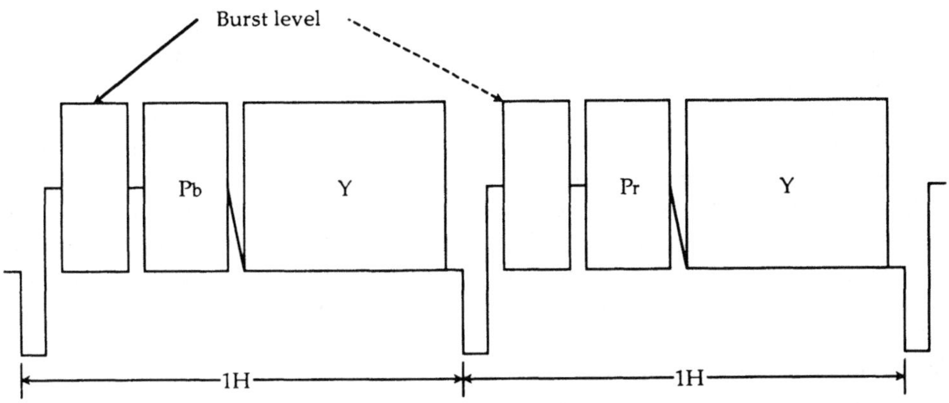

FIGURE A.13. TCI signal.

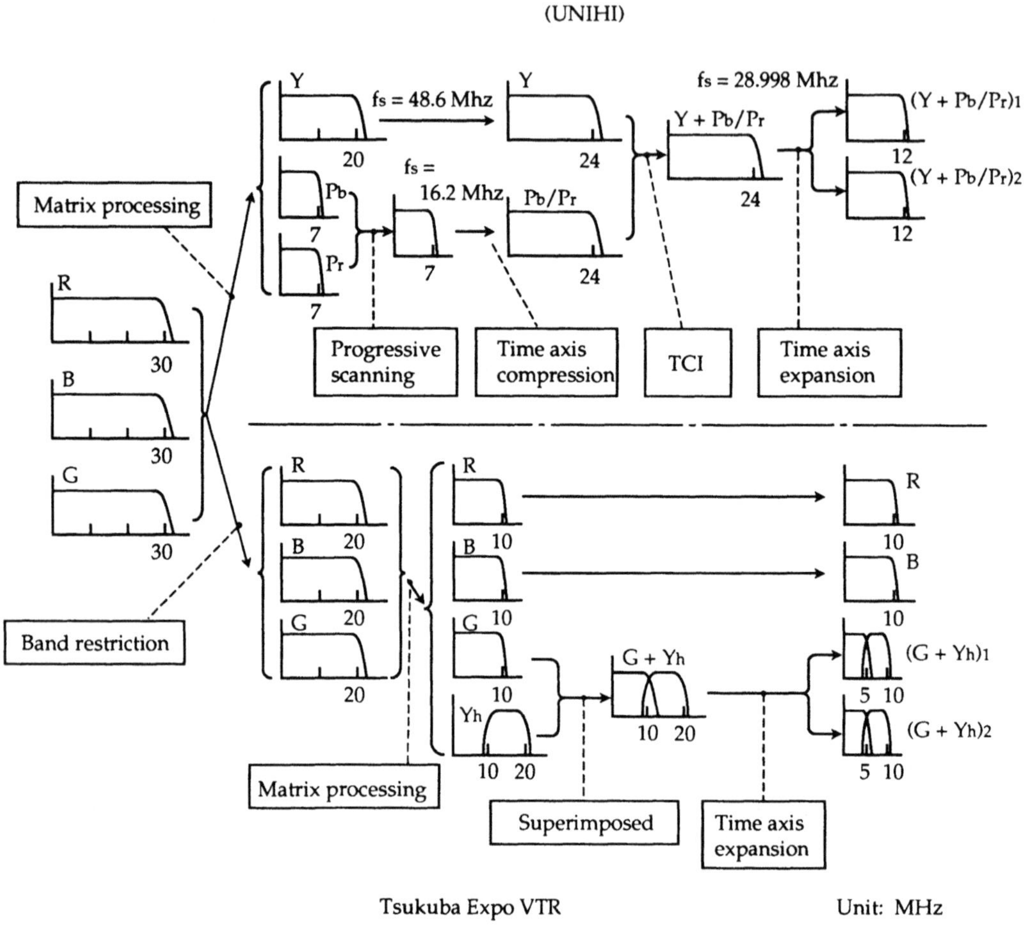

FIGURE A.14. Signal processing in Hi-Vision VTR.

the modulation frequency band. Figure A.16 shows the frequency spectrum during FM demodulation when demodulation is done by doubling and quadrupling the FM carrier frequency.

For demodulation, UNIHI has a demodulation system that quadruples the carrier frequency to improve image quality. While this method has been common knowledge, the circuits necessary for adequate performance had not yet been developed until NHK Laboratories developed a simple circuit that sufficiently reduces artifacts.[8] The circuit is shown in Figure A.17. This technology has been made available to manufacturers.

Seen from a different point of view, because this demodulator reduces the frequency of the FM carrier wave, the recording wavelength on the tape becomes longer, the CN ratio improves, and the SN ratio of the image improves. Because the amount of artifacts decreases, the low-pass filter need not be steep after demodulation, and the waveform characteristic can be improved.

Next, we will discuss the advantages of this demodulator. In Figure A.16, in addition to the amplification of the desired signal *D*, which must be demodulated, there exist incidental FM waves with carriers of $2f_c$ and $4f_c$, and the lower edge of their sideband wave component passes through an LPF and becomes the undesired signal amplitude *U*. That amount is expressed as a DU ratio (D/U) as follows:

$$\left(\frac{D}{U}\right)_2 = \frac{\beta B}{J_\gamma(2\beta)\{2f_c 2c - \gamma B\}} \text{(for doubling)} \qquad \text{(A.1)}$$

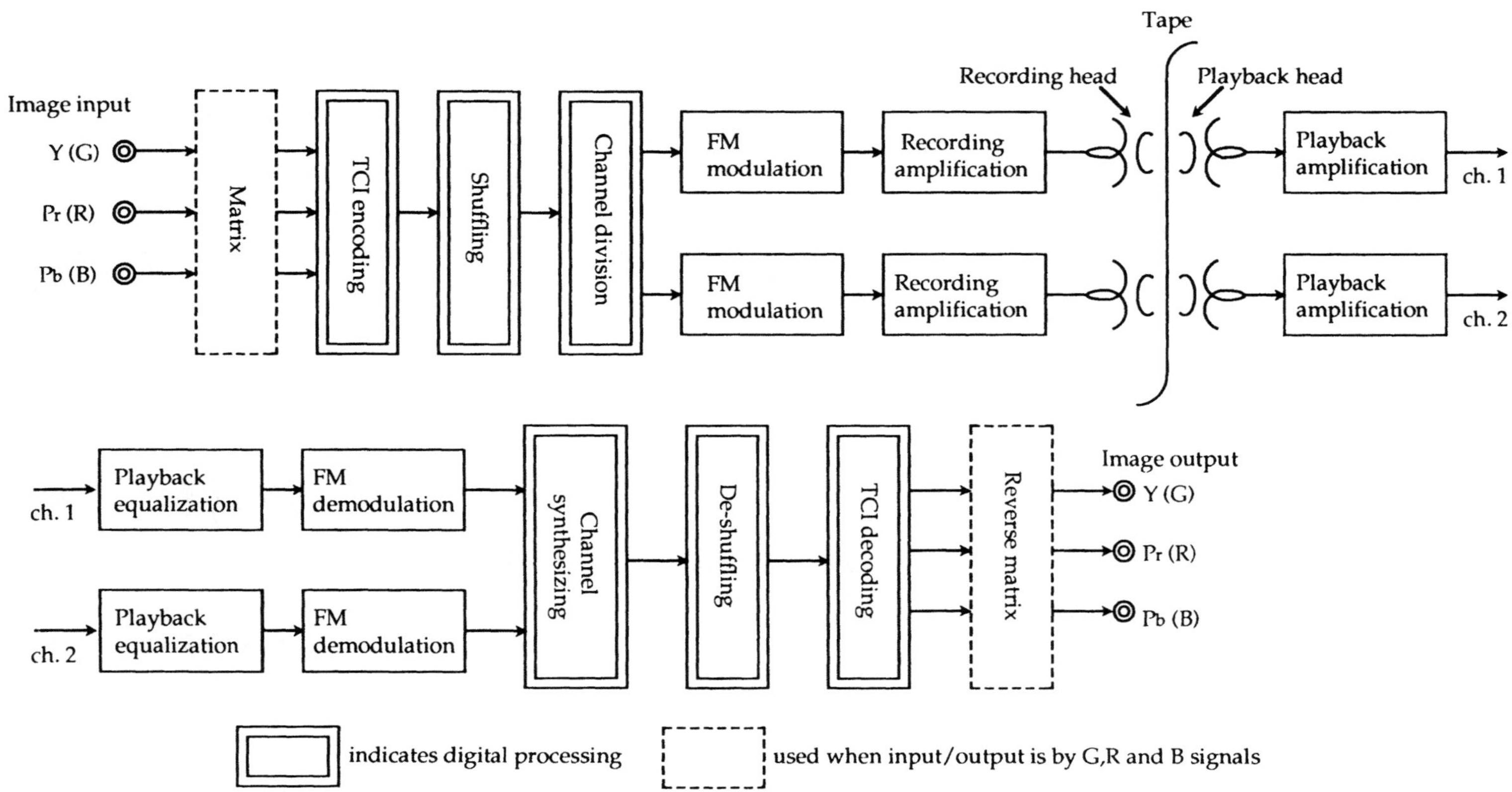

FIGURE A.15. Block diagram of video processing in the UNIHI system.

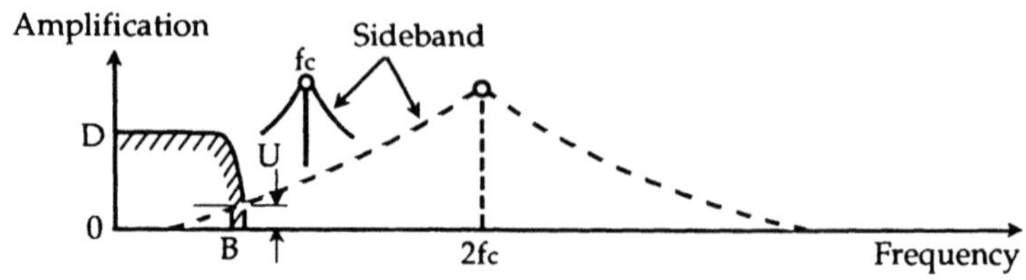

(a) Frequency doubling demodulation

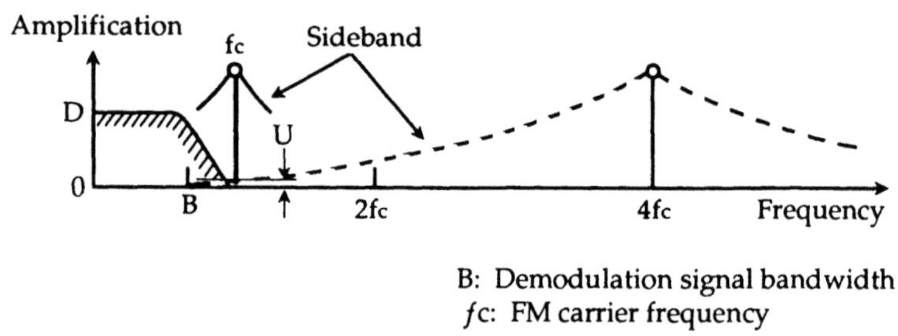

(b) Frequency quadrupling demodulation

FIGURE A.16. Frequency spectrum during FM demodulation.

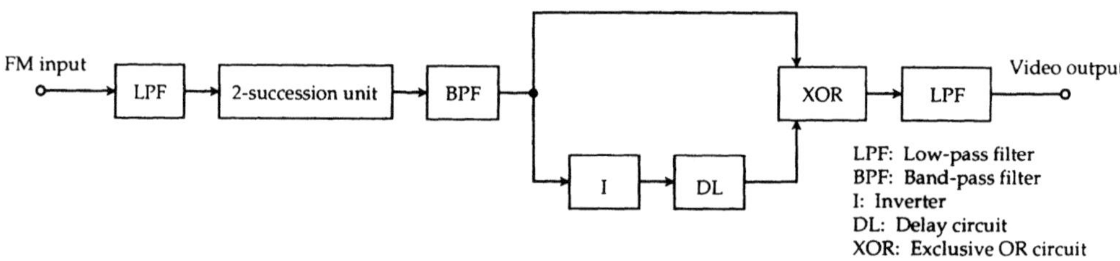

FIGURE A.17. Configuration of frequency quadrupling pulse counting demodulation circuitry.

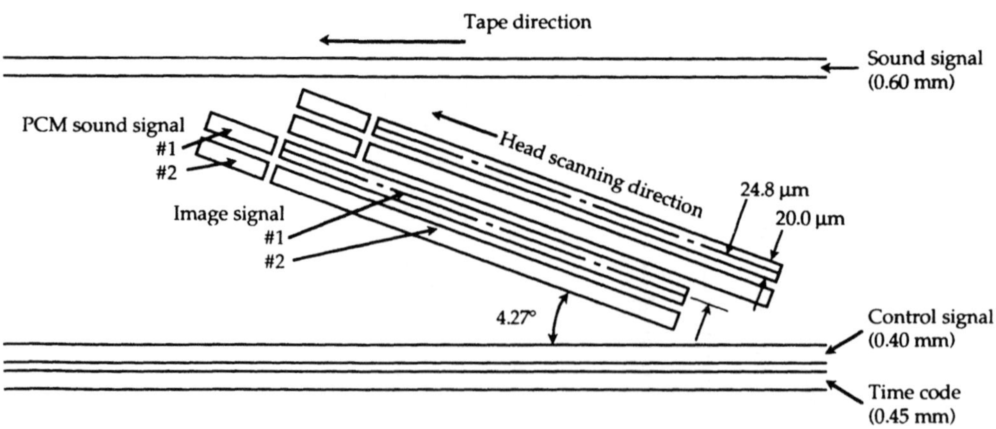

FIGURE A.18. UNIHI track pattern.

$$\left(\frac{D}{U}\right)_4 = \frac{\beta B}{J_\gamma(4\beta)\{4f_c 2c - \gamma B\}(\text{for quadrupling})} \quad \text{(A.2)}$$

where

β = modulation characteristic

$J\gamma$ = type 1 Besse function of γ order

γ = order of sideband

In the case of quadrupling, the generated carrier has a high frequency, and the order of the undesired sideband is also high and has a small amplitude. Thus the carrier frequency to obtain the same DU ratio f_c can be set low as shown in Figure 16(b). This allows the recording wavelength to be increased, and a range with a low output drop can be used. Then not only can the CN ratio of the playback signal be improved, but the demodulator output LPF can be set to have a smooth operating characteristic and the waveform characteristic of the output signal can be improved.

(5) PCM Audio Recording

The audio signal is digitally recorded, with the same heads as are used for video recording, on tape having the track pattern shown in Figure A.18. Two channels are accommodated on each track. Since playback and recording are both done with the rotating head, the DC component must be removed by using an 8-14 conversion as the modulation system. Digital recording of sound signals differs from video signal recording in that there is no correlation between lines or between fields, making it necessary to perform a thorough error correction. Based on experimental results, a double-Reed Solomon code (32,28,5),(28,24,5) system was adopted.

A.3 LSI MUSE DECODER

NHK has developed LSIs for a MUSE decoder with the goal of making Hi-Vision MUSE receivers more compact, lower in energy consumption, and lower in cost. The distinguishing features of the MUSE decoder developed at NHK are its large-scale circuitry, complex signal processing and high processing speed (maximum clock rate is 48 MHz).[13] The advanced technology used has not yet been seen in consumer electronics. Twenty-five types of LSIs have been developed for this MUSE decoder.[14,15]

A.3.1 Signal Processing in the LSI Decoder

We will explain the signal processing in the LSI decoder using the MUSE LSI block diagram in Figure A.19.

(1) Input Signal Processing Block

The MUSE signal first undergoes band restriction by an 8.1 MHz analog LPF, and then enters the input proccessing block. After clamping and sample holding (S/H), the signal undergoes A/D conversion. It is then split into two parts. One part enters the control signal generation block, while the other undergoes waveform equalization and compensation for distortion in the transmission path, and is again split in two. Of this second partitioning, one part is the main audio signal which goes to the audio processing block, while the other is the main video signal and undergoes nonlinear processing such as de-emphasis and transmission inverse γ.

(2) Control Signal Generating Block

In the control signal generating block the clock pulse used in the LSI decoder is reproduced, and the synchronizing signal and control signal used in MUSE video and audio decoding are separated. The clock reproduction is based on the frame pulse and horizontal synchronization. The subsampling phase, motion vectors, and other data are extracted from the control signal in a form usable to the decoder, and distributed to each block.

(3) Audio Signal Processing Block

The MUSE audio signal is interposed in the vertical blanking interval as 16.2 MHz ternary signals. Audio processing LSI 1 performs frequency conversion to 12.15 MHz, ternary-binary conversion, and de-interleaving to obtain a bit stream of 1350 b/s. Audio processing LSI 2 performs DPCM decoding. Following this, D/A conversion is performed, and 4-channel audio is obtained in the case of mode A, and 2-channel audio in the case of mode B.

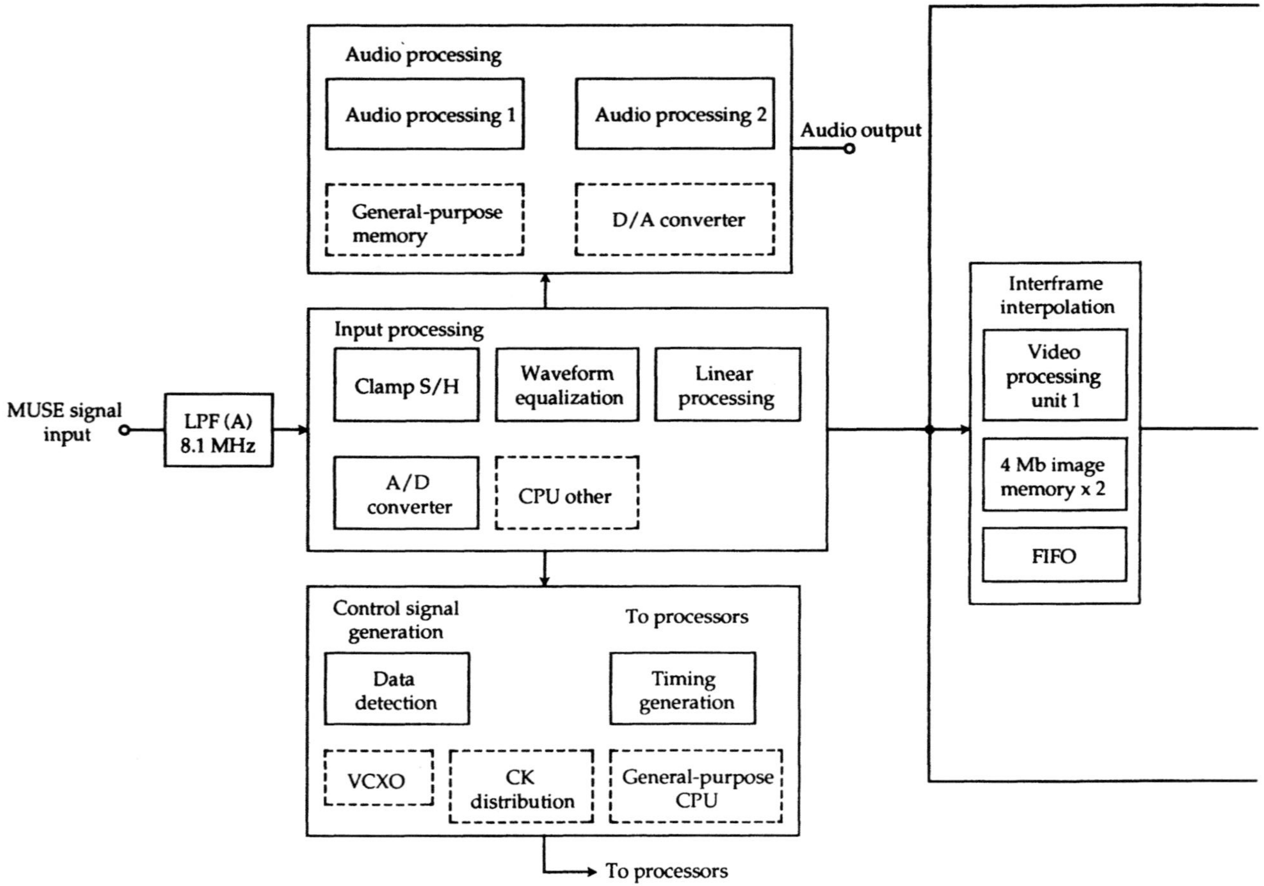

FIGURE A.19. LSI block diagram of MUSE decoder.

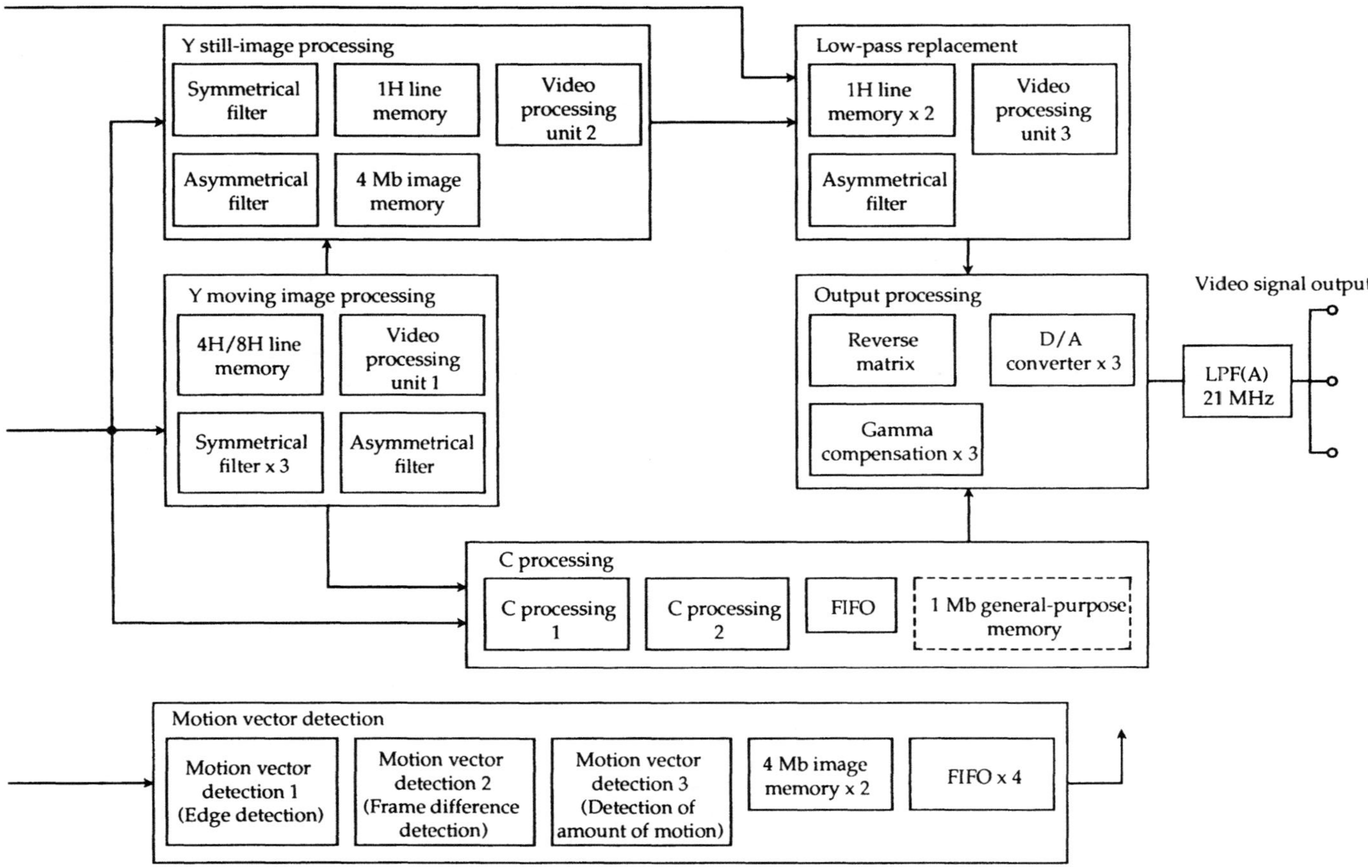

FIFO: first-in first-out register
VCXO: Voltage control quartz oscillator
CK: clock
S/H: Sample hold

Y: Luminance signal
C: Color signal

FIGURE A.19. (continued)

TABLE A.5. The MUSE LSI family.

Item	Count per set	Pins; package type	Operating frequency f (MHz)	Power use (W) at tuner input	Function
Symmetrical filter	4	64/SDIP	32.4	0.435	LPF
Linear processing	1	42/SDIP	32.4	0.385	Non-linear de-emphasis
Motion detection 1	1	64/SDIP	32.4	0.600	Edge detection
Motion detection 2	1	64/SDIP	32.4	0.600	Frame difference detection
Motion detection 3	1	64/SDIP	32.4	0.425	Y/C motion vector detection
Video process unit 1	2	64/SDIP	32.4	0.450	Frame interpolation, noise core
Video process unit 2	1	64/SDIP	48.6	0.625	Y mix (still image, moving image)
Video process unit 3	1	40/DIP	48.6	0.525	Low-pass replacement
Chroma process 1	1	48/SDIP	32.4	0.110	Time elongation, interpolation
Chroma process 2	1	42/SDIP	32.4	0.260	Progressive line decoding
Asymmetrical filter	3	48/SDIP	48.6	0.245	Sampling frequency conversion (32 MHz to 48 MHz)
Image memory	5	64/SDIP	32.4	0.590	Field delay (including vertical motion compensation)
Line memory (1H)	3	24/FLAT	32.4	0.290	Line (1H) delay
Low-capacity FIFO	6	44/QFP	32.4	0.385	Slight delay, horizontal motion compensation
Audio processing 1	1	84/QFP	16.2	0.205	Ternary to binary conversion
Audio processing 2	1	84/QFP	1.35	0.115	Sound decoding
Data detection	1	64/QFP	16.2	0.115	Control data detection
Timing generation	1	100/QFP	32.4	0.165	Control timing detection
Y4/C8 memory	1	40/DIP	32.4	0.690	Y 4H/C 8H delay
Reverse matrix	1	179/PGA	48.6	1.500	Y/C to R/G/B
Gamma correction	3	88/PGA	48.6	0.400	Display gamma correction
Sample hold	1	32/QFP	32.4	0.400	(Sample hold) clamp
A/D converter	1	48/QFP	16.2	0.400	(10-bit)
D/A converter	3	24/SDIP	48.6	0.255	(10-bit)
Waveform equalization	1	80/QFP	32.4	0.400	Waveform equalization filter
Total	46			(30W during operation)	

Packages: SDIP (shrink dual in-line package), QFP (quad flat package), PGA (pin grid array), DIP (dual in-line package), FLAT (flat package)

(4) Interframe interpolation block

In the interframe interpolation block, the interpolation of pixel signals is performed using signals from four fields. Two image memories, each 4 Mb in capacity, are used for this purpose. This block also performs noise reduction between two frames simultaneously along with the interframe interpolation.

(5) Motion Detection Block

The amount of motion in the MUSE decoder is obtained by dividing the linear mixing value of the change in level at the edge and the video level by the frame difference. Motion detection LSI 1 primarily detects the edge volume; LSI 2 detects the frame difference; and LSI 3 performs the division. The amount of motion is expressed in 4 bits, and two 4 Mb image memories are used for interframe interpolation of this motion.

(6) Y moving image processing Block

The Y moving image processing block repoduces the moving image luminance signal. From a signal that has undergone interframe interpolation, only one field of data is taken and then interpolated as the moving image. The interpolation is performed using a digital filter 7 pixels wide and 5 pixels high consisting of a line memory and, three symmetrical filters. The interpolated signals are divided into luminance signals and color signals. The color signals go to the C processing block, and luminance signals are directed to the Y still image processing block after undergoing a frequency conversion from 32 MHz to 48 MHz with an asymmetrical filter.

(7) Y Still Image Processing Block

In the Y still image processing block, a symmetrical filter is used to perform 12 MHz LPF processing on the signals which have undergone interframe interpolation, and an asymmetrical filter is used to convert the frequency from 32 MHz to 48 MHz. In video processing unit 3, after interfield interpolation is performed using the present field signal and another field signal delayed one field period using a 4Mb DRAM memory, the still image and the moving image are combined according to the amount of motion.

(8) Low Frequency Replacement Block

Since MUSE signals do not have aliasing signals in the low frequency range (0 to 4 MHz) of the luminance signal, information from the present field is sufficient to form the low frequency component of the luminance signal. When the low frequency portion of the luminance signal is formed from this data, it replaces the low fre-

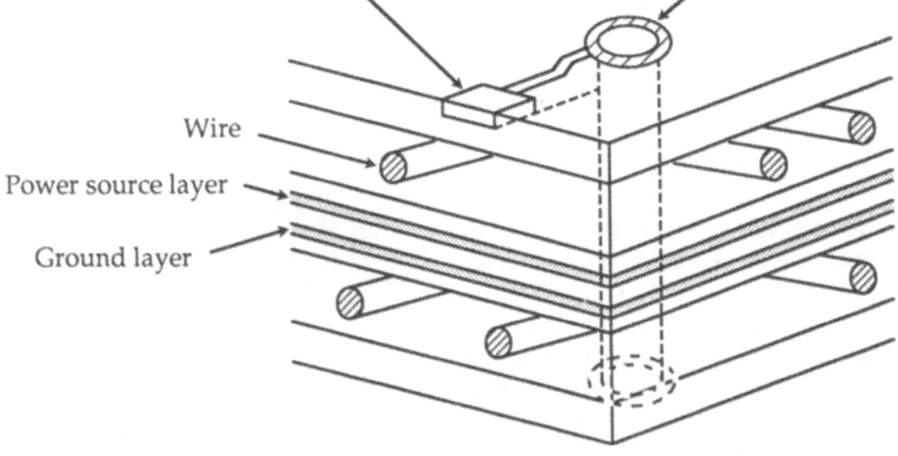

FIGURE A.20. Cross section of printed circuit board.

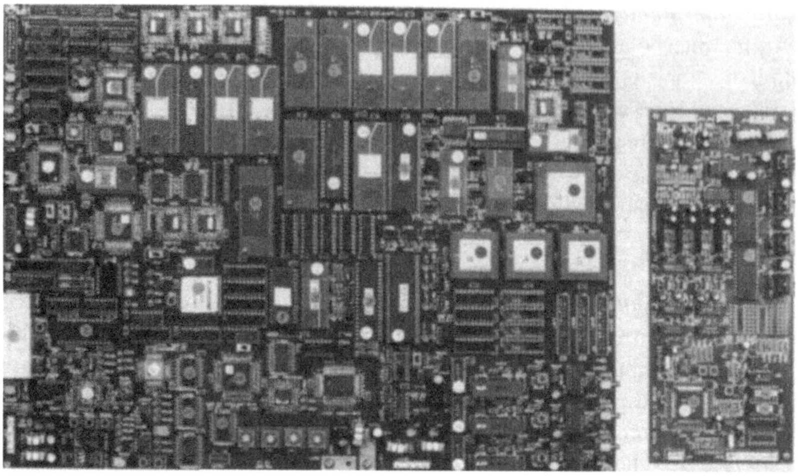

FIGURE A.21. Substrates of the LSI MUSE decoder (video on left, audio on right).

quency portion of the luminance signal that was decoded using the present field data. This processing is called low frequency replacement. For this reason, data which has not yet undergone interframe interpolation is necessary in the low frequency replacement block.

(9) C Processing Block
Reproduction of color signals is performed in the C processing block. Still image signals from the interframe interpolation block and moving image signals from the Y moving image processing block are interpolated by the still image filter and the moving image filter respectively, and are switched in 1-pixel units based on the amount of motion in the same way as are the luminance signals. After this, the sampling frequency of the color signals is converted from 16 MHz to 48 MHz, and the 48 MHz rate R-Y, B-Y signals are reproduced.

(10) Output Signal Processing Block
In this block, the R, G, and B signals are produced from 48 MHz-rate Y, R-Y, and B-Y signals using an inverse matrix, and undergo gamma processing for the display as well as signal enhancement. The R, G, and B signals pass through a 21 MHz analog filter and become the final video output after being returned to analog form by a D/A converter.

A.3.2 Hi-Vision Receiver

Table A.5 lists the 25 types of LSIs used in the MUSE decoders, including A/D converters, D/A converters, and field memories. Flat packages and shrink DIP packages are used for high-density mounting. Most of the LSIs are CMOS to achieve low energy consumption.

We designed and made a prototype substrate for a MUSE decoder having these LSIs. A multiwire substrate was used in the interest of shortening development time and making repair easier. Because shrink DIP, QFP, and standard DIP packages were intermixed, an additional layer was added to accommodate flat packages. Figure A.20 shows a profile of the board.

FIGURE A.22. LSI MUSE decoder component unit.

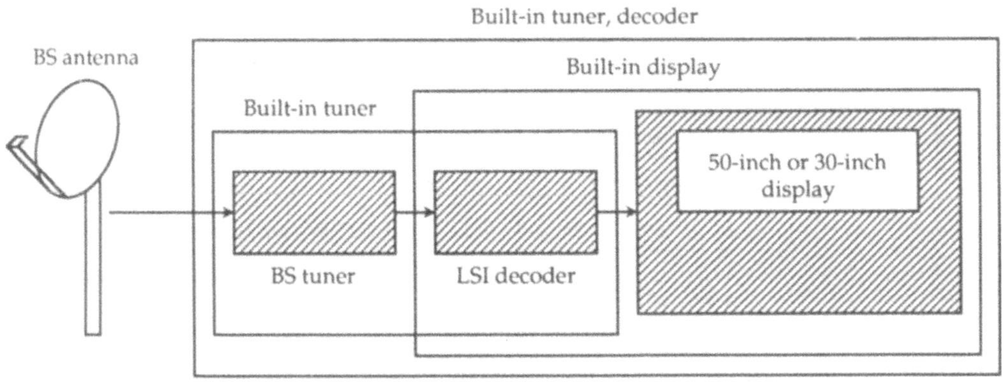

FIGURE A.23. Configuration of Hi-Vision receiver.

The analog section was positioned on the same board to consolidate the receiver's circuitry. To avoid intervention from the digital section, the power was divided up and the layout drawn so that the analog section comes to the edge of the board. The digital section was laid out to minimize wiring length. The clock is situated in the center of the board so that the distance to the ICs is evenly distributed. As a result of designing a high-density assembly board in this way, the board's video section is 40 cm × 30 cm, and the audio section is 23 cm × 10 cm. The assembled LSI board is shown in Figure A.21.

Figure A.22 shows a component type MUSE decoder, and which supplies power to the various boards. Its operation has been confirmed to be the same as with the prototype LSI decoder.

By developing LSIs for the MUSE decoder, we were able to fit both the decoder and BS tuner into our Hi-Vision receiver prototype for household use. The configuration of the Hi-Vision receiver is shown in Figure A.23.

A.4 CCIR HDTV STUDIO STANDARD RECOMMENDATIONS

At the 17th General Meeting in Dusseldorf, Germany from May 21 to June 1, 1990, the CCIR (Consultative Committee on International Radio-Communications) adopted recommendations on the long-standing issue of an HDTV studio standard.[16] Because the 16th General Meeting held in 1986 in Dubrovnik, Yugoslavia had been unsuccessful in adopting these recommendations, the CCIR worked diligently over the next four years, determined to have these recommendations adopted at the next general meeting. The recommendations are the fruit of these efforts. As shown in Table A.6, which details the progress in deliberations on the HDTV studio standard by the CCIR, in 1972 Japan proposed that the CCIR form a "Study Programme" for HDTV. The adoption of the recommendations eighteen years later thus marks a milestone for Japan.

A.4.1 Recommendations related to HDTV established at the CCIR

At the 1990 General Meeting of the CCIR, five recommendations related to HDTV were established, including the recommendations for a studio standard, as shown in Table A.7.[16–20] Each of these recommendations is important to the worldwide proliferation of HDTV. Here we will condense those recommendations which do not relate to the studio standard.

A.4.2 HDTV Studio Standard Recommendations

(1) Contents of Recommendations

The studio standard recommendations consist of 27 items, of which universal parameters have been set for 23. Tables A.8 through A.12 show

TABLE A.6. CCIR deliberation process for the HDTV studio standard.

Year	Event
1968	HDTV research begins at NHK.
1972	At the mid-term meeting, Japan proposes an HDTV "Study Programme."
1974	At the general meeting the question of HDTV research is formally adopted.
1976	At the mid-term meeting, the "HDTV Report" is made based on documents submitted by Japan.
1981	At the final meeting, an agreement is reached to promote research.
1983	At the mid-term meeting, IWP 11/6 is begun; "HDTV via Satellite" report is issued.
1985	At the IWP11/6 Tokyo Conference, a format conversion is demonstrated. At the final meeting, a 3-1 audio format is proposed.
1986	At the general meeting, there is deliberation concerning proposed recommendations on a worldwide 1125/60 standard. Recommendations are not established.
1987	At the mid-term meeting, Europe proposes a 1250/50 standard.
1989	(January) First proposals for recommendations made based on points commonly agreed upon at IWP11/6. (May) At the SG11 Special Conference the proposal for recommendations is revised. (October) After the final meeting, the colorimetry parameter remains to be studied.
1990	(March) At the IWP 11/6 Special Conference a colorimetry parameter is agreed upon. (May) At the general meeting, studio standard recommendations are adopted.

the values of the main parameters. Below are explanations of some of the parameters.

(a) Optoelectric Conversion Characteristic. The integrated optoelectronic conversion characteristic at the signal source, shown in item 1 of Table A.8, corresponds to the conventional gamma correction curve. This curve is precisely stipulated in the recommendations so that in studio post-production, gamma corrected signals can be accurately reconverted into linear signals for signal processing.

(b) Colorimetry Parameters. With regard to the three primary color chromaticity points of item 2 in Table A.8, the two concepts of the three tentative primary colors based on present display technology and future reference primary colors are incorporated.

With regard to the colorimetry parameters, in the process of the formation of the recommendations by IWP11/6 as well, there were opposing opinions, with some arguing for future systems and some arguing for systems which can be realized now. The parameters were brought together in this form as a result of those arguments. As shown in Figure A.24, red and blue employ the EBU chromaticity points and green employs a chromaticity point which is between the EBU and SMPTE chromaticity points. Future reference primary colors will attempt to widen the color reproduction range of the display and to improve the conversion between such media as film and color hard copies, so they are a subject for future investigation.

(c) Screen Characteristic. The 16:9 aspect

TABLE A.7. HDTV recommendations of the CCIR General Meeting.

Recommendation	Contents
HDTV basic parameter values for studio standard and international program exchange	Basic parameter values are recommended for the chromaticity points of the three primary colors and of reference white, aspect ratio, horizontal effective number of samples, luminance signal and color signal equation, and gamma curve.
Subjective evaluation method for HDTV image quality	Evaluation methods for determining HDTV image quality are recommended along with viewing conditions such as distance (three times the screen height), and brightness of the screen and background.
International exchange of electronically produced HDTV programs	It is recommended that the international exchange of HDTV programs which are electronically produced using television cameras and VTRs is done not by conversion to film but by electronic means with VTR tape.
HDTV conversion to film	When an HDTV image is recorded onto 35 mm film, the frame width should be the same as the ISO standard 35 mm film, while the frame height should result in an aspect ratio of 16:9.
Scanning range of 35 mm motion picture film in HDTV telecine	The film scanning range when 35 mm film is converted to HDTV by telecine should be based on the same range recommended for recording HDTV images onto film.

TABLE A.8. Provisions for optoelectric conversion.

Item	Parameter	Value		
1	Total optoelectric conversion characteristic at signal source	$V = 1.099\,L^{0.45} - 0.099$ $(1 \geq L \geq 0.018)$ $= 4.500\,L$ $(0.018 \geq L \geq 0)$ L: Brightness of object $(0 \leq L \leq 1)$ V: Corresponding electrical signal		
2	Chromaticity of three primary colors (CIE 1931) (Reference primary colors are explained in text.) Tentative primary colors based on present display technology	Color	Coordinates x	y
		Red	0.640	0.330
		Green	0.300	0.600
		Blue	0.150	0.060
3	Equal primary color signal chromatics $E_R = E_G = E_B$ (Reference white)	D_{65}	x: 0.3127	y: 0.3290

TABLE A.9. Provisions for screen characteristics.

Item	Characteristic	
	Parameter	Value
1	Aspect ratio	16:9
2	Effective samples per scan line	1920
3	Sample distribution	Orthogonal
4	Sample distribution and the number of effective scan lines are under study. Since these two are related, the effective samples per scan line may be reevaluated later.	

TABLE A.10. Provisions for scanning.

Item	Characteristic	
	Parameter	Value
1	Order of sampling scan	Left to right Top to bottom
2	Interlace ratio	See below
	The goal of the system is defined as progressive scanning, that is to have an interlace ratio of 1:1. With present equipment, an interlace ratio of 2:1 or low-pass processing with an equivalent sample rate may be used.	

TABLE A.11. Provisions for signal format.

Item	Characteristic	
	Parameter	Value
1	Luminance signal equation E_Y' System equation is based on present display technology and existing coding.	$E_Y' = 0.2125\ E_R' + 0.7154\ E_G' + 0.0721\ E_B'$
2	Color difference signal equation (analog) E_{PR}', E_{PB}' System equation is based on present display technology and existing coding.	$E_{PR}' = 0.6349\ (E_R' - E_Y')$ $E_{PB}' = 0.5389\ (E_B' - E_Y')$

Table A.12. Provisions for signal level and sync signal.

Item	Characteristic	
	Parameter	Value
1	Nominal level E_R', E_G', E_B', E_Y'	Reference black = 0 Reference white = 700 mV
2	Nominal level E_{PR}', E_{PB}'	± 350 mV
3	Sync signal format	Ternary bipolar
4	Sync level	± 300 mV All signals are synchronized.

ratio of item 1 of Table A.9 was supported fully by the member nations with no objections. A value of 1920 was supported for the effective number of samples for each sampling line (item 2).

However, as recorded in item 4 of Table A.9, there is room to reconsider this value in the future. For computer images, it is advantageous for processing if the sampling interval of the screen in the horizontal direction and the sampling interval of the screen in the vertical direction are equal. In this case, the sampling distribution is in square lattice form. When the effective number of samples per scanning line is 1920, the effective number of scanning lines becomes 1080 when this square lattice sampling

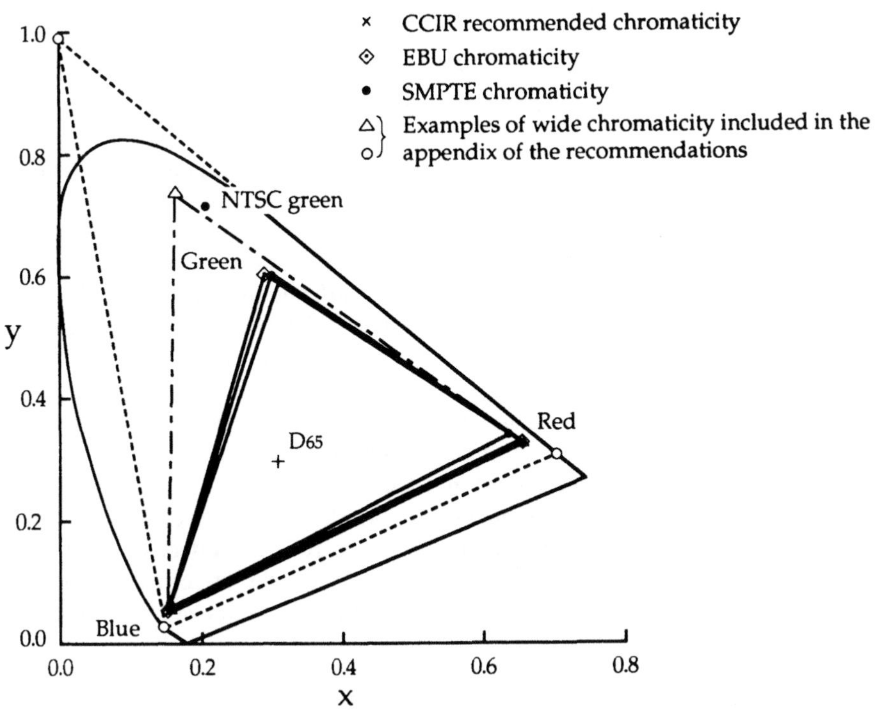

FIGURE A.24. Chromaticity points of three primary colors in HDTV.

distribution is adopted. Also, when the effective number of scanning lines is 1035 or 1152, the effective number of samples becomes 1840 and 2048, respectively. Therefore, the above is described in item 4 of the table.

(d) Scanning System. There are two stipulations for the interlace ratio in item 2 of Table A.10, one for equipment proposed for the future and one for present equipment. This is also a point where there were divided opinions between those who emphasize new future systems and those who emphasize specifications which can be realized now.

(e) Signal System. Item 1 of Table A.11 is the stipulation for the luminance signal equation. In this case, as in the case of the three primary colors of item 1 of Table A.8, both an equation for a system based on present display technology and existing coding methods, and an equation for a system based on future reference primary colors are being considered.

In the former system, which is based on present technology, the luminance level signal employs a method using R, G, and B signals which have undergone gamma correction. On the other hand, for systems which are based on future reference primary colors, research efforts are being intensified in methods such as constant luminance transmission. Table A.12 shows the parameter values for the signal level and synchronization signal.

Thus far we have outlined the main parameters. But there remain parameters on which agreement has not yet been reached. These are discussed below.[3]

(2) Significance of the CCIR Recommendations

The CCIR ranks these recommendations as very important achievements and as the first steps toward a uniform worldwide studio standard in the future. Also, examination of future recommendations related to HDTV will be carried out while utilizing sufficient data exchange with other international organizations such as ISO*, IEC** and CCITT***, since HDTV is expected to be applied not only in broadcasting but across a wide range of nonbroadcasting fields.

(3) Future areas for examination by the CCIR

The CCIR intends to continue its investigations as we move toward the realization of a uniform worldwide studio standard. Areas for future inquiry are listed below.

(a) Future Reference Primary Colors. Reference primary colors which will make a wider range of color reproduction possible will be studied, so three proposals have been mentioned in the Appendix as candidates for reference primary colors, as shown in Figure A.24. Two of these candidates use wider chromaticity points than the chromaticity points of the three primary colors which have been recommended. The other candidate will use the chromaticity points which have now been recommended in the future as well, but will attempt to expand the color reproduction range by using the lost portion of the R, G, and B image signals. In this case, if Y, P_R and, P_B signals are used, a wide color reproduction range can be transmitted without changing the contents of the present recommendations dealing with the signal level and the signal equation.

In addition, the question of whether a constant luminance transmission system can be employed will be examined.

(b) Effective Number of Scanning Lines and Field Frequency. These two parameters are very difficult to standardize. Therefore, there will probably be research on what methods could be used to standardize them in the future. This research will take time. Two possible methods proposed for future standardization are CIF (Common Image Format) and CDR (Common Data Rate).[21]

As shown in Figure A.25, CIF will attempt to standardize the effective number of samples per line and the effective number of scanning lines of the screen even in formats as different

*International Organization for Standardization
**International Electrotechnical Commission

***International Telegraph and Telephone Consultative Committee

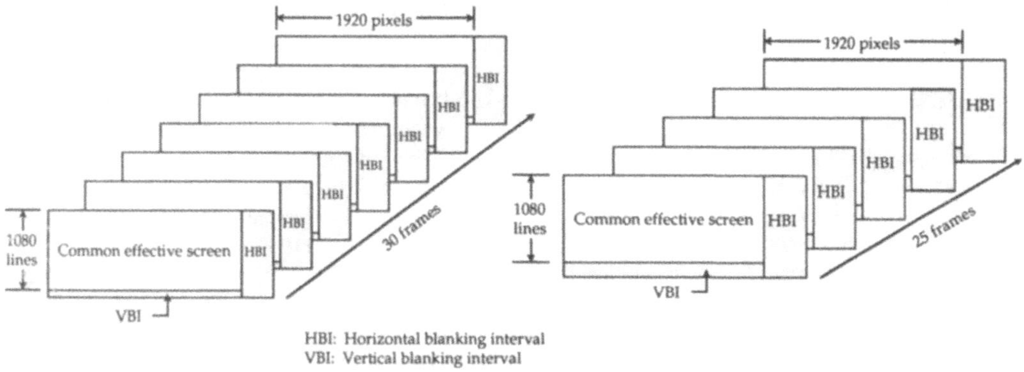

FIGURE A.25. Example of CIF (common image format).

as 1125/60 and 1250/50, so in the future it is also expected that an attempt will be made to standardize the number of frames per second. Even though CDR, like the digital specifications recommendations for present standard television (Recommendation 601), has multiple scanning specifications, it uses the same digital sampling frequency and data rate. This allows the studio digital instruments to change to the different specifications by means of a switch. CDR does not offer a clear route toward standardization of scanning parameters. Another idea that has been put forward is CIP (Common Image Part).[21]

Other items for study are digital parameters such as the quantization bit count, digital interface specifications, and the possibility of using bit rate compression in the studio. In addition, with regard to nonbroadcasting applications of HDTV, the common use of HDTV displays and computer displays will be investigated.

A.4.3 The Practical Application of HDTV in Japan

Since 1984, the High-Definition Television Committee of the Telecommunications Technology Deliberation Committee, Ministry of Posts and Telecommunications (MPT), has been looking into the matter of a domestic standard for a Hi-Vision satellite broadcasting system. The committee released its report at the same time that the CCIR issued their recommendations for an HDTV studio standard at their General Meeting in May 1990. In 1991, the MPT amended ordinances affecting Hi-Vision satellite broadcasting. NHK has been broadcasting in Hi-Vision eight hours a day on an experimental basis since November 1991 over a DBS back-up channel.

For Hi-Vision satellite broadcasting to develop smoothly in the future, low-priced receivers are absolutely essential. The first step in reducing prices has already being taken with the development of decoder LSIs for MUSE receivers. In the future, additional efforts to promote the diffusion of receivers will be needed, including further increases in the level of integration.

Meanwhile, steady progress is being made in the application of Hi-Vision technology in both program production and industrial applications. To take these applications to the next level, efforts are being made to reduce the cost of equipment such as VTRs and cameras. The establishment of the present CCIR studio standard recommendations will contribute to the production and application of Hi-Vision equipment, which will, in turn, reduce equipment costs and promote further diffusion.

Although agreement has not yet been reached on several basic parameters, the CCIR recommendations for an HDTV studio standard have been adopted. This achievement is highly regarded by participating nations as the widest attainable worldwide agreement at the present time. Remaining issues based on the foundation laid by these recommendations will continue to be investigated by the CCIR. Japan will continue to actively participate in and contribute to these studies.

REFERENCES

1. Sawada, "HDTV Lecture Signal Types and Transmission," *Journal of the Institute of Television Engineers of Japan,* (ITEJ), Vol. 42, No. 9, pp. 951–958 (1988).
2. Hirabayashi, "DPCM Coding Equipment for MUSE Digital Transmission," *Institute of Electronics, Information and Communication Engineers Japan* (*IEICEJ*) Spring Conference Proceedings, Vol. B-515 (1989).
3. Honda, "MUSE Signal 60 Mb/s Digital Transmission System," *ITEJ Technical Report,* Vol 4. No. 20, pp. 25–30 (1990).
4. Matsumoto, "120/140 Mb/s Compact Hi-Vision Coding," *International Communications Research,* No. 138, pp. 1–11 (October, 1988).
5. Yajima, "Configuration and Characteristics of 100 Mbit/s High-Efficiency Coding Equipment for HDTV," *IEICEJ Proceedings* (B) Vol. J73-B-1, No. 4, pp. 328–336 (1990).
6. Ito, "Research on HDTV High-Efficiency Coding Systems Using Vector Quantization," *Technical Report of IEICE,* Vol. 86, NO. 246, pp. 51–57 (1986).
7. Barbero et al., "Transmission System for HDTV Signals Based on Discrete Cosine Transform," *IEEE, ISCAS,* Vol. 4, pp. 1887–1890 (May, 1990).
8. Yajima et al., "HDTV/Standard-TV Compatible Coding Based on DCT," *IEEE, ISCAS,* Vol. 4, pp. 1895–1898 (May, 1990).
9. Ohtsuka et al., "Digital Bit-Rate Reduction Systems of HDTV," *IEEE, ISCAS,* Vol. 4, pp. 1903–1906 (May, 1990).
10. Iwamura, "Color Television Signal Transmission Systems," Patent No. 873435 (applied May 21, 1971).
11. Felix et al., "FM Systems of Exceptional Bandwidth," *IEEE Proceedings,* Vol. 112, pp. 1659–1988 (1965).
12. Kamijo, "High Image Quality FM Modulation Systems That Suppress Moire" *ITEJ Proceedings,* No. 6–9 (1989).
13. Ninomiya, Ohtsuka, Izumi, Gohshi, Iwadate, "MUSE Format Development," *NHK Technical Journal,* Vol. 39, No. 2, pp. 18–53 (1987).
14. Kokubun, Akechi, Kobayashi, Abe, Kobayashi, Kawashima, Ninomiya, "Development of LSIs for the Hi-Vision MUSE Decoder," *1989 ITEJ Proceedings,* No. 12–13, pp. 281–282 (1989).
15. Izumi, Goshi, Ozeki, Ninomiya, Mihashi, Omura, Yamaguchi, Kawashima, Takegahara, Suganami, "Development of LSIs for MUSE Receivers," *1989 ITEJ Proceedings,* no. 12–14, pp. 283–284 (1989).
16. CCIR Doc. 1007/11 (Rev.2), "Basic Parameter Values for the HDTV Standard for the Studio and for International Programme Exchange," (1990).
17. CCIR Doc. 1034/11, "Subjective Assessment Methods for Image Quality in High-Definition Television," (1990).
18. CCIR Doc. 1012/11, "International Exchange of Programmes Produced Electronically by Means of High-Definition Television," (1990).
19. CCIR Doc. 1024/11, "Recording of HDTV Images on Film," (1990).
20. CCIR Doc. 1022/11, "Scanned Area of 35 mm Motion Picture Film in HDTV Telecines," (1990).
21. CCIR Doc. 1009/11, "Future Development of HDTV," (1990).

Index

GPSR Compliance

The European Union's (EU) General Product Safety Regulation (GPSR) is a set of rules that requires consumer products to be safe and our obligations to ensure this.

If you have any concerns about our products, you can contact us on ProductSafety@springernature.com

In case Publisher is established outside the EU, the EU authorized representative is:

Springer Nature Customer Service Center GmbH
Europaplatz 3
69115 Heidelberg, Germany

Batch number: 10684222

Printed by Printforce, the Netherlands